SCIENCE EDUCATION RESEARCH IN THE KNOWLEDGE-BASED SOCIETY

Science Education Research in the Knowledge-Based Society

Edited by

Dimitris Psillos

Department of Primary Education,
Aristotle University of Thessaloniki, Greece

Petros Kariotoglou

Department of Primary Education of Florina,
Aristotle University of Thessaloniki, Greece

Vassilis Tselfes

Department of Primary Education,
Aristotle University of Thessaloniki, Greece

Evripides Hatzikraniotis

Department of Physics,
Aristotle University of Thessaloniki, Greece

George Fassoulopoulos

Department of Primary Education,
Aristotle University of Thessaloniki, Greece

and

Maria Kallery

Department of Primary Education,
Aristotle University of Thessaloniki, Greece

KLUWER ACADEMIC PUBLISHERS

DORDRECHT / BOSTON / LONDON

A C.I.P. Catalogue record for this book is available from the Library of Congress.

ISBN 1-4020-1412-0

Published by Kluwer Academic Publishers,
P.O. Box 17, 3300 AA Dordrecht, The Netherlands.

Sold and distributed in North, Central and South America
by Kluwer Academic Publishers,
101 Philip Drive, Norwell, MA 02061, U.S.A.

In all other countries, sold and distributed
by Kluwer Academic Publishers,
P.O. Box 322, 3300 AH Dordrecht, The Netherlands.

Printed on acid-free paper

Printed in the Netherlands.

TABLE OF CONTENTS

INTRODUCTION

We are pleased to present the ESERA 2001 Conference book, which is based on contributions submitted and presented to the Third International Conference "Science Education Research in the Knowledge Based Society" that was organised by the Department of Primary Education of the Aristotle University of Thessaloniki and held in Thessaloniki from August 21 to August 26, 2001.

The focus of the Conference was to discuss the scope, methods, outcomes and perspectives of research in science education in the context of the rapidly developing knowledge-based society. Some 450 researchers, teachers, and postgraduate students attended the conference. They came mainly from European countries, with a substantial proportion – some 20% – from countries outside Europe. While ESERA conferences reflect research carried out in Europe, they are increasingly becoming international events attracting researchers from all over the world.

A total of 220 works were presented in guest lectures, symposia, poster workshops, individual papers and poster sessions that took place during the conference along with alternative activities and informal meetings. All these works are already published in the Proceedings of the Conference (edited by D. Psillos, P. Kariotoglou, V.Tselfes, G.Bisdikian, G.Fassoulopoulos, E. Hatzikraniotis, M.Kallery).

For the publication of this book, participants who had presented their work to the Conference were asked by an open call to submit papers based on their work to be considered for publication in the Conference Book. All submitted papers were assessed, according to standard peer review procedure, by two independent reviewers, and forty-seven of them were accepted for publication in the present book.

The book is divided into six chapters. The first chapter includes papers focusing on certain theoretical approaches, research overviews and research methodologies. The second and third chapters focus on the teaching and learning of science, which, as in the previous conferences, attracted the wider interest of the researchers. Specifically, the second chapter deals with research on students' reasoning and understanding of science as well as learning in various contexts, while the third chapter includes papers which focus on the teaching and communication of science in a variety of contexts. Given the importance of new technologies in the knowledge-based society, a separate chapter, the fourth, is devoted to studies that are related to the development, use and integration of Information and Communication Technologies in Science Education. The fifth chapter focuses on science teachers' knowledge and aspirations as well as on the linking of research to teaching practices, an issue that is of particular concern to the researchers. The sixth chapter contains presentations of completed or current international research projects in which several groups in different countries are collaborating, which is a promising practice for the future.

Certainly, any classification contains elements of arbitrary placement. We felt that the one adopted facilitates the reading of the book and the issues that are under study for improving science education research in the context of the knowledge-based society.

We are grateful to the following distinguished colleagues who contributed to the revision of the papers and, thus, to the publication of this book. Without their help and patience this book would not have been possible. We want to thank them warmly.

Borghi Lidia
Clement Pierre
De Jong Onno
Duit Reinders
Fischer Hans
Fischler Helmut
Galili Igal
Gil Perez Daniel
Grea Jean
Ioannides Christos
Izquierdo Merce
Jimenez Aleixandre Maria Pilar
Kalkanis Georgios
Leach John
Lijnse Piet
Martins Isabel
Meheut Martine
Mikropoulos Anastasios
Millar Robin
Niedderer Hans
Osborne Jonathan
Pinto Roser
Ravanis Kostantinos
Russel Terry
Sassi Elena
Schwedes Hannelore
Sjoberg Svein
Soerensen Helene
Solomon Joan
Solomonidou Christina
Spiliotopoulou Vassiliki
Stavridou Eleni
Stavy Ruth
Tiberghien Andree
Tsaparlis Georgios
Tunnicliffe Dale Sue
Tytler Russel
Valanides Nikos
Viennot Laurence
Von Auschnaiter Stefan
Welzel Manuela
White Richard
Zoller Uri

We would also like to thank the following Greek and international institutions and enterprises whose grants of funds contributed substantially to the organisation of the conference, the publication of the proceedings and the present book:

The Hellenic Ministry of Education and Religious Affairs. The Hellenic Ministry of Culture. The Hellenic Ministry of Macedonia & Thrace. The Municipality of Thessaloniki. The Didaskaleio "D. Glinos", Department of Primary Education, Aristotle University of Thessaloniki. Fourier Systems International S.A. The University of Macedonia. The Prefectural Authority of Thessaloniki. Amaxotechniki S.A. The Hellenic Centre of Education Research. The British Council. The Hellenic Republic Association of Local Authorities of Thessaloniki. BOUTARIS Group. Kluwer Academic Publishers. The Goulandris Natural History Museum/Greek Biotope/Wetland Centre. "GLOBAL" Bookshop.

We would also like to express our thanks to Kluwer Academics for their help in publishing and distributing this book.

Finally, special thanks are extended to Mrs Sophia Fourlari, who undertook the difficult task of communicating with colleagues and making the camera-ready text possible. Her patience and skills were of great assistance in the smooth development of the book

The Editors

PART I

Theoretical and Methodological Approaches to Science Education Research

PRESIDENTIAL ADDRESS

What Can We Reasonably Expect Of Research In Science Education?

Robin Millar

University of York, UK

ABSTRACT

I want to focus on some questions which have always been important ones for science education researchers to consider but which are, I think, growing in prominence and therefore in importance. They are about the role and purpose of science education research. What is science education research for? What do we hope to achieve through the research, which we do? What can we reasonably expect of science education research? These are issues which cross-national boundaries and concern us all, whatever national context we work within. So although I will make some reference to recent debates about educational research in the UK, the issues that are raised are international.

1. RESEARCH AND PRACTICE

What can we reasonably expect of science education research? First it might be helpful to clarify who 'we' in that question are. I want to consider the question both from the perspective of people outside the research community who look to research to provide them with useful information and guidance, such as teachers or policy-makers, and from the perspective of people like us, inside the research community. It may be important to ask: do both groups have the same view on what we can reasonably expect of research, or are there differences of emphasis?

In the UK in recent years, there has been a high-profile debate about the value of educational research. The lecture, which started it off, was given by Professor David Hargreaves in 1996. In it he explored the question of whether teaching can be said to be a research-based profession. Hargreaves argues that at present it is not:

> Given the huge amounts of educational research conducted over the past fifty years or more, there are few areas which have yielded a corpus of research evidence regarded as scientifically sound and as a worthwhile resource to guide professional action. (p. 2)

He argues for changes in the way educational research is planned and funded and for an end to:

> second-rate educational research which does not make a serious contribution to fundamental theory or knowledge; which is irrelevant to practice; which is uncoordinated with any preceding or follow-up research; and which clutters up academic journals that virtually nobody reads. (p. 7)

D. Psillos et al. (eds.), Science Education Research in the Knowledge-Based Society, 3–8.

Of course, teachers are not the only potential 'users' of educational research. Some research relates to issues of policy rather than to classroom practice. A review of educational research in 1998 for the UK Department for Education and Employment concluded that:

> the actions and decisions of policy-makers and practitioners are insufficiently informed by research, … where research does address policy-relevant and practical issues it tends to:
> - be small scale and fails to generate findings that are reliable and generalisable;
> - be insufficiently based on existing knowledge and therefore incapable of advancing understanding;
> - be presented in a form or medium which is largely inaccessible to a non-academic audience; and
> - lack interpretation for a policy-making or practitioner audience. (Hillage et al., 1998, p. xi)

By 2000, we have the Secretary of State for Education arguing, in a major speech, that

> We need to be able to rely on … social scientists to tell us what works and why and what types of policy initiatives are likely to be most effective. And we need better ways of ensuring that those who want this information can get it easily and quickly. (Blunkett, 2000, 2)

These extracts that I have quoted are a tiny part of a major debate about the role of educational research in the UK, about the kind of research that should be done, and about whether or not it is worth funding from public resources. From conversations I have had with colleagues, I think that similar arguments and debates are taking place in many countries. Similarly the mechanisms for funding educational research in the European Union emphasise research to inform policy-making and support specific social and economic policies.

A key phrase in the debate about educational research in the UK is 'evidence-based education', or 'evidence-based practice' in education. More recently some people have begun to use the rather softer phrase 'evidence-informed practice'. These terms are chosen to make a deliberate parallel with the idea of 'evidence-based medicine'. In clinical medicine it is now widely accepted that doctors' decisions about the treatment to recommend for a particular patient should be based on evidence that it leads to better outcomes (or is more likely to lead to better outcomes) than other treatments – and not simply on traditional, or established ways of proceeding. Although many established procedures and treatments are not evidence-based in this sense, it is unlikely that any new treatment would be introduced unless it was supported by convincing evidence that it works better than the current one.

So can these ideas be applied to education? In particular, can they be applied to science education? Is it possible for science education research to tell teachers or policy-makers 'what works'? Can it tell us that one approach is better than another for achieving a particular goal? Are these reasonable expectations of science education research?

Before I address those questions let me first make a number of observations. First, we should all welcome, I think, this evidence that policy-makers are interested in research. Most of us hope that our work will be noticed, and that it will be useful. There is, of course, a corresponding danger that it is the policy-makers who set the agenda and define the framework within which the research questions are identified. Research has a role to play in challenging the current assumptions and questioning current practices, and few of us would be happy to restrict its scope to the issues that policy-makers currently define as central.

Some have argued that the focus on research that tries to tell us 'what works' is too narrow. In one recent article, Jenkins (2001) even questions whether all science education research has to be about improving practice. He comments that the chapters in the *International Handbook of Science Education* (Fraser &Tobin, 1998)

> seem to assume that science education as a field of activity is exclusively concerned with the practice of teaching and learning, together with supporting activities such as assessment, evaluation, and teacher education. Correspondingly, research in science education is about improving practice. (Jenkins, 2001: 11)

Jenkins argues for a view of science education research that would include historical and sociological studies of scientific institutions and practices and implies that 'understanding for its own sake' may be sufficient justification. It has always seemed to me, however, that the best studies of this sort are precisely those which carefully choose their subject matter and present their arguments in such a way as to illuminate current practices. I think most of us would see science education research as applied; its rationale, and the justification for doing it, is to gain insights that can contribute in some way to the improvement of practice. Its purpose is 'critically to inform educational judgements and decisions in order to improve educational action' (Bassey, 1995:39). The issue – which brings me back to the questions I raised a few moments ago – is not whether research should contribute to improving practice, but how it can contribute.

Although terms like 'evidence-based practice' may be new, the idea behind them is not. The goal of a 'science of education' goes back to the 19th century. Underlying it is the view that there are clear and definite answers to be found. Some have questioned this. The Canadian science educator, Doug Roberts, for example, argues that:

> There can be no such thing as a science-like 'theory of science education', that is, an explanatory theory with predictive capability. The reason is simple. The events of science education are unique [and] non-replicable. (1980: 65-6)

Another educational researcher, Michael Bassey, makes the same point:

> Teaching situations are so varied that it rarely, if ever, possible to say with certainty 'Do y instead of x and your pupils will learn more'. (1999: 48)

But does this mean that there is nothing that we can say? Are we satisfied with that conclusion? I'd like to quote a short passage from a chapter which Piet Lijnse wrote for a book that I edited about two years ago, along with Jonathan Osborne and John Leach, called *Improving Science Education*. Lijnse writes:

> I still remember my disappointment when, as a newly appointed didactician, I had to develop an innovative series of lessons to introduce quantum mechanics at secondary school. I turned to theories of education and educational psychology for help. However, hardly any such help appeared to be available. (p. 309)

I don't think we can simply put this down to the fact that the science topic is a rather unusual one, quantum mechanics. The same point could be made about almost any science topic. Just consider, for a moment, the science topic in your subject area that has received most attention over the past 25 years from researchers interested in teaching and learning. For the physicists it would probably be electric circuits, perhaps for the chemists particle ideas, and for the biologists photosynthesis. Let me call it X. If a teacher were to approach you saying: 'I have to teach X to my class of 14 year olds next month. What should I do to teach it as effectively as possible?', what specific advice could you give that you could honestly claim is based on sound research evidence? There is, Lijnse concludes, very little research about 'how to teach X', where X is any specific science topic, that gives clear and reliable guidance for practice. I think he is right. Often the kind of advice we would give is, if we are honest, based more on experience and 'practitioner wisdom' than on specific research findings. Yet teaching X is what science teachers do every day. It is hardly surprising, then, that research often seems to have little impact on their practice.

So, to answer the questions I posed earlier, yes, I think it <u>is</u> reasonable to expect science education researchers to address questions of the form:

Can we show that approach *x* works better than approach *y* for achieving goal *z*?

It would not, of course, be reasonable to look for an approach, which <u>always</u> works better than another, regardless of the learners or the context of the teaching, or the qualities of the teacher involved. 'Works better' has to be interpreted in a probabilistic sense: likely to result in better outcomes in more situations for more learners more of the time. And tackling questions of this kind does not imply that there is a 'best approach'. We can look for ways to improve on current practice, without necessarily believing that there is a single, best way to do it.

Not only do I think it is reasonable to address such questions, I would like to see more research in science education addressing them - not only because it would help to bridge the gap between researchers and practitioners, but also because it would lead to better research on teaching and learning in science, in two main ways.

First, it would require us to be clearer about goals. Two methods or approaches can only be compared if they are alternative ways of reaching the same goals. Of course the goals cannot be determined by research, but depend on our values and preferences; what we think is important to teach about science. Research cannot tell us what we ought to do; but once we have decided what we want to do, research may be able to help us learn how to do it effectively.

In order to compare two approaches we need not only to be clear about goals but about how we can measure, or detect, whether or not the students have attained them. This means agreeing about what we will accept as evidence that the goals have been attained. We would be making progress as a research community, I think, if there was more agreement about the tools we use for this sort of thing, so that every new study did not have to develop its own tools – with the result that findings are very difficult to compare with anyone else's. It is, for instance, interesting to see how the publication by some researchers in the United States of the Force Concept Inventory (Hestenes et al., 1992), a test for assessing students' understanding of some key ideas in mechanics, has led to a large number of independent but related studies, to discussion and improvement of the instrument, and to several studies testing specific ways of teaching to boost students' performance on this measure. Similar inventories are beginning to appear in other science areas. The OECD PISA project (Harlen, 2001), for example, will provide an operational definition of 'scientific literacy'. It is unlikely that everyone will agree with it, but it will advance the discussion of the area by providing a tool, however imperfect. All such tools raise questions and issues, and do not solve all the problems, but they can also facilitate and encourage programmes of related studies and enable us to make progress.

Second, it would require us to think harder about the implications for practice of our work, and try to spell out these implications more thoroughly. Many science education research articles end with a section, often quite brief, that discusses the practical implications of the work that has been reported. Two very common characteristics of these implications sections are that:

- The recommendations for action often seem to be based more on experience and common sense practitioner knowledge than on the research findings themselves.
- The work reported is rarely seen as part of a longer-term programme, which will test, whether the actions proposed would actually result in improvement.

There are, of course, some exceptions. In her book *Reasoning in Physics*, which I am delighted to say, has just been published in English (Viennot, 2001), Laurence Viennot discusses students' reasoning about a wide range of physics topics, based on their answers to written probes and interview questions. For each topic, the possible implications for teaching are then discussed, and some suggestions are offered. But she is careful to make clear that these have a different status from the research findings themselves:

> the teaching suggestions presented here, when given in detail, are set in boxes or placed in appendices. This is not because the author believes that pedagogical action is less important than fundamental reflection, but to avoid a confusion of genres. ... some pedagogical suggestions, though largely founded on prior analysis, are wagers of sorts; it has not yet been determined to what extent they are valid. (p. 12)

The 'not yet' in that final sentence implies that this is work that remains to be done, and is necessary to do. Instead of the word 'wagers', we might say that these suggestions are 'hypotheses'. The value of hypotheses is that they make clear what is being tested. They also of course involve taking risks. Stating your initial hypothesis and then testing it is riskier than simply saying that you are exploring events in an open fashion. As researchers, we should be more willing to risk being wrong – because that is the only way in which we will make progress.

I have argued in this talk that we need more research which tries to show that one approach is better than another for teaching something that we want to teach. But I have also, I hope, made clear that I do not think that <u>all</u> research in science education should be of that sort. The role of research is not only to tell us 'what works'. Some of the most valuable research studies have been ones that made people aware of problems in current practices. Research can inform practice in a range of ways that stop short of providing clear and definite answers: by providing the kinds of insights that enable us to see the familiar in a new way, by sharpening thinking, by directing attention to important issues, by clarifying problems, challenging established views, encouraging debate and stimulating curiosity (Nisbet, 1974).

I look forward over the next four days to hearing, discussing and thinking about many interesting studies of all these kinds.

REFERENCES

Bassey, M. (1995). *Creating Education through Research.* Newark/Edinburgh: Kirklington Moor Press/BERA.

Bassey, M. (1999). *Case Study Research in Educational Settings.* Buckingham: Open University Press.

Blunkett, D. (2000). Influence of irrelevance: Can social science improve government? London: Department for Education and Employment. Reprinted in *Research Intelligence*, 71, March 2000, 12-21.

Fraser, B. & Tobin, K. (1998). *International Handbook of Science Education* (2 vols.) Dordrecht: Kluwer.

Hargreaves, D. (1996). Teaching as a research-based profession: Possibilities and prospects. Teacher Training Agency Annual Lecture 1996. London: Teacher Training Agency.

Harlen, W. (2001). The assessment of scientific literacy in the OECD/PISA project. *Studies in Science Education,* 26, 79-104.

Hestenes, D., Wells, M. & Swackhammer, G. (1992). Force Concept Inventory. *The Physics Teacher*, 30, 141-157.

Hillage, J., Pearson, R., Anderson, A. & Tamkin, P. (1998). *Excellence in Research on Schools.* Research Report RR74. London: Department for Education and Employment.

Jenkins, E. (2001). Science education as a field of research. *Canadian Journal of Science, Mathematics and Technology Education,* 1 (1), 9-21.

Lijnse, P. (2000). Didactics of science: The forgotten dimension in science education research? In: Millar, R., Leach, J. & Osborne, J. (eds.), *Improving Science Education. The Contribution of Research* (pp. 308-326). Buckingham, Open University Press.

Nisbet, J. (1974). Educational research: The state of the art. In: Dockrell, W. & Hamilton, D. (eds.), *Rethinking Educational Research* (pp. 1-10). London: Hodder and Stoughton.

Roberts, D. (1980). Theory, curriculum development, and the unique events of practice. In: Munby, H., Orpwood, G. & Russell, T. (eds.) *Seeing Curriculum in a New Light* (pp. 65-87). Lanham, NY: University Press of America.

Viennot, L. (2001). *Reasoning in Physics.* Dordrecht: Kluwer. (Previously published in French as: Viennot, L. (1996). *Raisonner en Physique.* Brussels; De Boeck.)

CONSTRUCTIVISM IN SCIENCE EDUCATION: THE NEED FOR A CLEAR LINE OF DEMARCATION

Daniel Gil-Pérez

Universitat de València, Spain

ABSTRACT

Some voices have recently begun to question the constructivist positions, which have been considered the most important contribution of the last decades in science education. It could be thought then, that the "constructivist consensus" might just be a new fashion that would once again lead us back to the immovable reception model of science learning. This questions, at the same time, the idea of an advance in the field of science education towards the construction of a coherent body of knowledge.

However, these criticisms are frequently focused on the contributions of authors, such as Glaserfeld or Kelly, none of whom have worked in science education. We believe this is misleading, because it denies the possibility for science education to give way to a specific body of knowledge and reduces its theoretical bases to the *application* of external knowledge. The goal of this work is to analyse some of the current criticisms to the constructivist orientations *in science education*, without entering into other philosophical or psychological debates.

1. INTRODUCTION

After an impressive development throughout the last two decades, supported by a great amount of research and innovation, science education seemed to be becoming a new scientific domain. We are speaking, of course, of a development that, as in any other scientific field, has not had a linear character and within which have arisen and still arise fruitful controversies and more or less profound re-orientations. But this development has also shown real progress and convergence -in spite of many terminological and punctual differences- in the orientation of the process of the teaching/learning of sciences. This convergence is supported by a great amount of research and innovation that can be consulted in the large number of existing journals, and which has already made possible the publication of two *Handbooks* (Gabel 1994; Fraser and Tobin 1998).

This transformation of Science Education into a specific field of research and knowledge is usually associated with the establishment of what has been called an 'emergent consensus' about constructivist positions, considered by the AAAS as a real 'paradigm change'. However, some voices have begun to question these constructivist positions in science education, speaking, for example, of 'Constructivism *Deconstructed*' (Suchting 1992) or of the 'Rise *and Fall* of Constructivism' (Solomon 1994). These very different appraisals make Jenkins

D. Psillos et al. (eds.), Science Education Research in the Knowledge-Based Society, 9–17.

(2000) ask: 'Constructivism in School Science Education: Powerful Model or the Most Dangerous Intellectual Tendency?'

The goal of this work is to analyse some of the criticism that is being voiced and to study its implications for the development of science education as a coherent body of knowledge.

2. WHAT CONSTRUCTIVISM ARE WE TALKING ABOUT?

This debate demands, first of all, in our opinion, a clear line of demarcation, a clear answer to the question: What Constructivism are we talking about? At this respect, in the Editorial of a monographic issue of *Science & Education* dedicated to constructivism, Matthews (2000) reminds us that 'constructivism means different things to different researchers' and dedicates a whole paragraph to describing the 'Varieties of Constructivism'. This ambiguity is denounced as one of the main inconveniences of the idea of a 'constructivist consensus'. But, strangely enough, it is not taken into consideration when trying to '*deconstruct constructivism'* (Suchting 1992) or when announcing the *'fall of constructivism'* (Solomon 1994). In other words, all of us need to be accurate and precise in this debate, because there is a real danger of talking about different things. This is the reason we ask for '*a clear line of demarcation'*.

Let's consider, in the first place, Suchting's criticism. In his article 'Constructivism deconstructed', Suchting (1992) starts saying that constructivism is '*a doctrine which has for some time been very influential in thinking about education (...) associated especially with the name of its originator and principal exponent, Ernst von Glaserfeld'*. Without discussing the undoubted interest of such criticisms as Suchting's of von Glaserfeld's philosophical theses, we wish to point out that this debate has little to do with constructivist proposals *in the field of science education*. Particularly, von Glaserfeld cannot be considered its originator because his influence has been very small in the emergence of the constructivist orientation in science education. Effectively, the first references to von Glaserfeld in journals such as *Science Education, Journal of Research in Science Teaching, Studies in Science Education or International Journal of Science Education...* appear as late as in 1988! The same appraisal of the scarce influence of von Glaserfeld in science education can be obtained considering the very few references included in the two handbooks published till now (Gabel 1994; Fraser & Tobin 1998). So, to speak of von Glaserfeld as 'the originator' is an example of a serious failure of some current criticism: they 'aim at' a different target outside of the field of science education.

We can thus conclude that the debate put forward by Suchting and other authors *is not our debate*. We do not mean by this that there is no interest in studying von Glaserfeld's work and his possible contributions to the controversy concerning constructivist proposals in the field of science education. But we cannot accept a discussion assuming, as Suchting seems to do, that we are talking about constructivism 'in general' and that we are 'applying' von Glaserfeld's theses.

What is known as the constructivist consensus *in science education* has a very different origin: its origin is in many specific researches about the different aspects of science education: from concept learning, problem solving or practical works to evaluation or attitudes towards science... These researches have been undertaken to improve the poor results of the reception learning strategies, seriously questioned by research on, for instance, 'alternative frameworks' (Pfundt & Duit 1998). They have contributed, and continue to contribute, to a coherent body of knowledge, which supports the need to *implicate pupils in the (re)construction of scientific knowledge* (instead of just transmitting it) in order to make possible a meaningful and lasting learning (National Research Council 1996).

This is the reason why many researchers began to speak, with different terminologies, of pupils' *construction* of knowledge (Driver & Easley 1978), or of pupils' *production* of knowledge (Furió y Gil 1978), or of pupils' *generation* of knowledge (Osborne & Wittrok 1983). To sum up, this is why many of us have begun to speak of constructivism *in science education.* It is not the result, we must stress, of an ascription to philosophical or psychological constructivism.

Solomon's critique (1994) has, undoubtedly, a different character: she admits that constructivist approaches in science education have their origin in research about problems related to the science teaching/learning process. In fact Solomon associates the emergence of this trend to the publishing of Driver and Easley's article (1978) 'Pupils & paradigms: a review of literature related to concept development in adolescent science students'. But then Solomon affirms that, in the early 80s, 'it was found that what we can call 'the book of the theory' had been written nearly thirty years earlier by George Kelly (1955) a psychologist who studied patients locked away in the solitary world of the schizophrenic'.

Notice that Solomon does not say that Kelly's work *supported* the new ideas, but rather that it *constituted their theoretical base.* In our opinion, some of Kelly's ideas (Kelly 1955) can be thought provoking and *help* in the construction of a science education body of knowledge, but it makes no sense to merely apply them to our field. However, Solomon's critiques to constructivist approaches are centred on contributions made by Kelly and other authors such as, again, von Glaserfeld, none of whom work in science education. In particular, Solomon focuses on showing the limitations of Kelly's metaphor 'Every man his own scientist' assuming that constructivism was based, essentially, on the notion of *'the student as a scientist* '. But the idea of the student as a scientist is a metaphor that has also been criticised by science education researchers *from a constructivist point of view*, because it fails to correctly express what research has shown about the science teaching/learning process: actually, it is difficult to oppose the view that pupils *by themselves* cannot construct *all* scientific knowledge. But we do not think of pupils as practising scientists working in frontier domains: this metaphor, used by several authors has, of course, many limitations (Burbules and Linn 1991) and cannot give a useful view of how to organise pupils' work. A metaphor that contemplates pupils as *novice researchers* gives a better appraisal of the learning situation as a (re)construction of knowledge. Effectively, every researcher knows that when someone joins a research team, he or she can catch up quite easily with the standard level of the team. And that does not happen by verbal transmission, but through the treatment of problems in fields where his or her more experienced colleagues are experts. This is to say, through the participation in *guided researches*.

The situation changes, of course, when problems, which are new for every member of the team, are treated. In this case, the progress -if there is any- becomes slow and sinuous. The proposal to organise pupils' learning as a knowledge construction corresponds to the first situation, that is to say, to an *oriented research*, in fields very well known by the 'research director' (the teacher), and where the partial and embryonic results obtained by pupils can be reinforced, completed or even questioned by those obtained by the 'scientific community'. What is known as a constructivist approach to science learning responds to the characteristics of oriented research, a research where results obtained by different teams are steadily compared and where teams count on the feedback and help of experts. As Hodson (1992) has stated, 'Students develop their conceptual understanding and learn more about scientific inquiry by engaging in scientific inquiry, provided that there is sufficient opportunity for and support of reflection'. This synthesises many researches on science learning and mustn't be considered, we insist, the simple 'application' of von Glaserfeld, Kelly or any other philosophical or psychological doctrine.

On the other hand, this debate on what we understand by a constructivist approach in science education may help us to clarify what is the epistemological orientation of this approach, avoiding its incorrect assimilation to the theses of von Glaserfeld's radical constructivism. A clearer definition in this respect is undoubtedly necessary. In particular, special attention must be paid to what the history and philosophy of science show about how scientific knowledge is built. In fact, a powerful research line concerning spontaneous teaching conceptions about science (Bell and Pearson 1992; Désauteles et al 1993; Guilbert and Meloche 1993...) has shown that the understanding and taking into consideration by teachers of how scientific knowledge is constructed, appears as a conditio sine qua non –although not sufficient (Hodson 1993)– for really effective science teaching.

3. WHAT IS THE EPISTEMOLOGICAL ORIENTATION OF OUR CONSTRUCTIVIST APPROACH TO SCIENCE EDUCATION?

As Bell and Pearson (1992), among others, have pointed out, it is not possible to change what teachers usually do in the classroom (simple transmission of knowledge already elaborated) without transforming their epistemology, their conceptions about how scientific knowledge is constructed, their views about science. Effectively, teachers' spontaneous epistemology include many distortions and reductionisms acquired acritically by social impregnation that impede a correct orientation of science learning as a construction of knowledge. It is not a question, we believe, of engaging science teachers –not even science education researchers- in the nuances and subtleties of the epistemology of different authors. Fortunately, in spite of their differences, there is a common base in how authors such as Popper, Kuhn, Toulmin, Lakatos, Feyerabend, Laudan, Giere...conceive the nature of science, and it is this common base which must be enhanced, in our opinion, in order to organise pupils' learning as a (re)construction of scientific knowledge.

This common base can be expressed by the general rejection of an ensemble of distortions of the nature of science (McComas 1998). This is not just a question of denouncing once again the well-known extreme inductivism of many science teachers' conceptions. We have to pay attention to an ensemble of distortions, which support each other, as, particularly:

Extreme inductivism, enhancing 'free' observation and experimentation ('not subject to aprioristic ideas') and forgetting the essential role played by hypotheses making and by the construction of coherent bodies of knowledge (theories).

A rigid view (algorithmic, exact, infallible... dogmatic). The 'Scientific Method' is presented as a linear sequence of stages to be followed step by step and scientific knowledge is presented in its 'final' state, without any reference either to the problematic situations which are at its origin, its historical evolution, the difficulties overcome... or to the limitations of this knowledge, which appears as an absolute truth, not subject to change. The rejection of this rigid and dogmatic vision sometimes leads to an extreme relativism, close to the radical constructivist philosophical theses. A relativism both methodological ('anything works', scientific activity hasn't specific strategies) and conceptual (there isn't an objective reality which allows us to verify the validity of scientific constructions: the *only* base of scientific knowledge is the consensus of researchers). It is this extreme relativism, which is denounced by authors such as Suchting when they criticise constructivism. We agree, of course, with this criticism, but this has very little to do with what we call constructivism in science education.

An exclusively analytical vision which enhances the necessary fragmentation and simplification of the studies, but neglects unification efforts in order to construct wider bodies of knowledge.

A merely accumulative vision. Scientific knowledge appears as the result of a linear development, ignoring crisis and deep restructurings.

An individualistic and elitist view. Science appears as the activity of isolated 'great scientists', ignoring the role of co-operative work and of interactions between different research teams. No special effort is made to make science meaningful and accessible; on the contrary, science is presented as a domain reserved for specially gifted minorities, transmitting poor expectations to most pupils and falling into ethnic, social and sexual discrimination.

A socially 'neutral' view. Science is presented as something elaborated in 'ivory towers', forgetting the complex STS relationships and the importance of collective decision making on societal issues related to science and technology.

This spontaneous epistemology constitutes a serious obstacle to the renewal of science teaching in as much as it is accepted acritically as 'evident'. However, extending the constructivist approach to teachers' education (Tiberghien, Jossem and Barojas 1998), research has shown that it is not at all difficult to generate a critical attitude towards these commonsense views if teachers are given the opportunity to discuss possible distortions to the nature of science, transmitted by science teaching; the real danger seems to be the lack of attention to what is acquired by impregnation, without conscious reflection. This reflection is absolutely necessary in order to overcome simplistic tendencies to accept 'what has always been done' or to look for a new and more successful 'recipe'. We connect here to another kind of criticism to the constructivist proposals in science education.

4. CONSTRUCTIVIST PROPOSALS ARE NOT A RECIPE.

We shall refer now to some excessively simplistic and stereotyped interpretations of constructivist proposals such as "let's consider the pupils' previous concepts, let's provoke cognitive conflicts in them, and let's introduce the correct concepts". In fact, criticism of these simplistic views cannot be considered as a questioning of constructivist positions in science education; on the contrary, there is abundant literature on the subject in the field of science education and it has brought about a deepening of these constructivist positions. However, we must not forget that these strategies which today appear to us as simplistic formulas, were not presented by their authors under such a schematic mode (Posner et al 1982) and did mean a remarkable advance over other formulae which were even more simplistic as, for instance, 'let's explain concepts clearly and students will learn', or those naive proposals of 'learning by discovery' departing from the pupils' autonomous experimentation. These proposals -which some associate erroneously to Piaget, despite the fact that this author strongly rejected 'the myth of the sensorial origin of scientific knowledge' (Piaget 1971)- were criticised justly and on good grounds by many authors.

The greater efficiency of conceptual change strategies with respect to those of simple transmission of elaborated knowledge has been supported by much research carried out in different scientific fields. But it was soon noticed that certain alternative conceptions were resistant to instruction, even when this instruction was explicitly oriented to producing 'conceptual change' (Pfundt and Duit, 1998). To put it in other words: it became evident that the undeniable *progress* achieved by conceptual change strategies was still *insufficient*. From this situation arose an awareness about the need, among other things, to consider the *students' ways of reasoning*, overcoming conceptual reductionism (Gil-Pérez and Carrascosa 1985; Duschl and Gitomer 1991; Gil-Pérez 1996) and enriching conceptual change proposals.

There is still another aspect of conceptual change strategies which demands, in our opinion, re-examination: What is the sense of making pupils conscious of their ideas to immediately put them into conflict? In our opinion, the systematic confrontation of pupils' ideas with scientific ones can produce logical inhibitions. Effectively, it is easy to understand that a research is not undertaken to question ideas, or to produce conceptual changes, but to treat problems which interest scientists; problems which are treated, logically, with the possessed knowledge and with new ideas constructed in a tentative way. During this process the initial conceptions may suffer some changes or even be radically questioned, but this will never be the objective, which remains the solution of problems posed.

From a scientific point of view it is essential to associate knowledge construction with problems: as Bachelard (1938) stressed 'all knowledge is the answer to a question' and this is in radical conflict to conceptual change teaching strategies which take pupils' conceptions as a starting point. Furthermore, a basic scientific attitude in the treatment of problems is to take one's own ideas -even those 'most obvious'- as simple hypotheses that are necessary to question, conceiving other hypotheses. This can give another status to cognitive conflicts: they will not be seen as an external questioning of (incorrect) personal conceptions and reasoning (with its consequent affective implications) but just as a confrontation between different hypotheses.

For these reasons, the teaching strategy that seems to many of us more consistent with the characteristics of a scientific construction of knowledge, is to *organise learning as a treatment of problematic situations that pupils can identify as worth thinking about*. This strategy aims basically to involve pupils in the construction of knowledge, *approaching* pupils' activity to the richness of a scientific treatment of problems, including, among others:

- The discussion (a real discussion) of the possible interest and worthiness of the situations proposed, which gives meaning to their study and prevents students from becoming immersed in the treatment of a situation without having had the opportunity to form a first motivating idea about it.
- The qualitative study of problematic situations, taking decisions -with the help of the necessary bibliographic researches- to define and delimit concrete problems (an activity during which pupils begin to make their ideas explicit in a *functional way*). If we want pupils to really understand what they are doing, it is essential to begin with qualitative and meaningful approaches, as any scientist does.
- The invention of concepts and forming of hypotheses (occasion for using alternative conceptions to make predictions, i.e., using personal ideas in a functional way, not just answering questionnaires).
- The elaboration of possible strategies for solving problems, including, where appropriate, experimental designs to check hypotheses in the light of theory.
- The implementation of the strategies elaborated, and the analysis of the results - checking them with those obtained by other pupils and by the scientific community- that can produce cognitive conflicts between different conceptions (taking all of them as hypotheses), and requiring the formation of new hypotheses and the reorientation of the research.
- The application of the new knowledge in a variety of situations to deepen and consolidate it, putting special emphasis on the S/T/S relationships which frame scientific development, and orienting all this work to show the nature of a coherent body of knowledge of any scientific domain.
- The conception of new problems.

It is necessary to highlight that the orientations above –which are based on many researches in science education- *do not constitute an algorithm* that tries to guide the pupils' activity step by step, but rather they must be taken as general indications which draw attention to essential aspects concerning the construction of scientific knowledge not sufficiently taken into account in science education.

When we speak about constructivism in science education we are referring to these kinds of proposals, which are the result, we insist, of many convergent researches on pupils' alternative frameworks, problem solving, attitudes towards science learning, assessment..., and very particularly, on concrete scientific domains (Viennot 1996).

5. PERSPECTIVES

To conclude this analysis of current criticisms about constructivist approaches in science education, we would like to draw attention to the dangers of superficial readings and *vagueness* in the use of the term constructivism (Matthews 2000). A vagueness that, we add, makes it possible to qualify as 'constructivist' what one has always done ('I explain concepts and my pupils reconstruct them in their head'). This reading has, undoubtedly, its adepts: constructivism would be but an interpretation for learning and would have nothing to say about teaching. (To put it in other words: let's leave things as they are).

It is perhaps this vagueness (this grouping, under the constructivist umbrella, of simplistic recipes, of philosophic discussions far away from the precise science teaching/learning problems, of light interpretations allowing that anybody, whatever he does, can call himself 'constructivist' ...) which leads such authors as Osborne (1996) to speak of 'Beyond Constructivism'. In our opinion, however, the expression 'constructivist consensus' (Resnick 1983) is still useful *–if a clear line of demarcation is established with other meanings of 'constructivism'-* to highlight the basic convergence, within the field of science education, of many proposals that contemplate active participation of students in the (re)construction of knowledge (National Research Council's 1996). This convergence supports the idea of advance –not free of controversy as in any scientific field– towards the construction of a new science teaching/learning model able to displace that of the simple transmission/reception of previously elaborated knowledge and, finally, to advance towards the establishment of science education as a new field of knowledge. We must add that it will not be a simple task. Research has clearly shown that many teachers present serious resistance to adopting 'constructivist' positions –this is to say, to organising science learning as the (re)construction of scientific knowledge through an oriented research- and often introduce serious distortions. This sends us back to the problem of teachers' education and the need to involve them in the (re)construction of a science education body of knowledge (Tiberghien, Jossem and Barojas 1998). It is in this sense of implicating pupils *and teachers* in the construction of knowledge –overcoming the ineffective transmission/reception of this knowledge- that we, and many others, speak of constructivism in science education.

REFERENCES

Bachelard, G.,(1938). *La Formation de l'Esprit Scientifique.* Vrin, Paris.

Bell, B.F & Pearson, J.,(1992). 'Better Learning'. *International Journal of Science Education*, 14(3), 349-361.

Burbules, N & Linn, M. (1991). Science Education and Philosophy of Science: Congruence or Contradiction? *International Journal of Science Education*, 13 (3), 227-241.

Désautels, J., Larochelle, M., Gagné, B. & Ruel, F. (1993). La Formation À l'Enseignement des Sciences: le Virage Épistémologique, *Didaskalia*, 1, 49-67

Driver, R. & Easley, J. (1978). Pupils and Paradigms: A Review of Literature Related to Concept Development in Adolescent Science Students. *Studies in Science Education*, 10, 37-70.

Duschl, R. & Gitomer, D. (1991). Epistemological Perspectives on Conceptual Change: Implications for Educational Practice. *Journal of Research in Science Teaching*, 28 (9), 839- 858.

Fraser, B. & Tobin, K.G. (eds.), (1998). *International Handbook of Science Education*. Kluwer Academic Publishers, London.

Furió, C., & Gil, D,.(1978). *El Programa-Guía: Una Propuesta Para la Renovación de la Didáctica de la Física y Química.* Ice de la Universidad de Valencia.

Gabel, D.L. (ed.) (1994). *Handbook of Research on Science Teaching and Learning.* MacMillan Pub. Co. N.Y.

Gil-Pérez, D. (1996). New Trends in Science Education, *International Journal of Science Education*, 18 (8), 889-901.

Gil-Pérez, D. & Carrascosa, J,.(1985). Science Learning as a Conceptual and Methodological Change, *European Journal of Science Education*, 7 (3), 231-236.

Guilbert, L. & Meloche, D. (1993). L'idée de Science Chez des Enseignants en Formation: Un Lien Entre L'histoire des Sciences et l'Hétérogénéité des Visions. *Didaskalia*, 2, 7-30.

Hodson, D.(1992). In Search of a Meaningful Relationship: An Exploration of Some Issues Relating to Integration in Science and Science Education. *International Journal Of Science Education*, 14(5), 541-566.

Hodson, D. (1993). Philosophic Stance of Secondary School Science Teachers, Curriculum Experiences, and Children's Understanding of Science: Some Preliminary Findings. *Interchange*, 24(1&2), 41-52.

Jenkins, E.W. (2000). Constructivism in School Science Education: Powerful Model or the Most Dangerous Intellectual Tendency? *Science & Education*, 9, 599-610.

Kelly, G. (1955*). The Psychology of Personal Constructs.* Norton, New York.

Matthews, M.R. (2000). 'Editorial' of the Monographic Issue on Constructivism, Epistemology and the Learning of Science. *Science & Education*, 9, 491-505.

McComas, W.F. (ed.), (1998). *The Nature of Science in Science Education. Rationales and Strategies.* Kluwer Academic Publishers, Dordrecht.

National Research Council, (1996). *National Science Education Standards.* National Academy Press, Washington, Dc.

Osborne, J.F. (1996). Beyond Constructivism. *Science Education*, 80(1), 53-82.

Osborne, R. & Wittrock, M. (1983). Learning Science: A Generative Process. *Science Education*, 67, 490-508.

Pfundt, H. & Duit, R. (1998). *Bibliography: Students' Alternative Frameworks and Science Education.* Institute for Science Education, Kiel.

Piaget, J. (1971). *Psicología y Epistemología.* Ariel, Barcelona.

Posner, G.J., Strike, Hewson & Gertzog (1982). Accommodation of a Scientific Conception: Towards a Theory of Conceptual Change. *Science Education,* 66(2), 211-227.

Resnick, L.B. (1983). Mathematics and Science Learning: A New Conception, *Science*, 220, 477- 478.

Solomon, J. (1994). The Rise and Fall of Constructivism. *Studies in Science Education,* 23, 1-19.

Suchting, W.A. (1992). Constructivism Deconstructed, *Science & Education*, 1(3), 223-254.

Tiberghien, A., Jossem, E. & Barojas, J. (1998). *Connecting Research in Physics Education with Teacher Education.* I.C.P.E Book, International Commission on Physics Education.

Viennot, L. (1996). *Raisonner en Physique. La Part du Sens Commun*, De Boeck & Larcier S.A., Paris & Bruxelles.

OVERVIEWS OF THE RESEARCH PRESENTED AT ESERA 2001

[1]Richard White, [2]Hans E. Fischer, [3]Eleni Stavridou

[1]University of Monash, Australia, [2]University of Dortmund, Germany, [3]University of Thessaly, Department of Primary Education, Greece

ABSTRACT

In this chapter we reflect on the trends in research in science education that we saw in the papers delivered at the 2001 meeting of the European Science Education Research Association. Each reflection has a particular focus. Richard White compared the ESERA papers with the trends that he identified in his analysis of English-language research from 1975 to 1995, Hans E. Fischer commented on the papers in the light of developments in science education in Europe, and Eleni Stavridou summarised the papers on chemical education and on Information and Communication Technologies.

1. SCOPE OF THE CHAPTER

The 2001 ESERA Conference brought together scholars from six continents and many countries. The papers that they presented showed that research in science education is lively and diverse. What, though, of the future? Directions that research might take may be found from searching out trends within the diversity and identifying topics and methods that might merit attention and maintain the liveliness.

2. ESERA AND THE INTERNATIONAL ENGLISH-LANGUAGE RESEARCH; BY RICHARD WHITE

Although it was not published until October 2001, I wrote the manuscript on science education for the fourth edition of the Handbook of Research on Teaching in 1996. A part of the manuscript is an analysis of articles in the 1975, 1985, and 1995 issues of the American journals Journal of Research in Science Teaching and Science Education and the Australian journal Research in Science Education, and the 1985 issue of the European Journal of Science Education and the1995 issue of the International Journal of Science Education (the European Journal was first published in 1979, and became the International Journal in 1987). These snapshots of research at ten-year intervals revealed changes of sufficient scale to warrant the title that I gave the manuscript: 'The Revolution in Research in Science Education.'

D. Psillos et al. (eds.), Science Education Research in the Knowledge-Based Society, 19–30.

2.1 Types of Research, Methods of Data Collection, and Use of Statistics

The revolution in style of research is evident in a rise in popularity of descriptive studies, which were few in number in 1975 but had become the commonest form by 1995: they rose from a mere 9 of 137 articles in 1975 to 62 of 179 in 1995. Descriptive studies include classroom observations and probes of conceptions, in which there is no attempt, as there is in experimental studies, to measure the effect a change in an independent variable has on a dependent one, or none to measure the correlation between two variables.

Experiments and correlation studies continue to be done, but their fraction of the whole has declined. Also there has been a change in the nature of experimental studies. The more recent ones tend to involve longer interventions than those of the 1970s, to pay more attention to ecological validity, and to use a bigger range of dependent variables and more diverse methods of collecting data.

The much greater use of interviews and classroom observations is only partly due to the surge in descriptive studies. Many experiments and correlational studies now employ these techniques also. As a consequence, the number of articles that reported interview data soared from 3 in 1975 to 59 in 1995, and observations from 15 to 44.

The revolution in style also appears in the use of statistics. In 1975 it was either close to impossible to have published in JRST, Science Education, or RISE an article that did not include statistics - and inferential statistics such as analysis of variance or covariance or chi-square or t-test at that - or hardly anyone thought of submitting one. By 1995 many empirical studies reported no statistical analyses or summaries at all. Much of this shift is of course associated with the rise in descriptive studies, but there are even some experiments that involved no statistics.

The empirical studies reported at the 2001 ESERA meeting are consistent with the above trends. The conference saw many examples of descriptive studies; many of experiments, correlational studies, and descriptions that employed interviews, classroom observations, student artefacts, and other innovative measures; and many that relied on descriptive statistics or no statistics at all. There is no marked difference between the research reported at ESERA and the trends that appeared in my analysis of the four major English-language journals of research in science education.

It is important to think about these trends. How desirable are they? Should the revolution end, and the present styles become a stable, new orthodoxy? Or should we search for further improvements? My own view is that the revolution in research demonstrates existence of a lively, thriving, and thoughtful community of scholars, who are actively seeking to illuminate the complexity of learning and teaching of science and to build up a coherent body of theory plus principles and methods of effective practice. The end of the revolution would mark the death of that community. Fortunately, among the ESERA papers there are signs of further developments in the subtlety and power of insight of research methods. One such sign is the synthesis of the rigour of experimental design and statistical power with the richness of insight of descriptive studies, evident in research reported at the Conference that employed sophisticated statistical analyses of observational or interview data. A second sign is the number of longitudinal studies. Longitudinal studies have not previously been common, and it is a feature of the ESERA conference that so many were reported. These papers showed clearly the differences in insights that longitudinal studies are capable of providing from those from cohort studies.

2.2 The People Involved in Research

There are the people who do the research, and the people who are the subjects of it. In the four English-language journals that I analysed, in 1975 only 14% of authors were women. This had changed to 40% by 1995. Although women are now responsible for a greater proportion of research, equality is still some way off. My counts of audiences at plenary sessions of ESERA gave me the same figure of 40 percent women. This contrasted with the audiences at the conference of the European Association for Learning and Instruction, which took place in the week after ESERA, where more women than men were present.

The ESERA papers followed the trend that I identified in my analysis of journals for teachers to be increasingly often the subject of research, in which measures are made of their understandings of content and their attitudes and beliefs about the goals of education and for the learning of science

Another similarity of the Conference papers to the journals that I analysed is in the rarity of research on pre-school children. Of course research with very young children brings particular difficulties and requires special skills. Either these difficulties or the lack of the necessary skills have inhibited research in pre-schools. This is regrettable since research on alternative conceptions shows that children already form beliefs about the world before they reach elementary school. We need to know more about the foundations of these beliefs. I did not observe any studies at the conference in which pre-school children were the subjects, though there were several in which pre-school teachers or teachers-in-training were the subjects.

2.3 Content and Context

I am increasingly aware of the importance of content and of context as variables in education (see Fensham, Gunstone, & White, 1994, for a discussion of content in science). I have not done the analysis, but it would be interesting to see how many of the ESERA papers treated content as a variable, or subjected a piece of content to intense scrutiny. My impression, which could be mistaken, is that there were only one or two such papers.

In their attention to situated cognition, mathematics educators have shown more interest in context than have science educators. Situated cognition is only one aspect of context. Another is national differences. ESERA is particularly well-placed to promote comparisons of education from different countries and of different language groups. There is much of interest to be found in the way language affects conceptions of the world. It is encouraging that reports at the Conference included subtle studies that complement the mass surveys that IEA conducts.

National comparisons would go some way towards meeting my concern that despite ease of travel and communication educational research retains a parochial nature (White, 1995). Few of us (and I am as guilty as any) cite research in a language other than our own or English. We are geographically parochial also. It would be salutary to see how many of the chapters in this volume, other than those with Australian or New Zealand authors, have in their reference lists articles from the Australasian journal Research in Science Education, or whether any Asian journals are cited by people other than the scholars from Asia. Increased citing of research from countries other than our own or America, which of course is too prolific to ignore, would be welcome.

3. THE ESERA CONFERENCE AND THE DEVELOPMENT OF SCIENCE EDUCATION IN EUROPE; BY HANS E. FISCHER

Before I go into the details of this conference, I present a brief overview of the development in Europe of Science Education or Didactic of Science.

As politicians came to perceive that science and technology are important for developing a modern society, they encouraged the establishment of science education as a university subject. Countries that moved fastest in this now have their third or fourth generation of researchers in science education, while others still have their first.

Initially our discipline was closely linked with the science subjects themselves, and, accordingly, was oriented to design new classroom material, to reconstruct new content to use at schools and to develop curricula more reflective than before. Teaching methodology, teaching and learning processes, the effectiveness of the designed instruction or students' assessment and performance were mostly out of the debate. As a consequence science education did not develop as an independent discipline. It was dependent on the subjects and mostly former scientists were appointed to the chairs. As a result, research appropriate to the needs of the societies could not readily be established. Therefore, in Germany in the eighties many of those chairs were integrated into the subjects and were lost to science education. This development, in Germany at least, produced a disastrous failure to meet the needs of science teaching at schools and science education at universities. Nor has there been sufficient research in science education to help politicians to decide in a responsible way the design of the future educational system of their countries.

The deficits we experience now are not only related to developments in science education but also those of empirical pedagogy and psychology. Till the end of the eighties colleagues in those areas mostly investigated general questions of education and society, and school was only of interest because it was seen as a more or less closed system, appropriate for the study of social or educational effects. There was nearly no productive contact between science education and empirical pedagogy or psychology in Europe till 1990. We even did not understand each other.

A change occurred at the beginning of the nineties in Germany, France and other countries, similar to earlier trends in the US and UK. Science education, which I use in the sense of didactic of science, was discovered as science of knowledge transfer, learning, teaching, quality of education related to a subject and as problem of instructional design. In a meta-analysis of 1990, Wang, Haertel and Walberg (1990) identified 228 variables related to learning organized into 30 a priori scales within six broad categories of (1) State and District Variables, (2) Out-of-school Contextual Variables, (3) School-Level Variables, (4) Student Variables, (5) Program Design Variables and (6) Implementation, Classroom Instruction, and Climate Variables. In accord with this analysis, and drawing on Glaser & Bassok (1989), Snow & Swanson (1992) and the German National Council of Education (1971), the topics of didactic of science to describe and develop the quality of instruction can be characterised as follows:

1. The description of intended debit states and objectives related to the subject (analysis of competencies, definition of the teaching aims, content analysis, curriculum development).
2. The description of the actual condition of individuals relevant to the aims (analysis of the learning conditions).
3. The explication of the processes of transfer from the actual to the debit state (analysis of the learning processes).
4. The design of conditions to foster learning processes (instruction-design).
5. The description and the development of behavioural alternatives of teachers and students and their cognitive and motivational base (regulation of behaviour).
6. The development of procedures to test the success of learning processes or the effectiveness of learning environments (measuring and evaluation orders or systems).
7. The development of the school contents and materials according to the subject and results and principles of research in pedagogy, psychology and science education (didactical reconstruction).
8. The specification and evaluation of implemented learning environments (implementation and instruction-design).

The above categories of research can only be developed by a discipline related to either the subject or a larger domain of subjects and empirical pedagogy and psychology. The methodology has to be adapted from pedagogy and psychology and has to consider features of the subject matter. Therefore its character must be interdisciplinary and concentrated to solve domain-specific problems of educational quality without losing sight of potential generalization. It must rest on empirical evidence. It is fundamental to make statements about empirical evidence of the results of a research project to elaborate questions and aims. Therefore, our research must be more than a critical self-reflection. Compared to self-reflection we can solve our problems only on a higher level of abstraction to formulate research questions and on a high level of interdisciplinary linkage to find and develop adequate research methods.

Following these general remarks, I turn to the research projects presented at this conference.

Looking at the names in the two books containing the proposals, it seems that all relevant research work in the field of science education in Europe is represented. For this article the proposals were examined to characterize the theory and the methods

used (qualitative or quantitative) and to classify the projects roughly regarding quality and link between theory and methodology.
Related to the research question we can characterize the different approaches of this conference between descriptive-explanative and prescriptive-intervening.

3.1 Methods

We found a variety of different research methods, mostly standard methods like formal tests, questionnaires and interpretations on the basis of grounded theory or a system of categories. Sometimes qualitative and quantitative types of methods were used in parallel, without showing how to combine the different types of data and results. We found triangulation explicitly mentioned only in 4 cases of about 350. An often-mentioned idea about the advantage of mixing qualitative and quantitative methods can be seen in the following quote:

> "Using qualitative and quantitative complementary methods would lead me to a stronger position to be able to comment on the issues I set out to explore."

Confusion about the function of theory and methodology and a not very deep going understanding of an empirical analysis occurs in many projects. Quote:

> "All documents and student work was collected and analyzed in relation to the teaching schedule."

The qualitative style of empirical methods is seen in many studies as intuitive and not empirical and "case-study" often is nominated as a method of investigation and not as a type of qualitative studies which might be performed by different methods. Quote:

> "Our method is a case study. It consists in collecting data all along the teaching sequence."

Sometimes the meaning of a research method was not at all like it is seen in empirical research. Quote:

> "The participants' narratives were recorded in Word 2000, so as to be transcribed in a later stage, to ASCII and be stored in AQUAD five. After having interpreted, categorized and codified, we have inferred the main semantic nucleus."

Many of the case studies did not mention the problem of generalization. To the contrary, one or more general conclusions were drawn as results of most of the cases and not the development of hypotheses for further studies or a restriction to the case, as should be. Quote:

> "To investigate how students deal with familiar concepts in novel contexts a group of four students was given a problem requiring an understanding of the concepts of velocity, acceleration and force."

Some of the presented studies (about 30) can be characterized (and mostly done by the authors) as critical self-refection. Quote: "My aim was to increase my understanding of this teaching methodology in action".

3.2 Theory, Aims and Methodology

In many cases a theoretical model of the investigation was present. In most of these studies "constructivism", "communication", "interaction", "qualitative research" and so on represent a not very elaborated approach to answer a not very elaborated question of research which might be discovered as interesting in some cases. Quote:

> "A qualitative research was designed to deal with the interpretative approach of the study."

As a consequence of this poorly defined theoretical attempt there was often no link to be seen between theory and methodology. For example a constructivist background is explicitly mentioned but the study is based on a pre- post test without referring to learning processes or activities or it is only comprised of phenomenography on students' or teachers' misconceptions.

Many proposals did not show how the methods have been applied. For example "to analyze by video" does not say if grounded theory or a category guided coding scheme was used. In few cases even SPSS was mentioned as an empirical method to analyze the scales of a questionnaire.

Another crucial point of research is the formulation of the aims. Often methods were indicated but the aims and results did not refer to them. Accordingly, the research questions are not very elaborated. In many projects so many questions and aims are outlined that it might be doubted that any of them could be reached. On the other side, in very few projects, no research aim and no method at all were described. Moreover, in a lot of cases the method and the aim did not fit. To observe about 500 pupils and about 20 teachers by video, classroom observation and teachers' reports to find, quote: "The variety of representations of ideas to gravity", "the ways in which teachers succeed in helping children to become aware of their own representations", "the changes in children's representations" and to analyze "the impact of all these practices on the efficacy of learning outcomes or movement towards conventional science understandings of gravity" seems to be an ambitious challenge for two researchers using the above mentioned methods in a period of three years.

In only few cases of qualitative analysis was the inter-rater or cross-rater reliability mentioned, or in quantitative studies something like an alpha-value. Often the method was only vaguely described Quote: "the outcomes have been collated from experiences" or "the results are drawn from a detailed analysis of the videos".

Some of the presented studies were already some years old (before 1996) or did not refer to actual and relevant publications. Regarding the references it is striking that a large group of proposals mainly cites the author's own publications, which might be relevant for the project but does not give a positive impression about the authors' reception of ongoing research.

3.3 Résumé

To classify the quality of the proposals we had only a weak instrument and it was not validated at all, so the next statements are intended as personal comments that aim to initiate a discussion.

In my opinion, about 50% of the proposals provide a project of a high standard, suitable for an international conference like this. A remarkable amount of studies, for example most of the presented projects analyzing textbooks, can be classified as topics for master theses, while in some cases even this level was not reached.

On the other side about 20 of the proposed projects are funded either by national science foundations, ministries of education or the European Union, which might be an indicator of quality. Mainly the national science foundations in Europe and US are a kind of guarantee for a high standard of research. It occurred to me that some of these projects might have been downgraded or even rejected as not being research.

Thinking about the next conference I see me reading through three books of about 1300 pages, but with no increase of quality compared to this year. In my opinion we should use the two tools we have to foster the quality of research. There are the summer schools to present projects on PhD level even in early stages, which provide an opportunity to work on all the problems I have mentioned above. We should use the conferences as the second tool in between to present European research in science education on the highest possible level, which is easy to do because we have a lot of superb projects (about 150 in this conference) to present. Criteria for the quality of empirical research in science education that correspond with the needs of our European societies are already developed. They should meet the known standards of empirical pedagogy on the level of the European national science foundations.

4. RESEARCH ON CHEMICAL EDUCATION AND ON INFORMATION AND COMMUNICATION TECHNOLOGIES; BY ELENI STAVRIDOU

4.1 Chemical Education

During the Conference only 18 of the presented papers referred to Chemical Education. The majority of them concerned investigations of primary or secondary students' conceptions about chemical concepts, such as substance, chemical reaction, physical phenomena and chemical phenomena, transformations of matter, evaporation, freezing point, acids and bases, acid rain. In a few papers the use of particulate models of matter, molecules, etc. in explaining chemical phenomena and relating macroscopic and microscopic representations was also discussed. It is worth mentioning that almost all the papers referred to a small number of basic chemical concepts that appeared many times in the chemical education literature the last two decades.

In most cases, classic and well-established research methodologies were used, such as interviews and questionnaires, before, after, or during the teaching of chemistry concepts. In a few other papers, the use of relatively new methods, such as analysis of students' productions, of audio- or videotaped students' discussions, etc., seemed to be quite uncertain without a clear presentation of the results.

Among the 18 papers on Chemical Education, five were longitudinal studies lasting more than 2 or 3 years. The importance and interest of longitudinal studies are well known, as they offer the opportunity to have a clear and detailed view of the evolution of students' conceptions into specific learning or constructivist environments. These studies showed that children developing understandings are fundamentally situated both with respect to the phenomena and also their personal experiences. Depending on the phenomenon, some of the children use the molecule as a tool for understanding, while others do not.

Another four papers also reported research findings about students' conceptions. In three, the students were exposed to traditional teaching methods and in one case they were taught in a constructivist learning environment favoring collaboration and communication among students. In this constructivist study, interviews or questionnaires were used to detect students' conceptions, and to compare experimental and control group data. In the constructivist learning environment students' conceptions and understanding improved substantially, but the traditional teaching environment resulted in poor understanding, and the students' alternative conceptions did not change.

Three studies dealt with students' chemical problem solving. Data came from several sources: plans of operation, class observations, reports and group interviews, analysis of students' productions, conversations and interactions with the teacher, which were audio- and video-recorded.

Four papers dealt with university students' conceptions and difficulties. Various research methods were used, such as questionnaires, interviews, analysis of students' productions, audio- and video-recorded interactions. Two papers reported chemistry students' difficulties about the interpretation of graphs and the writing of organic reaction mechanisms. The other 2 papers concerned pre-service and in-service science teachers' learning difficulties.

4.2 ICT and the Conference

The label 'knowledge society' is a quite recent concept describing a new dimension of the structure of our post-industrial societies. For the first time in the history of humanity, every kind of knowledge and its sources can be close to every single person. But for the learner the old problem remains the same: how it is possible to catch other people's knowledge, to assimilate it, to use it, to learn from it. While science education researchers and teachers face the same old questions that the educators of the past, now the answer seems to be quite different as they more and more realize that the students and future citizens should develop new skills and appropriate competencies. They should be able to collaborate and communicate, to gather information and to select, arrange, manage and use it, and to learn in a lifelong process. These aims seem to be a new challenge for every educational system and shape to a new and very interesting research area.

According the overview of the papers presented in the Conference, 38 of them referred to the use of various types of modern technology in science education. More particularly, those papers can be classified into 4 different categories:

a. 10 papers reported Internet or Web-based projects,
b. 10 papers referred to the design and/or use of educational software,
c. 15 papers analyzed the use of educational software in labwork,
d. 3 papers were on other research areas.

a. Internet or Web-based projects

The design of new learning environments is a new challenge for every educational system. Over the world, researchers are constructing new and better approaches for the use of different kinds of modern technology to improve science teaching. In such new environments new teaching-learning materials are needed, both for the teachers and the students. Their design offers a considerable opportunity to put educational theory into practice and to evaluate their effectiveness. Designing technology enhanced science instruction for knowledge based society also means using the Internet to promote lifelong learning.

In this category, five papers described projects aiming at the implementation of ICT at national or local level, as in Norway, USA (WISE), Italy, Finland, and Scotland. In all these projects the main aims were the design and development of new learning materials and the improvement of science teachers' training (pre- and in-service).

Different uses of the Internet involving distance and lifelong learning are reported in 2 other papers:

- the Computer-Supported Educational Network for Collaborative Professional Development,
- the Web-based Multimedia System for a University Level Zoology Course.

As the majority of the above projects are quite recent, in most cases there is not a research methodology and the authors presented descriptions of the projects or a few qualitative results.

Three other papers discussed the problem of ICT use in schools. The innovative use of computers in the classroom leads to important changes to the traditional roles of all the partners involved in the teaching and learning process. Pupils should be active and responsible partners in the construction of their own knowledge, working either individually or in small groups, while teachers should be conceivers and designers of pupils' learning activities, and pupils' guides and assistants in the learning process.

It is evident that the simple presence of computers in the classroom could not result in such radical changes concerning teachers' and pupils' roles. Teachers' prior practices and routines influence changes teachers make in their classroom to accommodate technology. In fact, teachers tend to modify the technology to fit their teaching styles rather than modify their teaching style. Educational reformers and researchers face the problem of how to train teachers to be effective when using ICT in their classrooms.

According to these 3 papers the teachers are not prepared to use ICT in their classroom and their needs for support, guidance and training are important.

b. Design and/or use of educational software

The design and development of quality educational software with clear pedagogical or didactical ideas is another big challenge for science education researchers. Seven of the presented papers are research reports in which multimedia applications and other educational software and material are specifically designed and developed on the basis of students' or pupils' previous ideas, conceptions, conceptual difficulties and needs, in order to help them construct scientific knowledge. For example, constructive software was developed for the computer-aided teaching and learning of sound, geometrical optics, light and colors, holography, mechanical interactions and Newton's 3rd law, the erosion of the Earth's surface.

In 3 other papers, the main question was whether or not the representation of a traditional model of teaching has changed and whether new models of teaching have been build in the context on ICT. The real problems raised by teachers was how to integrate and generalize the use of multimedia in classroom practice, and how to manage the transition from a rather traditional pedagogy to a teaching based on a project pedagogy. Teachers who have taught with multimedia have developed a new image of their job, a new relation to knowledge and a new way of teaching with fewer teachers' interventions and more responsibility on the students' part.

The papers of this category presented a variety of research methods, as personal interviews, questionnaires, tests, as well as analysis of audio and video-recorded discussions, students' written productions, and observations of students.

c. Use of educational software in labwork

This category comprises an important number of research papers in which computers and software (already existing or specifically designed and developed - simulations, spreadsheets), and other material is used in students' labwork, in order to facilitate the making of sense of manipulations and observations of natural phenomena in science lessons. As scientists are familiar with labwork, it seems quite evident that the use of ICT applications in the labwork resulted in more research papers than other categories. In some papers constructivist hypotheses were explicitly mentioned, while other ones seemed to be more science content oriented. In two cases the computers were used in the lab, only as a tool facilitating the communication between students.

Some research questions were:

- how to link theory to practice, the real phenomenon under study and its model,
- does the complementary study of real and simulated experiments result in better students' learning outcomes,
- what kind of difficulties science teachers have and in which degree they adopt the use of ICT (such as spreadsheets) in labwork.

The research methods were questionnaires, tests, analysis of audio and video-records, students' written productions, observations of students, concept maps before and after teaching, control-experimental group. Results included:

- ICT use promotes sense-making discussions among groups of students.
- The use of simulations and modeling tools may result in better learning effectiveness in linking theory to experiments

- Project based instruction leads students to having more responsibility for their own learning.
- ICT use changes the role of the curriculum, the teacher, the learner and the school.
- Many students are more familiar with the use of ICT than their teachers.

4.3 Some concluding remarks and perspectives

The use of ICT in science education is leading to the development of new learning environments and new learning materials as well as to radical changes concerning the teaching practices. A small but quite important number of researchers applied the theoretical and methodological tradition of the science education research and gave shape to this new research area concerning the use of ICT for the improvement of science education. Constructivist ideas are well established and form a solid theoretical background for the majority of the researchers.

The overview of the papers presented in the Conference showed that science education research is going across a transitional phase concerning both its research object and its methodology. Many researchers are more and more interested in the reality and complexity of the real classroom -with or without ICT- and they investigated the process of learning itself and not only its outcomes. For this kind of research different tools were used, such as analysis of audio- and videotaped discussions, of students productions, etc, but it seems that these methods are not well established, in comparison to more traditional tool, such as pre- and post-tests, questionnaires and interviews. So, more attention should be paid to the improvement of these methods as well as to the development of new technological research tools (appropriate software) for the study of the student-computer interaction and the evolution of students' conceptions into these new learning environments.

REFERENCES

Deutscher Bildungsrat (1971). *Deutscher Bildungsrat, Strukturplan für das Bildungswesen*. Stuttgart.

Fensham, P., Gunstone, R., & White, R. (Eds.) (1994). *The content of science*. London: Falmer.

Glaser, R. & Bassok, M. (1989). Learning theory and the study of instruction. *Annual Review of Psychology*, 40, 631-667.

Snow, R. E. & Swanson, J. (1992). Instructional psychology: Aptitude, adaptation, and assessment. *Annual Review of Psychology* (43), 583-626.

Wang, M. C., Haertel, G. D. & Walberg, H. J. (1990). What infuences learning? A content analysis of review literature, *Journal of Educational Research*, 84 (1), 30-43.

White, R. T. (1995). International scholarship and the AERA. Educational Researcher, 24(6), 19-21.

ON THE METHODOLOGY OF 'PHENOMENOGRAPHY' AS A SCIENCE EDUCATION RESEARCH TOOL

[1]P. Buck, [2]M. J. Goedhart, [3]W. Gräber, [2]W. H. Kaper, [4]T. Koballa, [5]C. Linder, [6] F. Marton, [7]H. Schwedes, [8]V. Spiliotopoulou, [9]N. L. Tsagliotis, [10]M. Vogelezang,

[1]*University of Education, Heidelberg,* [2]*University of Amsterdam,* [3]*IPN, Kiel* [4]*University of Georgia, Athens, USA,* [5]*University of Uppsala,* [6]*University of Göteborg,* [7]*University of Bremen,* [8] *Patras Technical and Vocational Teacher Training Institute,* [9]*University of Crete,* [10]*Scholengemeenschap Het Assink of Haaksbergen*

ABSTRACT

The contributors to this symposium have performed phenomenographic research in different projects. Their epistemology as well as their philosophy of science did more or less differ. It was the scope of this symposium to discuss whether this can be an obstacle or may lead to fruitful debates. After brief descriptions on the methodology of the research performed, 3 questions were discussed: (a) Is it appropriate to "borrow" and use phenomenography methods in non-phenomenological research contexts? (b) Is it justifiable to substitute the term 'experience' used in phenomenography by 'understanding', or 'conception'? (c) How can we secure validity of phenomenographic research?

1. INTRODUCTION, RATIONALE

For teaching concepts in which the *quality* of understanding natural phenomena as well as the understanding of (normal) scientific thought is important to the international science education community. The focus may be on coherent, contiguous and personal insight or on a system of results of Normal Science. If it is accepted that individuals produce personal insight[1], then methods of educational research are needed that trace back and describe such ways of individual understanding. For this purpose, *phenomenography* can be taken as an accepted research method. It was developed in Sweden in the late 1970s out of a Husserlian philosophical background in the context of psychological research, more precisely in a context of learning as coming to experience the world, or aspects of the world, in particular ways (cf. Marton & Booth, 1997). Researchers in other countries have independently developed similar research approaches (ten Voorde 1977, Buck 1993, 1996).

[1] More about the pedagogical issues in this connection cf. (Buck et al, 2002)

D. Psillos et al. (eds.), Science Education Research in the Knowledge-Based Society, 31–41.

Phenomenography should actually not be regarded as an independent method but rather as being embedded either in psychology (learning theory) or in pedagogy (educational philosophy). I it can be (and is) "borrowed" from outside the underlying phenomenological philosophy, for instance within a constructivist research design. Both the existence of similar methods and their use outside phenomenological research settings give rise to methodological questions. It was the purpose of the symposium from which this paper was drawn, to focus on methodological questions related to a) how data were collected and interpreted, b) how generalizations have been validated, and c) whether or not terms such as 'conception' or 'understanding' can be substituted for Marton's term 'experience'.

In the discussion Marton & Booth's description of the idea of phenomenography (Marton & Booth, 1997, chapter 6) was taken as a referent. All individual contributions during the symposium, either delivered as oral contributions from the panel or as statements from the symposium audience during the symposium, were anchored in this idea.

2. OBJECTIVES, DATA COLLECTION AND VALIDATION[2]:

Phenomenographers typically take a given phenomenon as a starting point and seek to find out how a person experiences this phenomenon. A current way to do this, as described by Marton & Booth (1997), is to conduct a two-phase interview. In the first phase the phenomenon is introduced in an "open and concrete way", usually by asking the interviewee to perform a certain material task. The second phase is then devoted to letting the interviewee formulate and reflect on his experiences. Marton stresses that the interview is a learning situation, both for the researcher (who learns how other people interpret the phenomenon) and for the interviewed person who is possibly expressing his experience in new ways. This situation is structurally similar to the situation the van Hieles used to promote a level change (van Hiele-Geldof, 1957, van Hiele 1986).

Wolter Kaper and *Martin Goedhart's* research was based on van Hiele's level-theory of concept formation in mathematics. It was adapted by ten Voorde and collaborators in three projects 'understanding chemical substance at secondary schools' (ten Voorde, 1977), 'understanding ionic crystals' (Joling, 1993) and 'understanding energy' (Kaper, 1997). Both phenomenographers, as well as "van Hiele-level" researchers, are looking for a small number of qualitatively different ways – ways of experiencing in the case of phenomenography, ways of using words (geometrical or chemical or physical) in the case of van Hiele level research. They might possibly be looking from different perspectives at more or less the same thing.

Michiel Vogelezang reported on a case study of a pupil learning geometry in the context of "level raising education" by van Hiele and then proceeded to chemical education on melting and solidifying. In the classroom situation presented, the students were forming a certain conception of melting and freezing points. He de-

[2] For the synopses of the contributions discussed here cf. symposia 5, 6 and paper session 30 in Psillos et al. (2001)

scribed how the *teacher* was learning what he can expect the students' own experiences to be at this moment of their learning biography. He used this instance as a close-up picture of what phenomenographic researchers do: They look carefully (i.e. they apply hermeneutic procedures) upon the underlying conceptions and ask questions based on their awareness of the way in which learning progresses and what is constitutive in a given notion of a given phenomenon. Accordingly, it is the hermeneutic skill of the researcher that makes the results valid.

Cedric Linder and *Delia Marshall's* study dealt with tutors' approaches to encourage introductory physics students at the University of Western Cape, South Africa to widen their 'experience of variation' as part of their learning to solve problems. The study covered 5 groups of students working in a "physics co-operative learning setting" (Heller & Hollabaugh, 1992) of 3-5 students on introductory Newtonian mechanics tutorial problems.. The groups of students met regularly in a venue wired with high optical resolution and magnification video cameras.

The study had two consecutive parts. Both parts involved 2 tutors working with 4 groups of students (the same groups of students). For the first 4 weeks of the course the tutors basically responded to student questions. After this (part 2) the tutors were shown how to use Schön's three coaching models as intervention strategies to introduce variation into the students' learning experiences.

The recordings were done at both the macro level (group interaction level) and the micro level (what individuals in the group wrote down and discussed what they wrote). The tapes were then analysed for evidence (1) that the tutors had recognisably begun using Schön's coaching models (cf. Schön 1988, Buck et al, 2002) as a way to encourage the students to search for variation as they struggled to make sense of why and what to do and (2) that discernible learning then began to take place by making use of systems of variation. It is mainly this second concern that is of interest in the context of phenomenographic research: Physics tutors will have to use a combination of approaches when they intervene in group work situations. The kind of intervention which they decide to adopt will be determined by how well they are able to judge where the students are in terms of making sense of a problem and what their difficulties are. It is the researcher's crucial skill that is required by phenomenography.

Hannelore Schwedes and *Lydia Murrmann's study* dealt with the ways in which 10-year-olds' understand light, shadow and vision. They interviewed children and videotaped classroom lessons and complemented their data collection with pre- and post-intervention paper & pencil tests. The detailed analysis of the transcripts of the interviews and lessons started with the reconstruction of meaning of individual statements and actions. For each child, then, lists of reconstructed meaning (understanding) were formulated that refer to coherent interview passages. The categories were refined and finally revised until they covered the whole variety of reconstructed ideas expressed in this research setting.

Vassiliki Spiliotopoulou's research, as far as the *perspective of phenomenography* is involved, aims to describe the essence of all the ways in which the phenomena can be experienced and searches for the appropriate descriptive tools.

Some 280 children, aged 6 to 16 years, in 8 schools in the Merseyside area in England were asked to express their thinking as produced not only from their immediate experience in particular instances but also from their *overall* experience of the world, through their actions, imagination, speculations, theorisation and inferences. The main tool in her study was a questionnaire-grid, in which students had to answer fundamental questions about 22 entities, such as spoon, earth, sugar, man, stars, space, atoms, water, universe etc. In addition she asked questions that solicit children's experience of different phenomena, like life's duration, movement, existence of forces. The question-entity pairs create different situations about which children were asked to state their ways of experiencing it (Spiliotopoulou, 1997).

A detailed and refined consideration of the children's answers was followed by a synthetic approach through the use of systemic network analysis (Bliss, Monk & Ogborn, 1983). The answers to every question for all entities and for all children were studied, thus leading to a very large number of possible combinations expressed in the triplets: "entity – question - aspect of response". The study proceeded by constructing a network based on the abstraction of the triplets and by recontextualising the produced categories. This procedure was repeated until a certain degree of certainty was reached that the decontextualised categories of the network are good descriptions of all existing triplets for each question and for all entities.

Marton's phenomenographic perspective (1981) suggests, "we would deal with both the conceptual and the experiential, as well with what is thought of as that which is lived". This objective requires consideration of the wholeness of the system. Not only the notion of elements should be considered, but also of modes of connection, the structure and organisation, the levels of structure and the control and function of the system. This is quite a complicated task and there is no way for a study to approach *integrally* all the aspects involved. Spiliotopoulou pointed to the synergy produced by combining the phenomenographic and the systems' theory approach.

Tom Koballa and *Wolfgang Gräber* carried out two studies to understand how prospective teachers conceptualise science teaching as part of their university experience. One study was conducted in the US and the other in Germany, the first one using phenomenography as a research method to construct a structure of categories of student conceptions of science teaching and the second, described as "partially" phenomenographic, involved a "search for predetermined categories" (Ashworth & Lucas, 1998, p. 420). The phenomenographic research study involved 7 students who were interviewed in a semi-structured format to elicit descriptions of their conceptions of science teaching and learning. The interview transcripts from these participants were treated as a whole. From these transcripts, 53 utterances were identified as revealing understandings related to science teaching. These utterances were then coded independently by two researchers into 12 and 17 categories, respectively. Examples of the coding categories generated by the two include "linking class activities," "covering content," "considering students' misconceptions," "using hands-on activities," and "making content interesting." These categories were then discussed by the two and revised by eliminating and merging categories and by grouping other categories based on perceived linkages. This sequence was repeated until a set of categories that both researchers agreed represent the data was constructed. Throughout the data analysis process, the two researchers worked independently to examine the transcripts and construct categories and jointly to discuss the categories and negotiate their re-construction. The outcome of the process was the proposal of a categorical system with qualitatively distinct, conceptual categories. It was by this partially independent and partially negotiated research procedure that Koballa & Gräber undertook to secure some validity of their phenomenographic research. The conceptions generated are nevertheless "provisional descriptions" (Entwistle, 1997).

Nektarios Tsagliotis, in an attempt to view conceptual change *within* Marton & Booth's Theory of Awareness (1997), elicited and compared children's conceptions on mechanical energy. Within a *second-order perspective* (Marton & Booth, 1997, p. 117-121) focusing on the descriptions of phenomena as they appear to the person, 13 children were interviewed before and after a teaching intervention on mechanical energy. The Interview-About-Instances technique was used (Gilbert, et al., 1985). The same set of eight depicted instances was presented and discussed with the children in both cases. These instances pictured everyday activities (like a man pushing a heavy box, children playing on the seesaw, swinging, going down the slide) or natural phenomena (like a stone rolling down a hill slope) etc. The interviews were analysed on two levels. At the first level the *Pre-Intervention Interviews* and the *Post-Intervention Interviews* were analysed separately in order to elicit a variety of qualitatively different conceptions about each picture before and after the teaching intervention. At the second level the elicited conceptions were compared within the context of each picture in order to identify differences in discernment, simultaneity and variation at deeper layers of awareness. This is an attempt to reveal the dynamics of conceptual change. Within a dynamic approach to conceptual change, children appear to experience and discern some varied features of energy, in terms of characteristics and aspects that come to the fore and remain in focus within the particular context of the depicted instances of the interview cards. As Marton (1990) has argued, *"within the internal dynamics of a conception of something, a restructuring is taking place and one meaning develops into another"* and this appears to be a gradual and on-going procedure. In this sense, if a person's awareness is the experienced world and learning can be seen as a change in the structure of awareness, then conceptual change can be seen as a change in the ways of experiencing various phenomena in terms of changes in discernment, simultaneity and variation of deeper or more advanced layers of understanding these phenomena, within a person-world relationship.

Peter Buck, in an attempt to come to objective phenomenographic descriptions, developed a method, which he called "chaining of [non-] understanding". It makes use of non-understanding situations of someone's understanding (i.e. his explanation). Once we can describe what is *not* understood and why, we may understand, what else *has* been understood.

Thus by chaining non-understanding to understanding and thereby gaining insights about the ambiguity of the situation (i.e. reaching meta-understanding, in other words by nesting three explication-demanding situations one into the other) we may reach enhanced interpersonal understanding of the whole. This is accomplished by creating plausibility (viability) of every step of understanding. In a formalized scheme (cf. fig. 1) he depicted his general understanding of what is taking place in phenomenographic research and applied it to case studies on the phenomena of floating and sinking and of distillation.

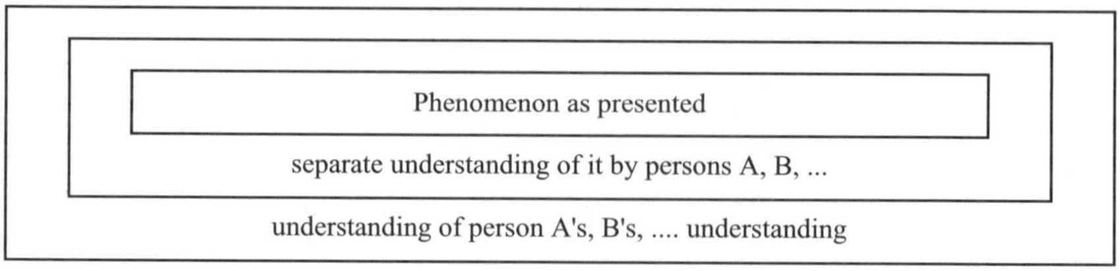

Figure 1. Chaining [non-]understanding

3. MARTON'S IDEA OF PHENOMENOGRAPHY AS DEPICTED BY BUCK

During the symposium Buck produced the following depiction of Marton's procedure of phenomenography (fig. 2): Marton & Booth state: "The main task.. for our researcher/learner is to discern the internal structure and the intertwined meaning of the object of research [i.e. understanding of the {phenomenon} dealt with by the researcher R or the testee $T_{1,...n}$] " (1997, p. 133); this is depicted by double arrows in the ovals of fig. 2. It is the *experience* of T that is compared to the *experience* of R, the latter being molded by his/her professional training. The experience is expressed in words; this *shift of quality* being depicted by rectangles.

Buck saw some need for terminological revisions (bearing their origins in his underlying concern of educational philosophy, (cf. Buck et al., 2002). He insisted on a distinction between *'experience'* which is inaccessible, and *'understanding'* which is the essence of any dialogue, be it a dialogue between human beings or be it the "dialogue" between a person and a phenomenon (also here the person puts questions and the phenomenon "answers"). In this general dialogue situation there are three qualitatively different "stages" of understanding, indexed with 1, 2 or 3: 'Understanding$_1$' in Buck's terminology is synonymous to Marton's 'experience', whereas 'understanding$_2$' is of different quality. It is a relationship ("⇔") between persons or a person and a phenomenon. 'Understanding$_3$' is again of a different quality. It is *generalized* understanding, thus abstracted from the persons that have been maintaining their understanding$_2$. It is a "thought phenomenon" now.

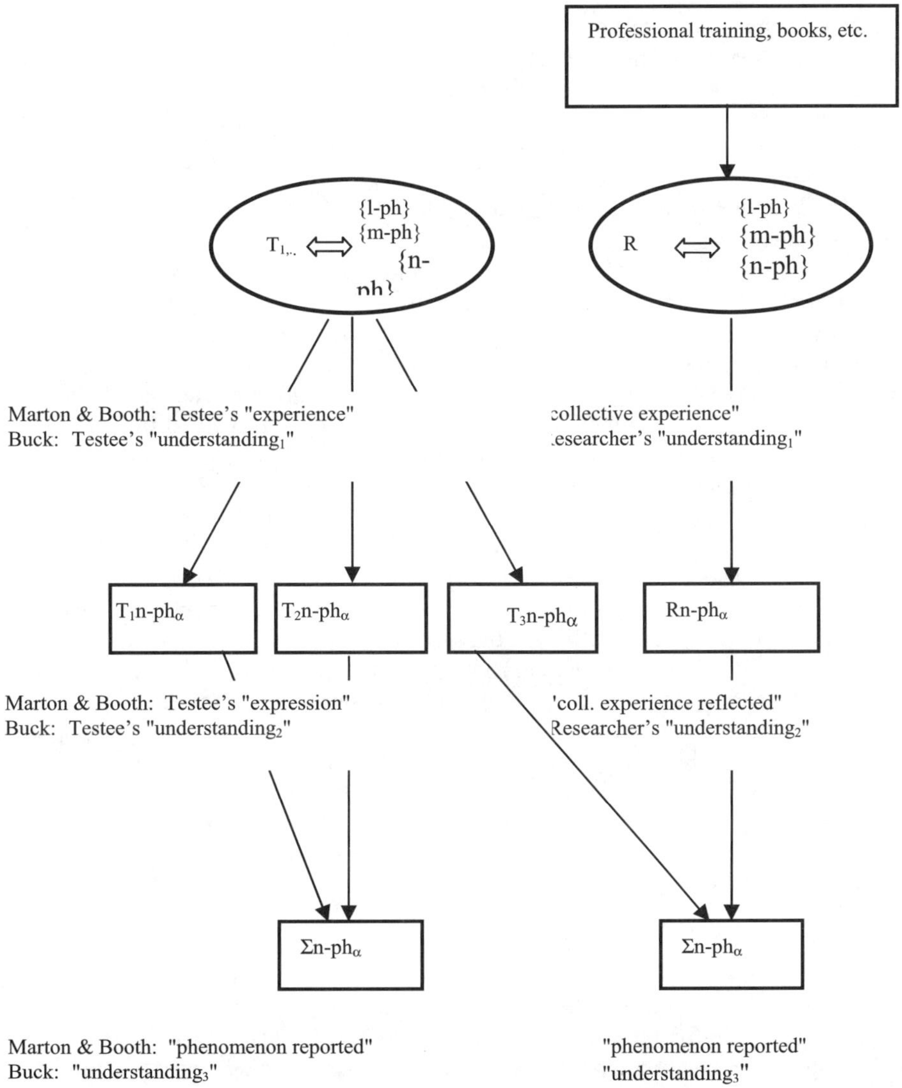

Figure 2. The procedure of phenomenography as understood by Buck

4. DISCUSSION

The question of sample taking, data collection and data processing was not really a matter of dispute. The discussion, instead, focused on science philosophy implications: The compatibility of Systems Theory and Phenomenography was at stake in

Spiliotopoulou's contribution. Both seem to complement each other in a favourable way. The research methods developed by the van Hieles and ten Voorde and Buck seem to have structural commonalities with Marton's Phenomenography. The differences lie in the appraisal of the nature of the phenomenon described: "experience" or "text" or "conception"? Mainly three questions were discussed during and after the symposium:

1. Is it appropriate to "borrow" phenomenography in non-phenomenological research contexts?
2. Does it matter if we substitute Marton's term 'experience' by 'understanding' (Buck) or 'conception' (Tsagliotis, Koballa & Gräber)?
3. How can we secure validity of phenomenographic research?

The answers have to be quoted individually: *Wolter Kaper* insisted that he did not exactly borrow the research method of phenomenography. The van Hiele's independently devised their method of distinguishing qualitatively different ways of using subject-related language. Only after careful study we can conclude that both methods have much in common and are perhaps based on equivalent assumptions. Secondly, on the basis of van Hiele-level research he argued that a change in students' use of language shows the emergence of new objects-of-experience (like ‘chemical substance’), therefore of new ways of experiencing. Therefore a change in ‘conceptions’ has been related to a change in students' experience. However, that does not mean that both terms share the same meaning.

Hannelore Schwedes' research context is the analysis of learning processes in physics education. Her basic epistemological assumption is the paradigm of constructivism. To her constructivistic research is compatible with phenomenographic research although they are based on different philosophies. Both in her research and that of Marton’s are the way that learners experience phenomena they are asked to deal with in physics instruction the central element of analysis. The use of phenomenography gives a special new turn to constructivist approaches: The idea of preconceptions in students’ minds can be given up and there is no need for postulates about the structure of the brain. Learning, in this context, amounts to a change in the experience of phenomena, related to the physics curriculum.

The power of Marton’s term ”experience” lies in its wholeness of this conception. It describes a process that unites perception, cognition and emotion in doing something. Therefore it cannot be substituted by the term conception or understanding. Understanding is part of the process of experiencing, and conceptions may be constructed during this process.

The objects of study in phenomenography are experiences. To experience implies mentally to discern the object from its context, i.e. constitute a phenomenon. This means it is experienced as something particular and distinct. By this discrimination on the one hand the context is constituted as context or ”outer horizon” and on the other hand the intentional object or the ”inner horizon”. The conceptions outer and inner horizon (terms from phenomenology) indicate parts of the structure of experience and are at the same time elements of the concrete experiences.

Tom Koballa joins in here: It *is* acceptable to borrow the research method of phenomenology for use in a non-phenomenological research context because the essential character of the research is to ”question the way we experience the world” (van Manen, 1990, p. 5). The focus of phenomenography is on particular phenomena, such as science teaching or learning, and how they are experienced. Different from psychology, the focus is not on remembering, problem-solving or decision making. Phenomenography and phenomenology both share human experience with phenomena as the object of research. However, they have different purposes. Phenomenology focuses on how people experience the world. In contrast, phenomenography tends to focus on ways that people experience the world that enable them to cope with it more or less efficiently (Marton & Booth, 1997). This extended purpose helps to distinguish phenomenography from phenomenology, but does not preclude the use of the methods of description, interpretation, and critical analysis that have long been associated with phenomenological approaches to research.

Further supporting the use of phenomenography is that it combines the explanatory power of both individual and social constructivism. Rather than focusing on the learner's active role in the construction of knowledge or the influence of culture on knowledge development, phenomenography "transcend[s] the person-world dualism imposed" by the two respective schools of individual and social constructivism and considers the person and world as one (Marton & Booth, 1997, p. 12). This enables exploration of the world that people experience.

Conceptions, on the other hand, *Nektarios Tsagliotis* continues, can be viewed as systems or categories of explanation or as "categories of description", which are not seen as individual qualities, but can be used in facilitating the "grasp of concrete cases of human functioning", in a person-world relationship. As Marton had mentioned during the symposium, the 1981 paper on *phenomenography* was subtitled "describing *conceptions* of the world around us" (emphasis added) just as Tsagliotis had argued in the synopsis of his paper (cf. Psillos et al, 2001, p. 121): "This approach depicts *conceptions* as being characterisations of categories of description....". Thus, he agrees with Marton that we ought not to make a distinction between immediate experience and reflected conceptual thought, since they both may carry different meanings and understandings [or non-understandings] of a certain phenomenon. He also agrees with Koballa's view that "people construct conceptions and come to understanding based on their own lived experiences". Tsagliotis also argued that "the notion of multiple conceptions can be seen as a challenge to a theory of conceptual change, which assumes conceptual stability and does not focus on the dynamics of awareness (Pong, 1999). If multiple, varied conceptions are context dependent, it appears more important to be able to recognise a context, discerning some of its features into focal awareness, and in this sense evoke an *appropriate* conception, in terms of *conceptual appreciation*, delimited by the particular context (cf. Linder, 1993)."

Vassiliki Spiliotopoulou bringing the systems theory approach and phenomenography together argues: The term "experience" is used in a meaningful way in the phenomenographic context and it turns to be a distinctive way of arguing about human learning. Marton & Booth (1997, p.122), by claiming, that experience is neither a mental, nor a psychological, nor a physical entity, but rather "an internal relationship between the subject and the world" and that "an experience is of its essence non-dualistic", draw a line between the analytical approaches in phenomenographic research and traditional constructivist ones

It does not seem that any conflict exists between the epistemological foundations of phenomenography and the *use* of terms like 'understanding', 'thinking' or 'conception'. Because, when we say "we can experience something as something thanks to the two basic capabilities: (a) we can discern entities and aspects, and (b) we can be focally aware of a few entities and aspects simultaneously" (Marton & Booth, 1997, p.123), do we not refer to the human capability of thinking or to the human need to understand? The essence of the point is that the focus chosen for the analysis and the discernment of aspects in the data for producing categories of experiences are of a different nature from the ones produced from non-phenomenological research contexts, which are usually of psychological or mental nature.

Peter Buck argues against these positions (and *Wolter Kaper* agrees to it): Although experience (understanding) is in the focus of phenomenography, it is not accessible as such but only after it has gone through the metamorphosis of becoming a *script*, i.e. it is *language* that we do research on. Making phenomenographic descriptions is hermeneutics almost like philology: it is interpreting texts. It *matters*, if we change the term: by using 'understanding' we state more precisely what we are dealing with. As to the validity of phenomenographic research: we are facing exactly the same problems hermeneutic research is facing. Whether or not to "borrowing" a phenomenological method by, say, a constructivistic approach: As it is the primary starting point of any research enterprise that matters, in such cases phenomenographic results are turned into constructs, so they are swallowed by the underlying and dominating epistemology.

5. CONCLUSIONS

A one and a half hour symposium of course cannot lead to a more elaborate discussion, nor can it lead to recommendations for future and further research. What the discussion boils down to is the question of what the object of research is. For example take the case where the object of research is what different meanings of various phenomena learners might develop and how these meanings are developed. Given that this is the object of research we might use similar research methods and come up with – at least partially similar – results, the interpretations of which might differ at a deeper level. This is certainly true as far as the question is concerned, what different meanings of a certain phenomenon might learners develop. The very same findings may originate from a constructivist or from a phenomenographic perspective, however the accounts of the results may differ.

If our focus is "differing meanings" it seems legitimate not to make distinctions between immediate experience and reflected conceptual thought, for instance, as both may – or may not – carry the same meaning of a certain phenomenon.

In order to secure validity we have to identify recognisable "critical features" which discriminate between different meanings of a certain phenomenon in an intersubjectively reliable way and we have to demonstrate that "the critical features" are not only critical when it comes to distinguishing between different categories of meanings, but *are also pedagogically relevant* in the sense of being powerful tools for understanding the difference between more and less successful cases of learning.

REFERENCES

Ashworth, P. & Lucas, U. (1998). What in the 'world' is phenomenography? *Scandinavian Journal of Educational Research*, 42 (4), 415-431

Bliss, J., Monk, M. & Ogborn, J. (1983). *Qualitative Data Analysis for Educational Research*. London: Croom Helm.

Buck, P., (1993). Läßt sich Verstehen beobachten? (Can understanding be observed?) In: Ulrich, W. & Buck, P. (eds), *Video in Forschung und Lehre* (pp. 211 – 227). Weinheim: Deutscher Studienverlag.

Buck, P. (1996). Über physikalische und chemische Zugriffsmodi (On physical and chemical modes of approach). *Zeitschrift für Didaktik der Naturwissenschaften,* 2, 25-28

Buck, P., Genseberger, R., Linder, C., Marton, F., Vogelezang, M. & Woyke, A. (2002). *European Science Education Philosophies in which 'phenomenography' as well as a 'pedagogy of awareness' plays a key role,*(Symposium at the ESERA Conference 2001 Thessaloniki; in this volume

Entwistle, N. (1997). Introduction: Phenomenography in higher education. *Higher Education Research & Development*, 16 (2), 127-134

Gilbert, J.K., Watts, D.M. & Osborne, R.J. (1985). *Eliciting Student Views using the Interview-About-Instances Technique.* In: West, L.H.T & Pines, A.L. (Eds.) *Cognitive Structure and Conceptual Change* (pp. 11-27), London: Academic Press.

Heller, P. & Hollabaugh, M. (1992). *Teaching problem solving through co-operative grouping. Part 2 Designing problems and structuring groups*. Am J. Phys. 60 (1992) 637-644.

Joling, E. (1993). *Chemie in gesprek*. Amsterdam: Thesis publishers (thesis).

Kaper, W.H. (1997). *Thermodynamica leren onderwijzen* (Teaching an learning thermodynamics). Amsterdam: Thesis publishers (thesis). Summary at http://www.summerschool.dk/esera/home.html

Linder, C.J. (1993). A challenge to conceptual change, *Science Education,* 77(3), 293-300

Marton, F. (1981). Phenomenography-Describing Conceptions of the World around Us. *Instructional Science*, 10, 177-200.

Marton, F. (1990). The phenomenography of learning - a qualitative approach to educational research and some of its implications for didactics. In: Mandl, H., De Corte, E., Bennet, N. & Friedrich, H.F. (Eds.). *Learning and Instruction* (Vol. 2.1, pp. 601-616). Pergamon Press,.

Marton, F. & Booth, S. (1997). *Learning and awareness*. New Jersey: Lawrence Erlbaum Ass.

Pong, W.Y. (1999): The Dynamics of Awareness. *Paper presented at the 8th EARLI conference*, Göteborg, August 24-28

Psillos, D. et al. (eds)(2001). *Proceedings of the 3rd Internat. Conf. On Sience Education Research in the Knowledge Based Society*, Thessaloniki, ISBN 960-243-578-X

Schön, D.A. (1988). Coaching reflective teaching. In: P.P. Grimmett & G.L. Erickson (eds.), *Reflection in teacher education*. New York: Teachers College Press.

Spiliotopoulou, V. (1997). *The Cosmologies of children, 6-16 years old*. Unpublished Ph.D. Thesis, Department of Education, University of Patras, Greece.

ten Voorde, H.H. (1977). *Verwoorden en verstaan* (Putting to word and understanding). 's Gravenhage: Staatsuitgeverij (thesis).

van Hiele-Geldof, D. (1957). De niveaus in het denken, welke van belang zijn bij het onderwijs in de meetkunde [Levels of thinking which are important in teaching Geometry]. *Ped. Studien*, *XXXII,* p. 290f

van Hiele, P.M. (1986). *Structure and Insight, a theory of mathematics education*. Orlando: Academic Press.

van Manen, M. (1990). *Researching lived experience: Human science for an action sensitive pedagogy*. New York: State University of New York Press

CONVERSATION THEORY AND SELF-LEARNING

Rufina Gutierrez

Science Education, Fundación Castroverde, Madrid. Spain, Grup de Recerca TIRE, Science Education Dept., UAB, Barcelona, Barcelona, Spain

ABSTRACT

Recent empirical data available in Science Education literature show a relationship between the capacity to generate *self-explanations* of phenomena and/or problematic situations, (explanations offered by subjects from themselves, without incorporating new information coming from any external source), and *self-learning* (learning generated by subjects from themselves, without incorporating new information coming from any external source). The empirical data are related to a number of situations and contexts, and to different age groups (children, secondary students, and university students). To generate self-learning the usual means is a conversation. This conversation can be established either within him/herself, with a person, with a written text, with a problematic situation, with a computer. In this paper we will propose a theoretical framework to interpret this issue, and we will analyse practical cases to illustrate the validity of that model.

1. BACKGROUND, AIMS AND FRAMEWORK

The empirical data that show the relationships between self-explanations and self-learning are acknowledged by researchers. When faced with this evidence, two main positions are held:

a) To take advantage of this relationships, introducing teaching practices favouring self-explanations, putting the subject in situations which stimulate this capacity; or
b) To search for theoretical models which allow explaining how and why these two terms (i.e. self-explanations and self-learning) are related.

In teaching practice, it has been used: to favour learning transference (Brown and Kane 1988); in assessment (Duschl and Gitomer 1997); as a teaching method (Brown and Polincsar 1989); in problem solving (Chi et al 1989, Bielaczyc et al 1995); in the production of analogies and explanations (Brown, 1992); in the acquisition of new categories (MacGregor 1996); in the construction and reconstruction of mental models (Chi et al 1994, Gutierrez 1994).

Difficulties arise when we take a closer look at the interpretative frameworks of the empirical data which are, for the most part, fragmentary, inspired in different psychological models, which range from Cognitive Science and Artificial Intelligence (Bielaczyc et al. 1995, Brown and Kane 1988, Chi et al. 1989, Chi et al. 1994, Gutierrez 1994), to purely social ones (Resnick et al. 1993), and a variety of piagetian and vygotskyan constructive positions in between. For example, to explain the efficacy of self-explanation, there exists a hypothesis which claims that it

D. Psillos et al. (eds.), Science Education Research in the Knowledge-Based Society, 43–49.

contributes to the delimitation of a "zone of proximal development" which facilitates learning (Brown and Palincsar 1989). Another hypothesis suggests that self-explanation generates the learning of procedural rules, which lead to greater acquisition in problem solving (Chi and Bassok 1989, Bielaczyc et al. 1995). However, the authors agree that *" we do not understand the mechanism underlying self-explanations"* (Chi et al. 1989, p. 169).

1.1 Conversation Theory

Nevertheless, there is a model that, in our opinion, can give a coherent explanation of how and why self-explanations generate self-learning. We are referring to the *Conversation Theory* proposed by G. Pask (1975, 1976a). The aim of this paper is to show the utility of Pask's theory to accomplish this goal.

Pask develops his theory in the context of Artificial Intelligence. The question he wants to answer with his theory is as follows: *what are the minimum conditions for a cognitive system (machine or human being) to be able to learn?* Starting with this question, he describes the basic structure, which gives to the system the capacity for an autonomous learning. A further question posed by Pask is the following: *how is it possible to preserve what has been already learned?* And he postulates, taking advantage of biological data, the need of a mechanism of *self-reproduction*, which allows for a *cognitive replication of itself* to the system. The minimum structure required for a cognitive system in order to accomplish these two conditions is defined by Pask as a *Conversation.* Thus, cognitive systems are Conversations (in Pask technical terms), which learn and preserve what has been learned telling or "conversing" to themselves. This process is obviously carried out unconsciously by the subjects.

1.2 Self-explanations and self-learning

As a consequence of his theory, Pask considers that if a cognitive system learns by means of self-replication (maintaining a Conversation with itself), it also can learn by replicating in others cognitive systems, namely, maintaining a Conversation with them. According to this, self-explanation (cognitive self-replication) constitutes the essence of the learning process, and it is so when it is carried out both within itself and when it is made for others. In this case we are referring to a subject who is giving an explanation to others about his/her learning contents. This very act of explaining to others (explaining *in* others, Pask would say) constitutes an act of learning (self-learning) for the subject. At this respect, he writes:

> "There is a theory of learning and teaching together and that is all. Results from many studies support this point of view as, also, does the evidence of commonsense. So, to be dogmatic (but with confidence) learning implies teaching and teaching implies learning. Sometimes, the teacher and learner responsible for the joint process are obvious (a student at a desk and another person wearing an academic cap and gown). Sometimes, the teacher and the learner are not so obviously distinct and turn out to be unexpected but, once indicated, intuitively plausible entities" (Pask 1975, p 33).

1.3 Self-learning and Teachback

As mentioned before, the self-replication mechanism within ourselves is not conscious. Thus, Pask designs a strategy for interviewing, to facilitate the process of gaining access to the content of cognitive system, called Teachback. It essentially consists on the following steps: A) the interviewer asks the interviewee an explanation *(teach me)* of what he/she knows about a topic, object of the interview. B) the subject comes up with an explanation of it. C) the interviewer explains to the subject (teachback) his/her understanding of what the subject has just explained. D) if the subject recognises in the explanation given for the interviewer what he was intending to say, the interview goes on; if not, the interviewer keeps asking and telling back to the subject what he is understanding about the subject's explanations, until the subject considers that the interviewer understood what he/she wanted to say.

The differences between teachback interviews and other, more classical, form of interviews used previously in educational research are presented by Pask himself (1976a p19-28, 1976b p13-15), in the following terms:

A *vygotskyan interview* is carried out between a *subject,* who is interviewed (the interviewee) and an *experimenter.* The interview is mediated by a physical artefact, visible to both. The experimenter poses a problem that the interviewee must solve. The questions asked are *how and why* type questions, and the explanations are about the functioning of the mediating artefact. The dialogue is normally prepared in such a manner that the interviewee is unable to answer some of the questions posed by interviewer and then he will ask for help. The interviewer will then give him the information necessary for him to be able to answer the questions correctly. In this way the participants can reach a consensus explicitly, and this consensus is open to an external observer. *-In the teachback interview the interviewee is not provided with information.*

In the *piagetian interview* a physical mediation is also used. This could be operated by the interviewee. But in this case, if certain keys are not available (for example, information about the psycho-evolutionary levels), consensus will not be reached on many occasions. And for an external observer this consensus, if it exists, will always be ambiguous. *The objective of the teachback interview is not to diagnose the cognitive capacities of the interviewee.*

In teachback interview there is never asymmetry between the interviewer and the interviewee; the only essential mediation is language; consensus is inherent to the process and always explicit; and it is always open to the scrutiny of an external observer.

These characteristics of the teachback interview ensure that the questioner's interpretation of the knowledge that the subject communicates to him is exactly what the latter considers to be a fair reflection of his thought (Shaw 1979, Johnson 1985, Johnson and Johnson 1986, Bliss and Ogborn 1987).

2. MODE OF INQUIRY

The Teachback technique for interviewing has been successfully used for eliciting implicit thinking by several authors (Johnson1983, Serrano 1992, Gutierrez and Ogborn 1992, Gutierrez 1994, Solsona 1997, Perea 1999). Moreover, in our work (Gutierrez and Ogborn 1992, Gutierrez 1994) we observed that teachback interviewing not only permitted us to gain access to the implicit thinking of students (as the other studies shown), but also that the content of the explanations offered by the subjects was increasingly rich throughout the interview process. The research was carried out with ten Spanish students, five aged 13-14 and five aged 17-18. The first group had had only an elementary introduction to Newtonian mechanics; the second had all chosen Physics for university entrance, and had had extensive courses in newtonian mechanics. Students were shown three comics picture, each showing a different everyday motion, and were asked to explain the three motions, and make graphical representations of what they were intending to say in the meanwhile, as a mean to help their explanations. The role of the three comics/situations were to test if the newly acquired knowledge while self-explaining one of them were used when self-explaining the others (robustness). Interviews were approximately one hour long, and they were tape-recorded and transcribed. The analysis of the protocols was carried out using categories taken from de Kleer and Brown's mechanistic mental model (de Kleer and Brown 1983).

3. OUTCOMES

The results of the analysis allowed us not only to characterise the different mental models implicit in the students' explanations, but also how the models stated by the students were progressively enriching in the amount of entities they contain and in the quality of relationships between the entities they expressed. All the subjects of our sample were adding elements and expressing more complex relationships while they were explaining the functioning of the different physical systems. And they did so not because they received additional information, but as due to the very act of explaining (self-explaining) to the interviewer their own thinking. Thus, we could say that these self-explanations produced self-learning, in this case by means of teachback. The complete account of the results can be found in Gutierrez 1994, and further elaborations in Gutierrez and Ogborn 1997, Gutierrez and Pinto 1997, and Gutierrez 1999.

3.1 Examples from students' graphical representations

As we said before, students were producing graphical representations of what they were intending to say while explaining their thinking to the interviewer. As an example, in FIGURE 1 we copy the representations produced by three students from the older group, JAC, JUC, and OSC, while explaining the movement, shown in one of the comics. In this comic a ball resting in a plank is set in movement by applying one instantaneous force to the ball. The ball moves on top of the plank until it reaches the edge, and falls down. Figures A were drawn when the interview started, and Figures B at the end. The enrichment of the second set when compared with the first is apparent.

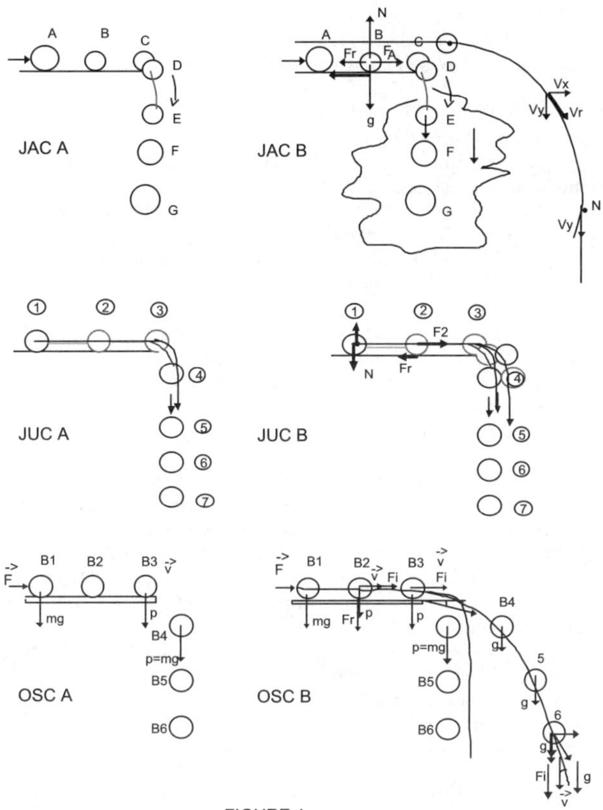

FIGURE 1

Figure 1

4. CONCLUSIONS AND IMPLICATIONS

The relevance of *conversation* (with one or another technique) to produce self-learning is a relatively new issue in Science Education literature. Conversation Theory offers a coherent theoretical framework for explaining and interpreting the empirical data available, which can be helpful in order to open new perspectives of research. For example, we have recently shown how teachback can promote conceptual change (Gutierrez 2001). But the real incidence in didactics of what research can contribute will depend on teachers' willingness to accept methodological changes, quite a revolutionary idea for many of them, but completely natural for their pupils: to converse in order to produce self learning.

REFERENCES

Bielaczyc, K., Pirolli, P., L. & Brown, A., L. (1995). Training in self-explanation and self-regulation strategies: investigating the effects of knowledge acquisition activities on problem solving. *Cognition and Instruction, 13* (2), 221-252.

Bliss, J. & Ogborn, J. (1987). Knowledge elicitation. En: Ennals (ed*), Artificial Intelligence, State of the Art Report, 15*. Pergamon Infotech. Oxford.

Brown, A., L. (1992). Design experiments: Theoretical and methodological challenges in creating complex interventions in classroom settings. *The J. of the Learning in Sciences, 2,* 141-178.

Brown, A., L. & Kane, M., J. (1988). Preschool children can learn to transfer: Learning to learn and learning from example. *Cognitive Psychology, 20,* 493-523.

Brown, A., L. & Palincsar, A.,S. (1989). Guided, cooperative learning and individual knowledge acquisition. In: Resnick, L. B. (ed), *Knowing, learning and instruction*. LEA. Hillsdale, NJ, p 393-451.

Chi, M., T., H., Bassok, M., Lewis, M. T., Reimann, P. & Glaser, R. (1989). Self-explanations: How students study and use examples in learning to solve problems. *Cognitive Science, 13* (2), 145-182.

Chi, M. T. H. & Bassok, M. (1989). Learning from examples via self-explanations. In: Resnick, L. B. (ed), *Knowing, Learning and Instruction* (pp. 251-282). LEA. Hillsdale, NJ.

Chi, T., H., M., Leeuw, N. de, Chiu, M., H. & Lavancher, C. (1994). Eliciting self-explanations improves understanding. *Cognitive Science, 18*, 439-477.

De Kleer, J. & Brown, J., S. (1983). Assumptions and ambiguities in mechanistic mental models. In: Gentner, D. & Stevens, A. L. (eds), *Mental Models* (pp. 155-190). LEA. Hillsdale, NJ.

Duschl, R., A. & Gitomer, D., H. (1997). Strategies and challenges to changing the focus of assessment and instruction in Science classrooms. *Educational Assessment, 4* (1), 37-73.

Gutierrez, R. (1994). *Coherencia del pensamiento espontáneo y causalidad. El caso de la dinámica elemental.* Doctoral dissertation. Univ. Complutense. Madrid. CD-ROM Edition: Servicio de Publicaciones. Univ. Complutense. Madrid, 2001.

Gutierrez, R. (1999). Aprendizaje autónomo. Un modelo Conversacional. In: Gutierrez, R. et al., *Aspectos didácticos de la Física y Química (Física)* (pp.11-36). 8. ICE Univ. de Zaragoza. Zaragoza.

Gutierrez, R. (2001). Mental Models and the fine structure of Conceptual Change. In: R. Pinto and S. Surinach (eds), *Physics Teacher Education Beyond 2000* (pp. 35-44). Elsevier Editions. París.

Gutierrez, R. & Ogborn, J. (1992). A causal framework for analysing alternative conceptions. *Int. J. of Science Education, 14* (2), 201-220.

Gutierrez, R. & Ogborn, J. (1997). The process of Conceptual Change. Paper presented at *ESERA First International Conference.* Roma.

Gutierrez, R. & Pinto, R.,(1997). Mental models of physical systems: from description to explanation. In: Oblak, S., Hribar, M., Luchner, K. & Munih, M. (eds), *New ways of teaching Physics* (pp.381-384). Proceedings of the GIREP-ICPE International Conference . Board of Education of Slovenia. Lubliana (Slovenia).

Johnson, L. & Johnson, N., E.,(1986). Knowledge elicitation involving teachback interviewing. In: Kidd, A. (ed*), Knowledge elicitation for Expert Systems*. Plenum. London.

Johnson, N., E. (1983). *Elicitation and representation of children's arithmetic knowledge*. Doctoral dissertation. Chelsea College. University of London.

Johnson, N., E.,(1985). Varieties of representation in eliciting and representing knowledge for IKBS. *International J. of System Res. and Information Science, 1*, 69-90.

MacGregor, J., N. (1996), Incremental category learning without external information: An algorithm for category-opening internal learning (COIL). *British Journal of Developmental Psychology, 87,* 81-103.

Pask, G. (1975). *Conversation, Cognition and Learning*. Elsevier. Amsterdam.

Pask, G. (1976a). *Conversation Theory. Applications in Education and Epistemology*. Elsevier. Amsterdam.

Pask, G. (1976b). Conversational techniques in the study and practice of education. *British J. of Educational Psychology, 46,* 12-25.

Perea, A, (1999). *Razonamientos de estudiantes de Secundaria sobre los procesos de cambio.* Doctoral dissertation. Univ. Autónoma de Barcelona.

Resnick, L., B., Salmon, M., Zeitz, M. C., Wathen, S., H. & Holowchak, M. (1993). Reasoning in Conversation. *Cognition and Instruction, 11* (3-4), 347-364.

Serrano, T. (1992). *Desarrollo conceptual del sistema nervioso en niños de 5 a 14 años. Modelos mentales.* Doctoral dissertation. Univ. Complutense. Madrid.

Solsona, N. (1997). *L'emergència de la interpretació dels fenòmens químics.* Doctoral dissertation. Univ. Autónoma de Barcelona.

Shaw, M., L. (1979). Conversational heuristic for eliciting shared understanding. *Int. J. Man-Machine Studies, 11,* 621-634.

ANALYSIS OF VIDEO DATA OF SECONDARY SCHOOL SCIENCE STUDENTS

Laurent Jeannin

UMR GRIC, équipe COAST, Université Lyon2, France

ABSTRACT

This research work has three origins. Firstly, in the European project "Labwork in Science Education", researchers developed a method of video analysis of labwork in real time (Niederer et al, 1998). Secondly, in our research team, video data of students working during teaching sessions was collected for several physics teaching sequences at the upper school levels. Thirdly, the "Kronos" software was used to analyse the video data, being devoted to treating observational data from a chronological perspective. Videos appear to be very useful and fruitful "instruments for taking data" in didactics. However their use raises a lot of theoretical and methodological questions, to which this paper aims to contribute: how and why should one analyse a video?

1. BACKGROUND, AIMS AND FRAMEWORK

1.1 Introduction

This study is situated in the setting of a research study on teaching and learning about sound at the level of the first year of the upper secondary school (15-16 years).

Most of the research studies in physics education show that physics learning — and conceptual understanding in particular — takes place over a long period of time, which can run over weeks and years. At the same time, recent studies, even with different theoretical frameworks, show that understanding students' learning requires going deeper in our knowledge of students' activities all along the teaching sessions. On the other hand, the advances in video technology allow us to use it easily. Videos appear to be very useful and fruitful "instruments for taking data" in didactics. However their use raises many theoretical and methodological questions. This paper is a contribution to this questioning. It presents a case study of the use of video data consisting of *directly coding* the videos when they are being observed, without using a written transcription as an intermediary (Jeannin, 2000). The videos have been recorded in a class where the teacher makes use of a sequence conceived by a group comprising teachers, researchers and Inspectors (Vince, 2000). Different previous studies in didactics (Buty 2000, Pateyron 1997, Roth 1998, Welzel 1998) used the video. Then, in the setting of the project European Labwork in Science Education, an analysis of strips video has been finalised: the CBAV-method, (Niedderer and al, 1998). Currently, several software permit the analysis of videos: Media-tagger (Max

D. Psillos et al. (eds.), Science Education Research in the Knowledge-Based Society, 51–59.

Planck Institute), Themecoder (Magnusson, 1993), Mac Shappa 1.0.3 (Sanderson P., 1994) and Kronos 2.4 (Kerguelen, 1997). We chose to work with Kronos 2.4 for its accessibility, the quality and the functionality of its graphic representations and for its modularity in the construction of categories of analysis.

1.2 Theoretical framework

Our study is focused on the construction by the students, of the meaning of the knowledge involved in the teaching situation. We consider that "*our knowledge of the world is mediated by our interaction with it*". On the one hand, this statement reflects a common criticism made by Soviet psychologists of theories that treat humans as if they were passive receivers of stimuli; on the other hand, it reflects a criticism of idealistic theories that treat humans as if they created their knowledge of the world by conscious reflection." (Wertsch, 1979, p. 38). In this study, it is not stated that conscious reflection does not play an important role in learning but that interaction, whatever it is, plays a role in students' construction of meaning of their surroundings. To go from the concept of interactions to the observable, relevant to being coded from the videos with the chosen software (Kronos), the concept of "milieu", derived from the theory of situations of a French didactician of mathematics is used (Brousseau 1998, Veillard, 2000). The "milieu" is constituted by the elements with which the student interacts; a given social and material environment can lead to different surroundings according to the actor that interacts with this environment. For a student in the observed class, the "milieu" may be constituted by the other students, the professor, the furniture, the experimental material, etc...We limit ourselves to three types of interactions: interactions with the material elements of the environment, reading and writing, and social interactions. We present below how we specified the observables that could characterise these interactions, before discussing the relevance of the data obtained.

1.3 The teaching sequence

The teaching sequence included six laboratory sessions (1h 30 each). In the analysis of the teaching content, Vince (2000) presents four models involved in this content in relation with experimental domains: (1) sound chain with the associated concepts of emitter, milieu, receptor, (2) the model of the vibration to create sound with the associated concepts of vibration (related to back and forth movement), frequency and period, amplitude; (3) particular model for the sound propagation with the concepts of vibration of the milieu and displacement of the vibration, no displacement of matter, variation of pressure, displacement of the particles (vibration); (4) model of the dynamic of sound (displacement and velocity of sound, reflection and absorption of sound) (p. 119).

2. METHODS AND SAMPLES

2.1 Methodology

The use of videos as data, together with their processing, involves several steps, both in the case of transcribing and in the case of direct coding: (1) videotaping; (2) particularly in the case of direct coding, digitalisation and compression in order to be able to get the files of the videos on a computer (3) choice of the codes of transcription or choice of categories and the corresponding observable; (4) data treatment after coding to get new data. Here we will present and discuss the type of data that are obtained with software such as Kronos. Firstly, in order to limit the difficulty of analysis and to be able to better understand how the students construct a meaning of the situation, a same pair of students were videotaped all along the teaching sessions with a fixed camera and two lapel microphones. We are aware that the video analysis will only allow us to go a step further and not to achieve the whole reconstruction of the meaning that students construct. We chose to analyse the video through an *a priori* and an *a posteriori* analysis. The *a priori* analyses deal with the interactions involved in each of the questions, on their duration. The analysis of a teaching sequence (6 times 1h30mn) requires decomposing the sequences into easily analysable parts. The question of how to split the teaching situations into meaningful parts was raised. In our case, the labwork sheets, for each 1h 30 session, proposed a series of questions. It was decided to deal with each question if the students have taken the question into account that is the case most of the time. Then from the six-labwork sessions, 50 questions were obtained, *a priori.*

2.2 Elaboration of categories and observables of coding

As it was introduced, there were three main categories corresponding to interactions: social interactions and interactions with the material elements of the surroundings, categories corresponding to writing and reading, and finally another type of category dealing with the content analysis. In this elaboration, three steps are necessary: (1) to select the elements of the *milieu* which (*a priori*) can play a role in the construction of the meaning of the situation from the point of view of the knowledge involved in this situation; (2) to define the relevant students' actions; (3) to determine the corresponding observables. Associated with these "visuals reference marks", we will take in account the different temporalities that some actions can have. Indeed, it is necessary to take a "visual reference mark" that is not limited in time (too short or too long), and to take in account the possibility of overlapping of such "*visuals reference marks.*"

2.2.1 Interactions with the elements of the material milieu

The study of the teaching practice in the observed class leads us to select some elements belonging a priori to the milieu with which the students may interact. These elements are: (1) all objects of the experimental device, in particular the low frequency generator (LFG), the oscilloscope, the loud speaker, a microphone, a tambourine, a tuning fork; (2) the laboratory sheet given by the teacher at the beginning of the session, (3) the notebook where each student has to write her report.

Concerning "handling the material objects", and the gestures corresponding to taking perceptual information (in particular hearing, touching, and seeing), we made the following choices in relation with the knowledge involved in the session to take into account the membrane of the loudspeaker and the microphone. To categorise the "**touch of membrane of the loudspeaker/microphone**", the "visual reference mark" is the following: the student puts her hand on the membrane. To categorise "**to look at an event**", the "visuals reference marks" are more complex. We made the choice to categorise the events that attract the students' attention. Then, our principal criterion is based on the insistence of the student's look. To categorise "**hearing**", we made the same choice as previously. Therefore, the "visual reference mark" is: the student "pricks up his/her ear" toward the loudspeaker. Concerning "handling the material objects", it is necessary to select very specific actions. We always make a distinction in relation to the knowledge involved in the teaching-learning situation. Therefore, we chose to distinguish *manipulation with the material objects* with respect to the **LFG**, **the oscilloscope**, the tuning fork or the tambourine. So for the oscilloscope and the LFG, we take as "visual reference mark": the student manipulates the buttons frequency and/or amplitude for the LFG. **To manipulate oscilloscope**, we take as "visual reference mark": the student manipulates buttons of regulating. The actions on the tambourine and the tuning fork produce a sound directly and we categorize them by: **to make a direct action**. These "visuals reference marks" are: the student hits on the skin of the tambourine or the tuning fork with sticks and/or with fingers. In the case of reading and writing, the visual reference marks are defined in the same way as previously. With respect to reading, a difficulty appears concerning the complexity of the visual reference marks. We can assume that the student is not always going to read with his sheet in the hands and the gaze fixed toward this sheet. But an element remained invariant in this phase of reading: the gaze. Indeed, the gaze is always directed towards the laboratory sheet to be read. Therefore, we take as a "visual reference mark" **to read the laboratory sheet**, the moments where the student's look is fixed on the laboratory sheet. **Reading measurement** and **reading the notebook** will be processed in the same way as for reading the laboratory sheet.

With regards to written answers, the associated actions are well determined. The student must have a pencil and his notebook ready to write. His writing can be with words and/or drawings. Then, we made the choice to take into account these two possibilities: **to write** and **to draw**. The "visual reference marks" are: the student takes his/her pencil and writes or draws

2.2.2 Social interactions

Social interaction is a complex phenomenon where several dimensions — cognitive, social and cultural — interact in a spatio-temporal setting, historically and situated culturally, labelled by rules and by values. The interaction can be considered to be a privileged place for the construction and emergence of social, cognitive and linguistic forms. Different authors recall the three shapes of containers of the interaction: (1) the interaction as a primordial place of formation and transformation of social bonds, (2) the interaction as a place and means of acquisition via coordination and the exchange during located social activities, (3) the interaction as a prototypical place for the use of the language. The link between these three features is *the language*. This language is configured by the developments of the speakers, adjusting with each other and adapting their activity in context. The possible link between these activities and discursive expertise resides in the phenomenon of *deixis*. There can be verbal deixis ("it", "him", ...) and gestural deixis, that often occurs by means of pointing to an entity using the index finger. We will take exclusively in account the phenomena of gestural deixis: **to show with the finger**. The "visual reference mark" is: the student shows with the finger an element or an event of the situation. Through this definition of social interactions, the verbal or gestural exchange between protagonists is essential. The inventory of the possibilities of exchanges of the student brings us: a student can exchange with his partner (**E1 and E2**), with a student external with his dyad (**E1 and other**), with the teacher (**E1 and teacher**) or only (**E1**). The "visual reference marks" associated with these different categories are auditory. On the other hand the reference mark for E1 is more complex. Indeed, a student can speak alone in a way that may or may not give rise to an exchange. Therefore, the reference mark for E1 is a student who speaks alone and whose speech does not engage in the continuation of an exchange with another student.

2.2.3 Contents verbalisations

We will use a categorisation distinguishing the verbal productions that deal with: (1) the elements of a theory which correspond with the explanatory system of modelling; this theory is not necessarily correct of the point of view of physics, (2) the elements of a model that consist in formalisms, or in qualitative and quantitative relations involving some physical quantities (or generally of variables) and (3) terms referring directly to the material objects or perceived events (Tiberghien, 1994, 1997). We regroup the first two elements under the terms: "the world of the theory/model" and the third in "the world of the objects/events". To be able to code the content verbalisation, we carried out an *a priori* analysis of the lexical fields in relation to the knowledge concerned in the various categories: theory/model, objects/events and relation theory/model and objects/events. These lexical fields must take in account the evolution during the sequence of labwork. This lexical field enabled us to obtain "auditory reference marks".

2.3 Setting up of the analysis video protocol: Kronos 2.4

The software Kronos 2.4 makes it possible to treat mainly statements of direct observation but also statements of measurements and verbal communications. It proposes a mode of coding of the data by assignment of attributes intended for interpretative analyses, which enables real time coding of specific events, which in turn generate a state over the duration. Thus coding must be made from observables corresponding to events, which are easy to characterise in the video by avoiding making judgements as much as possible. Our analysis grid is made up of the elements describing the interactions with the elements of the material milieu, the speakers and the verbalizations contents describe by three modelling levels. These will be inserted in the first document generated by the software, heading: protocol of description. This software allows synchronisation and a simultaneous viewing of the video and Kronos 2.4. This association of the video and the software does not allow us only one simple and effective coding, but also a temporal synchronisation between the elements of the video which one codes and data of observation that we obtain by the software. Once our analysis of the video finished, we can very easily obtain various types of information: graphs, statistics, etc....To illustrate the data obtained by the treatment of the software, we present (1) the *a priori* analysis of a question of the second labwork session, (2) the data obtained and (3) their analyses.

2.4 A priori analysis

This labwork session n°2, entitled "the propagation of the sound in air, the temporal period of the source of sound", presents the relation between frequency and low-high values. It draws upon the concepts of transmitter, milieu of propagation, receiver and vibration (presented in paragraph 3; Vince, 2000). We present the question (2.c) that is representative of the greater variation that we could observe between the *a priori* and *a posteriori* analyses. We note that we could observe five important significant variations out of the 50 questions gathered in six labwork sessions. For information, it was requested from the students to place a lit candle in front of the membrane of a loudspeaker connected to a LFG and to vary the frequency of the generator of zero at a frequency corresponding to a sound. These three questions are: 2.a: To name the milieu of propagation of the sound involved in this experiment; 2.b: What do you observe and hear at very low-frequency and higher frequency?; and 2.c: Can you deduce the way in which the milieu of propagation behaves when a sound is emitted by the loudspeaker? To answer the question 2.c, the student is not obliged to reason, because the elements of the previous questions are sufficient to produce the answer. The exchanges that the students can have will be based on the behaviour of the milieu of propagation from the point of view of theory/model; in addition, they should make some relations between theory/model and objects/events. We summarise our a priori analysis below in table 1.

Table 1. Forecasts of the actions for the question 2.c

Socials interactions	E1 and E2	E1 and others	E1 and professor	E1	Show	
	Yes	No	No	No	Yes	
Material	Touch of membrane	Look an event	Hearing	Manip LFG	Manip oscilloscope	Make action
	No	No	No	No	No	No
Read and write	Read the laboratory sheet	Read a note book	Read a measure	Write	Draw	
	Yes	Yes	No	Yes	Yes	
Contents	Theory/model	Objects/events	Relation			
	Yes	No	Yes			

3. RESULTS

We have statistics at our disposition for each code of our analysis grid, for the totality of the labwork sessions or each question, and the diagrams of states represented in ordinate axes, our grid of analysis and time in the abscissa.

3.1 The diagram of states

We obtain this type of diagram for each labwork session or for each question for each student. This diagram allows a general visualisation of the role of the different interactions. For example, figure 1 shows three moments of resolution by student E1. (1) E1 reads then takes some perceptible information while handling and calling upon the teacher, then (2) writes then reads, exchanges in terms of relation between theory/model and objects/events with his partner, reads again, writes and (3), exchanges in terms of theory/model, handles again, shows and charges in terms of relation with other students longer than with the teacher.

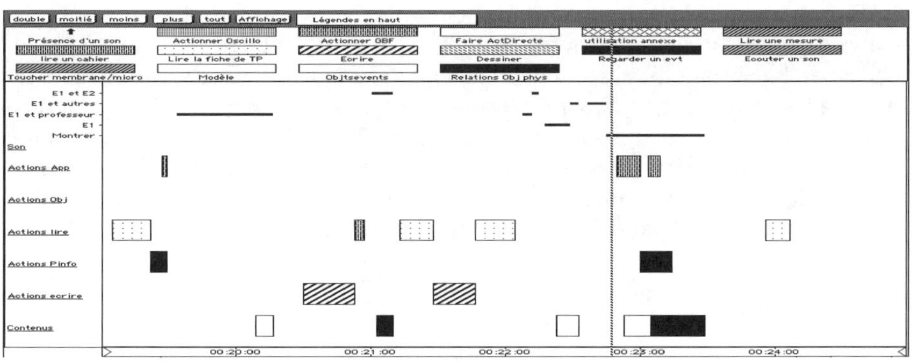

Figure 1. Diagram of states of the question 2.c

3.2 *Overall picture: Labwork n°2*

Table 2. Percentages of some actions during the labwork session n°2. (100%= duration of the question)

TP2	1.1	2.a	2.b	2.c	1.2.a	1.2.b	1.2.c	2.1.a	2.1.b	2.1.c	2.2
E1 and E2	22%	28%	43%	4%	17%	30%	23%	31%	6%	44%	35%
E1 and teacher	3%	0%	0%	15%	0%	5%	10%	0%	35%	0%	20%
Manipulate LFG	11%	18%	10%	6%	0%	0%	11%	39%	0%	0%	0%
Read labo. sheet	25%	20%	27%	20%	6%	20%	5%	13%	25%	16%	26%
Write/draw	31%	0%	28%	14%	43%	25%	10%	47%	0%	44%	40%
Look an event	29%	33%	14%	7%	0%	0%	0%	0%	0%	4%	2%
Duration (mn)	11:08	02:49	04:59	05:07	07:48	03:19	11:06	01:47	01:54	03:06	08:00

Table 2 gives us a general view of the student's activity during this labwork session. The first line presents the questions and the first column gives elements of our analysis grid. It also permits us to see the overlapping of the actions. For example, for the question 2.b, if we make the sum of the percentages, we get 150%. It permits us to also see the different proportions of the actions. For example, the questions 2.c and 2.1.b present the smallest percentages in the category **E1 and E2**, but strongest in the category **E1 and teacher**. But, they also permit us to compare the labworks. For example, we could observe that the students exchange, on average 25% of the time during the 50 questions.

3.3 *Comparison between a priori and a posteriori analyses*

Figure 1 and table 2 show what happens effectively, and a shift with our a priori analysis appears. Figure 1 shows that the students handle a first time with the teacher and a second time with other students. The table 2 permits us to have a fast quantitative view on the labwork session. It permits us to detect quickly the gaps between the *a priori* analysis and what happens effectively, but it does not permit us to go in the details of the scenarisation of the resolution. Figure 1 facilitates the visualisation of the sequentialisation of the labwork in actions, but it does not allow us to fully understand the students' construction of meaning.

4. CONCLUSIONS AND IMPLICATIONS

Our work was concerned with the relations between the activity of the student and teaching sequence on sound phenomena. Our first preoccupation was to describe a characterisation of the influence of the teaching situations on learning, using an analysis of video based on the study of the activity of the learners. We provide some reflexions on the status of this new type of data and the new questions that it raises.

This approach, once put into effect, allowed us to identify more precisely the relevant actions for learning. With this remark, we add that this first processing using the video can make it possible to deal with a very large corpus (several hours) and to offer some means of "zooming" on moments which seem to us relevant, which could then be transcribed in more detail. This study provided information concerning the duration of the different types of actions. The students speak together on average around 25% of the total time of the labwork session. The students read (70% of the total time of reading during labwork session) quickly (reading time lower than 30s). Our study also allowed us to give an account of the alternation, the succession or the overlapping of some actions. But this work and the results obtained, also give us new questions, notably: how can such results be used to understand the students' co-construction of meaning? Must one consider the gestures and the speech as a single system or dissociate them? But these questions depend on the choice of the gestures to be coded. Should we only take in account the gestures that seem *a priori* relevant to us in relation to the learning? Should we also take scientists' gestures or communication into consideration?

REFERENCES

Brouseau, G. (1998). *Théorie des situations didactiques*. Grenoble, La pensée sauvage.

Buty, C. (2000). *Etude d'un apprentissage dans une séquence d'enseignement d'enseignement en optique géométrique à l'aide d'une modélisation informatique*. Thèse, Université Lyon2.

Jeannin, L. (2000). *Mémoire de DEA: Etude de TP de physique en classe de seconde à partir d'une méthodologie d'analyse de bandes video en temps reel*. Université Lyon2.

Kerguelen, A. (1997). The Kronos Software : a tool for work activity analysis. *Proceedings of 13th Triennal Congress of International Ergonomics Association* (pp. 240-247). Volume 7, Tampere, Finlande, 29 juin-4 juillet.

Magnuson, M.,S. (1993) *THEME USER'S MANUAL: with notes on theory, model and pattern detection method. 86 pages*. Human Behavior Laboratory, University of Iceland.

Niedderer, H., Buty, C., Tiberghien, A. et al. (1998). *Category Based Analysis of Videotapes from Labwork (CBAV)-the Method and Results from four Case-Studies*. Working Paper 9 from The European Project Labwork In Science Education (Targeted Socio-Economic Research Programme, Project PL 95-2005), 41 pages.

Pateyron, B. (1997). *Mobilisation des savoirs dans la formation professionnelle*. Thèse, Université Claude Bernard-Lyon1.

Roth W.,M. (1998). Learning process studies : examples from physiques. *Int. J. Sci. Educ*, 20(9), 1019-1024.

Sanderson, P.M. (1994). *Exploratory Sequential Data Analysis: Software*. Engineering Psychology Research Laboratory Technical Report (EPRL-94-01). University of Illinois, Urbana, IL.

Tiberghien A.,(1994). Modeling as a basis analyzing teaching. Learning situations. *Learning and instruction*, 4, 71-87.

Tiberghien, A. (1997). Learning and Teaching : Differentiation and relation. *Research in Science Education*, 27 (3), 359-382.

Veillard, L. (2000). *Rôle des situations professionnelles dans la formation par alternance : cas des élèves-ingénieurs de production de l'ISTP de Saint-Etienne*. Thèse, Université Lumière Lyon2.

Vince, J. (2000). *Approches phénoménologique et linguistique des connaissances des élèves de seconde sur le son*. Contribution à l'élaboration et l'analyse d'un enseignement et au développement d'un logiciel de simulation.Thèse Université Lumière, Lyon2.

Welzel M. (1998). The emergence of complex cognition during a unit on static electricity. *Int. J. Sci. Educ*, 20(9), 1107-1118.

Wertsch, J., V. (1979). *The concept of activity in soviet psychology*. Amonk, New York: M.E. Sharpe, Inc.

LONGITUDINAL STUDIES - PROVIDING INSIGHT INTO INDIVIDUAL THEMES IN SCIENCE LEARNING AND STUDENTS' VIEWS OF THEIR OWN LEARNING

Gustav Helldén

Kristianstad University, Sweden

ABSTRACT

The present paper reports on findings from an analysis of twenty-three students' conceptualisation of ecological processes and the students' reflections over their understanding of these processes. The students were interviewed eleven times from 9 to 15 years of age about conditions for life and decomposition, and the role of the flower. Characteristic individual themes in the students' conceptions can be followed year by year. At 15 and 19 years of age, the students listened to audiotapes of what they said respectively as 11- and 15-year olds. I asked each student to comment what they heard and to describe how they thought their understanding had developed. The students were asked to describe how they had developed their understanding. Through analysis of the students' descriptions of themselves as learners, it was possible to discern their different views of learning in relation to fundamental ecological processes.

1. BACKGROUND, AIMS AND FRAMEWORK

In order to create teaching situations during which students' ideas can be challenged, science educators must know more about the development of students' conceptions. Short-term studies cannot capture the whole picture. Despite these needs, very few studies of science learning have been longitudinal in nature (Arzi, 1988; White, 2001). Such studies can also provide the researcher with the opportunity to study the nature of the learning process, learning pathways and the influences on student learning. In order to learn how to help students to develop deeper understandings and more scientifically accurate explanations of some ecological processes I carried out a longitudinal study of students' conceptualisation of conditions for life, decomposition, and the role of the flower in plant reproduction. The students were interviewed eleven times about these processes (Helldén, 1995, 1998, 1999).

Already at an early stage of the research project, there were features in the students' descriptions that reappeared year after year. These features seemed to result from personal experiences that were hidden from the researcher but had become a part of the students' episodic memory (White, 1988). Such features could be of importance for the description and interpretation of the students' conceptual development. Why not ask the students themselves about such features? Therefore, I

D. Psillos et al. (eds.), Science Education Research in the Knowledge-Based Society, 61–68.

decided to let them as 15- year-olds listen to audiotapes and make comments of what they said at 11 years of age and as 19-year-olds what they said at 11 and 15 years of age. The students also showed great interest in what they said during earlier interviews.

The aims of the present research paper are to study

- How it is possible to identify and describe individual themes that have emerged from analysis of interviews with students during a longitudinal study;
- How the process of review of earlier interviews can contribute to the interpretation of individual themes in the development of ecological understandings;
- The different views students have concerning the development of their learning.

2. DESIGN AND PROCEDURE

I, like many other researchers in science education, have found that clinical interviews can give in depth information on students' thinking about natural phenomena. Over the course of my study I interviewed the same twenty-three students concerning ecological processes on eleven different occasions from grade 2 (9 years) to grade 8 (15 years) and at 15 and 19 years of age about their views of the development of their understanding.

To challenge the students' ideas about the conditions needed for life, we grew plants in sealed transparent boxes. I started the interview with the question: "What do you think the plant needs to be able to grow in the box with glass lid pasted on?" Another question was: "What do you think will happen to the plant in the box if we plant it there and glue the lid on?" For my interviews about decomposition, I had soil, brown leaves and litter on a table in front of the students. The opening question was in this case: "What do you think will happen to the leaves on the ground in the fall?" Later during the interview about decomposition, I also asked the students: "What makes the leaves fall from the trees in the autumn?" For my interviews about the flower's role in plant reproduction, I showed the students different kinds of flowers.

After the students had been interviewed about the ecological processes at 15 years of age, each student listened to audiotapes of my interviews with them at 11 years of age. At 19 years of age the students listened to audiotapes from the interviews at 11 and 15 years of age. During these interviews I asked the students to comment on what they said in the earlier interviews, to describe how they thought they had developed their ideas and what they thought had been of greatest importance influencing their ideas.

3. ANALYSIS

All interviews were audiotaped and transcribed verbatim. Concept maps were constructed from the interview transcripts according to a design that was developed at Cornell University. Ausubel's theory of meaningful learning had important implications for the analysis of the interview data (Novak, 1998). By comparing concept maps and transcripts from the interviews with each student through the years, I have identified personal developmental features in the students' conceptions. I have also analysed interviews with the students as 15- and 19-year-olds after they had listened to earlier interviews with them in order to see if they could identify such personal features in the earlier interviews with them and what views they had of the development of their understanding.

4. FINDINGS

As a result of my analysis of the interviews over the years, I have found that individual students' assimilation of different concepts could be shown to depend upon their prior conceptualisations of ecological processes. Through tracing their thinking over many years it has been possible to identify personal themes in the development of the students' understanding that would be impossible to identify by probing their conceptions at a single time.

4.1 Personal themes in the development of students' understanding

4.1.1 About conditions for life

Some of the children's advances in understanding conditions for life in the sealed transparent boxes could be traced to classroom processes. In order to explain to the 10-year-old-students what was going on in the sealed boxes, the teacher introduced the concept of the water cycle. Many children described individual cycle models as themes during the subsequent interviews to explain how the plants in the boxes could get water, air, oxygen and carbon dioxide. However, much of their knowledge seemed to relate to more personal and earlier experiences.

Already at the very first interview with Oscar about conditions for life, he argued that the plants needed creepy crawlies in the soil. He expressed the same argument in the interviews at 11, 13 and 15 years of age. In fact this idea helped him at 15 years of age to explain how the plants in the sealed box could get carbon dioxide.

When Sofia explained why the plants could survive in the box, she always described in an illustrative way how the water cycle could provide resources for many of the plants' needs. She always mentioned the concept of dew as part of the cycle like in the following interview segments.

Sofia at 11y

> 'Well, first you poured in water. Then in the morning, there sort of will be dew on the leaves and then, when it is a little warmer in the box, it rises up towards the floor. And then it is raining down when there is too much.'

Sofia at 15y

> 'You had watered the soil before you planted them and had sealed the box. The plants absorb the water. Then there is dew on the plants that evaporates. It rises. Therefore, it is condensation there. And then it runs down back into the soil.'

4.1.2 About decomposition

Also in the interviews with Anders, we can follow a personal theme as a part of his efforts to explain the decomposition of leaves on the ground. He always referred to composting in a quite detailed way as we can see in the following segments from two interviews with him.

Anders 11y

> 'Well, they have dried. Then they haven't had enough water and then ..... You put many other things there, you can also put eggshells and so on the compost heap. And they stay there and rot in some way. It takes a long time, it takes just about four years to get real and very nice soil.'

Anders 15y

> 'It is like composting. You have some soil on the bottom and then it stays there becoming smaller and smaller bits. I think it's the same with the leaves, that they are mixed up and then there'll be more and more of it. I don't think that there'll be soil but I think there will be small, small bits that is mixed up with the soil, then you say that it has become soil.'

4.1.3 About the role of the flower and defoliation

Some students continued to express anthropomorphic explanations as personal themes year after year, especially concerning the role of the flower in plant reproduction. Oscar was one of those who preferred to express anthropomorphic and human-centred formulations but demonstrated a development towards a more 'scientific explanation' at the age of 15 y. He explained the role of the flower at 11 years of age in the following way.

Oscar at 11y

> 'In order to attract animals that can then suck nectar from inside there so that they can reproduce or perhaps they can boast to the other flowers and make themselves beautiful and so on in the same way as women powder themselves and so on. If you have them in a garden, perhaps you water them very much more cos they are so pretty.'

Several students explained defoliation through the years by using anthropomorphic formulations like Oscar at 9 years of age: *'It doesn't get any water. Or it has no muscles left to be able to stay on the branch'*. When I analysed the interviews about defoliation I found it possible to identify continuity in the development from anthropomorphic ideas towards more scientific explanations. I could also identify another continuity in the descriptions of defoliation in several students' statements. They changed from more 'leaf-centred' ideas at 11 and 13 to more 'tree-centred' at 15 years of age as in Oscar's case.

Because of the longitudinal design of the research project, it has been possible to identify personal themes in the students' conceptions of some ecological processes. The personal themes can be of a structural nature, a way of explaining a phenomenon as in the students' descriptions of the cycles in the sealed box or as in Oscar's case when he used anthropomorphic formulations to explain the role of the flower in plant reproduction.

4.2 The students' views of the development of their understanding

4.2.1 About their ideas.

When the students as 15- and 19-year olds listened to the audiotapes of their earlier interviews which laid out for them the history of their thinking, they could in many cases identify personal contexts that continued to persist as themes in their thinking through the years. Initially, I asked the students to make comments on the ideas they expressed in the interviews and why they said as they did. Nearly all the students appreciated what they had learnt at school about the ecological processes but still they referred to personal out of school experiences that had been of great importance for the development of their understanding. Here follow some examples of the comments the students expressed after they had listened to the interviews with them.

When Sofia heard how she referred to the concept of dew in her explanations of conditions for life in the sealed transparent boxes, she claimed: *'It's from childhood. The dew has always fascinated me. It is unbelievable beautiful.'* Then she said that she had wonderful experiences of dew when she spent a couple weeks every summer at a relative's summerhouse. Also Anders could at 19 years of age trace back his explanations of decomposition to experiences from childhood. When Anders heard his descriptions on the tape recorder at 19 years of age he smiled and said: *'We had a neighbour who carried out composting in a special way. I liked to be there together with him. The man even put eggshells and coffee grounds on the compost. I remember the first time I was there. And he asked me to empty a bucket with coffee grounds and some eggshells on the compost heap. I was confused. I think I was 7 years old.'*

The students often related their way of describing the ecological processes to concrete personal experiences that occurred between the ages of 5 and 10. A majority of the students claimed that early experiences are of great importance for the development of their understanding. Early episodes in childhood seem in many cases to have such an important influence on their future descriptions that the students used this as a part of their explanations of a phenomenon that we can follow as themes in the interviews.

Some students found their anthropomorphic formulations a bit strange. Others expressed an appreciation of the way they explained the phenomena. Oscar clearly expressed anthropomorphic and human-centred ideas to explain the role of the flower and the defoliation at 11 years of age, though he used the anthropomorphic ideas as analogies rather than as explanations. There is continuity in Oscar's accounts from anthropomorphic ideas towards more 'scientific' explanations at 15 years of age. At 19 years of age after he had listened to earlier interviews with him, he claimed that the anthropomorphic features in his descriptions about defoliation and the role of the flower could be a result of his mother's way of explaining the phenomena to him. She often used anthropomorphic descriptions when explaining different phenomena. He said that he of course knew as a boy that the leaves did not have muscles. It was a way of explaining why the leaves fall down from the trees in autumn. This resulted in anthropomorphic formulations that I can follow as a consistent theme year after year. Also in other cases, he referred to his parents as important for the development of his understanding of ecological processes.

4.2.2 About their conceptualisation.

The longitudinal data allowed students to develop and describe insights into their own conceptual history and also opened up a meta-conversation about their learning. At 19 years of age, twelve students only recognised a minor change in their understanding of ecological processes between 11 and 15 years of age after they had listened to the interviews with them. They did not correct the answers they gave four years earlier, but rather extended their explanations. Some of them were more aware of the way they talked about things than their understanding of the processes. Even if they did not recognise any change they expressed positive feelings about learning. *"Cos when you were a child you didn't understand that much but learnt still more. You didn't understand anything and were happy to learn."*

Eleven students in the group expressed recognition of an obvious change in their understanding of the phenomena. Six students in this group had a more descriptive than reflective view of learning. They talked more about addition of new facts than analysing their thinking. Anders belongs to this group of students. He was fascinated by the recognition of how he described some details about decomposition in the same way as four years earlier: *"I said the same today. It is still several years later. You don't remember what you said and although it is somewhere 'in the back of your head'."* Although he found himself use the same expressions as four years earlier, he described a change in his thinking about the decomposition of leaves on the ground.

Five of the twenty-three students were able to analyse and reflect over their earlier ideas. All the students in this group also articulated as 15-year-olds a more complete description of the ecological processes. They willingly discussed the reasons why they said as they did and how they formulated their explanations. More or less, these students made comments on their earlier use of analogies between a plant's needs and a human being's needs. Sofia made the following comment on her explanation as an 11-year-old: *"You can often compare them even if the seed hasn't any human thoughts. You can see yourself how you develop. The seed grows too. It's like ..... the seed grows and I grow.*" The students in this group emphasised more than the other students the importance of building upon earlier experiences and knowledge. They described an active and constructivist view of the learning process. For them learning could start with reading a book, listening to others and catching some words. A couple of them described how they caught ideas piece by piece and then put them together. Perhaps these students, because of a more reflective and metacognitive view of their learning, could also have developed a higher ability to integrate new perspectives in their thinking. They tended to use a richer language with colourful and anthropomorphic formulations.

5. DISCUSSION

Through the long-term study of students' conceptualisations it was possible to identify personal contexts and continuity as recurrent themes in the students' explanations of the different phenomena. Even if the explanations changed through the years towards more or less complete explanations, it is in several cases possible to identify core ideas that can be followed by analysis of the interviews with the same child from 9/10 to 15 years of age. These core ideas can be of structural nature, a way of explaining a phenomenon like several students' use of a cycle model to explain how the plants could get necessary resources or Oscar's use of anthropomorphic formulations to explain defoliation and the role of the flower. It has also been possible to identify individual themes that concern the content of the students' statements. Sofia always used the concept of dew as a part of her explanations. Anders talked year after year about eggshells when he explained composting.

The students showed great interests in their earlier statements, a great ability to interpret what they said – ie. why they said what they said, and to identify key life experiences that shaped their understandings. Most of the students were rather critical of what they said in the interviews and sometimes expressed astonishment over the poor development of their understanding. However, six of the students demonstrated a more reflective, metacognitive view on the nature and development of their learning. When they heard what they said in the earlier interviews, they realised that they at that time had not always understood the concepts they mentioned. One student argued that she used words without knowing what the words really meant and therefore missed the general picture. Some students described how they snapped words from what they heard from others, mostly at school and at home, and how they had formed a general picture by putting together pieces of puzzle.

At 19 years of age, some of the 'reflective' students claimed that the anthropomorphic features in their earlier descriptions about defoliation and the role of the flower could be attributed to the influence of adults' attempts to explain the phenomena to them. They could say that they knew as children that plants did not have feelings nor the leaves muscles. The anthropomorphic features were a way of explaining why plants had colourful flowers and why the leaves fell down from the trees in autumn. The 'reflective' students also showed a greater ability to use language as a tool to support their efforts to understand the phenomena and express their thoughts.

Other students demonstrated a limited ability to reflect on their learning. They recognised that their learning was influenced by their prior experiences but they described the change in their explanations through the years as a process of addition of information. Some students said that they knew more, simply as a result of being older. Others described a learning process by which new facts were fastened to their minds, but not in any way that changed their cognitive structure. Even if they recognised a change in the explanations they gave at age 11 and 15, some students were unable to trace the change in their understanding to any identifiable experiences. They did not know why their understanding had changed, and they often gave the same explanation to the phenomena as they did four years earlier.

When they were asked to comment on what had been of greatest importance to the development of their understanding, the students with a reflective ability could as 19-year olds trace back such features to experiences they had made together with teachers, neighbours, playmates or family members. Even if the students appreciated what they had learnt at school, they described experiences outside school, often in the early ages that contributed to their understanding.

6. CONCLUSIONS

Experiences from the present research project have shown

- That even if there was a substantial conceptual development, there was also a very strong element of personal context and continuity in the students' thinking about the ecological processes
- That the long-term study of students' conceptualisations has made it possible to identify aspects of their personal biographies which had important implications for their learning
- That students did not mostly replace one understanding with another one but widened their ranges of ideas or increased their repertoires of ideas
- That the possibility of listening to previous interviews allowed the students to identify important precursors to their understandings, many of which related to social situations
- That the students with a reflective and metacognitive ability had a richer language and a greater success in analysing earlier interviews with them

The longitudinal design makes it possible to gain insights into students' individual learning and thereby contribute to the development of a more complete description of students' learning in science.

REFERENCES

Arzi, H., (1988). From short- to long-term: Studying science education longitudinally. *Studies in Science Education,* 15, 17- 53

Helldén, G. (1995) Environmental Education and Students' Conceptions of Matter. *Environmental Education Research.* 3, 267-277.

Helldén, G. (1998). A longitudinal study of students' conceptualisation of ecological processes. Paper presented at the Annual Meeting of the National Association for Research in Science Teaching in San Diego.

Helldén, G. (1999). A longitudinal study of pupils' understanding of conditions for life, growth and decomposition. In: M. Bandiera, S. Caravita, E. Torraca & M. Vincentini. *Research in Science Education in Europe.* Dordrecht: Kluwer Academic Publishers.

Novak, J., D. (1998). *Learning, creating, and using knowledge: Concept maps as facilitative tools in schools and corporations.* Mahwah, NJ: Lawrence Erlbaum Associates.

White, R., T. (1988). *Learning in science.* Oxford: Blackwell.

White, R., T. (2001). The revolution in research on science teaching. In V. Richardsson (Ed.) *Handbook of Research on Teaching.* New York: Macmillan.

CHANGING REFERENTIAL PERSPECTIVE IN SCIENCE CLASSROOM DISCOURSE

Eduardo F. Mortimer, Flávia Maria Teixeira Santos

Faculdade de Educação, Universidade Federal de Minas Gerais, 31370-901 Belo Horizonte - MG, Brazil, e-mail: mortimer@dedalus.lcc.ufmg.br

ABSTRACT

This paper discusses some methodological issues that emerge in constructing data from classroom transcripts. Based on a 45 minutes lesson of a high school chemistry teacher, we distinguish between data that can be constructed from this lesson alone and data that only become visible if you look at the sequence of lessons in which this one is included. We explicitly discuss the methodological procedures to break up the 45 minutes transcript and obtain episodes to be analyzed. The focus of the analysis is on meaning and meaning making and we will emphasize the shifts in referential perspective, from empirical to theoretical, and will relate this to the shift in the characteristic of the "text" being produced, from dialogical to univocal.

1. BACKGROUND, AIMS AND FRAMEWORK

Research in science education has traditionally identified contents of ideas, inspired by physiological research on conceptual development. More recently, a number of studies, in what could be called a "discursive turn" in science education, has changed this focus in highlighting the process of building concepts in social interactions and how they change through negotiation (Edwards and Mercer 1987, Ogborn et al 1996, Scott, 1998). In this new tradition, concept is equated to meaning (Vygotsky, 1987). Therefore, the focus is on meaning and meaning making. Meaning is viewed as polysemous and polyphonic, created in the interaction and then internalized by the individual. The process through which data is obtained is itself viewed as a process of meaning making and this has a methodological consequence, i.e., the data is constructed in the interaction and not viewed as something out there, which is objectively accessed and described.

The aim of this paper is to contribute to the theoretical and methodological debate in this emergent tradition. We selected a 45 minutes lesson of one high school chemistry teacher that was observed and video-recorded during the entire academic year. In presenting the data, we explicitly refer to the steps through which we break up the data and obtained an episode for analysis. This episode will be analyzed using a methodological framework we have been developing (Mortimer, 1998, Mortimer and Machado, 2000, Mortimer and Scott, 2000, Scott and Mortimer, 2002, Mortimer and Scott, in press) for describing the flow of discourse and meaning in science classrooms. In this paper we present an example of how this

D. Psillos et al. (eds.), Science Education Research in the Knowledge-Based Society, 69–78.

methodological framework can be used to highlight the process of changing referential perspective in classroom discourse, from empirical to theoretical referentiality, which we consider to be one of the biggest challenge for teachers that adopts approaches in which interaction and dialogue between teacher and students are seem as playing an important role in the construction of scientific meanings by the students. We consider that the process of meaning making in science classroom is complex and multifaceted and that is impossible, in the scope of one paper, to address all this complexity. In this sense, here we are going to restrict ourselves to illustrate how the framework we have been developing can be used to analyze the process of change in referential perspective.

The framework we are going to use in this paper is theoretically rooted in a sociocultural tradition that build on ideas of Vygotsky and Bakhtin to describe meaning making as mediated by cultural tools, such as social languages, speech genres and other semiotics devices (Wertsch, 1998). In previous papers (Mortimer, 1998, Mortimer and Machado, 2000), one of us made an attempt to describe the discursive patterns in a science classroom, through the analysis of different "voices" (Bakhtin, 1981), which interact in the dialogues. We also tried to describe the alternation between persuasive and authoritative discourse, covering two distinct functions: that of generating new meanings and that of transmitting, consolidating or reinforcing meanings already shared by the whole class. These features, inspired in Wertsch's work (1991), can be summarized using scheme 1, reproduced from Mortimer and Machado (2000).

To convey meanings	To generate new meanings
⇓	⇓
Univocal	Dialogic
⇓	⇓
authoritative	internally persuasive
⇓	⇓
sender and receiver codes coincide	Text is a thinking device and not only a passive link between sender and receiver
⇓	⇓
Demand allegiance and not the free appropriation of words	Seeks counter-words; it's shared
⇓	⇓
A single voice	Multiple voices
⇓	⇓
I-R-F evaluative	Chains of interaction with elicitation and elaboration

Figure 1. Two functions of texts

The "texts" produced in classroom will carry a tension between univocal and dialogic function, where the latter will be more appropriate to contexts in which new meanings are being constructed while the former will be more appropriate to those contexts in which already shared meanings are being used or reviewed. An important feature of this functional dualism of texts is that they can be thought as being dialogic or univocal independent of being uttered individually and interactively. What make a text functionally dialogic is the fact that it express more than one voice, more than one conceptual horizon, and not the fact that is was produced by group of individuals or by an individual alone. This leads us to another dimension that should be explored, i.e., if the talk is interactive, in the sense of allowing for the participation of other people, or non-interactive. In this dimension we should add that, independent of being interactive or not, classroom talk can be lead by the teacher, as most of it use to be, or lead by the students. It is a quite big demand on science teachers that they should be able to sustain students talk and that students should be able to lead talking not only in working in small groups but also in broader classroom discussion (Mortimer and Scott, in press).

In this paper we are going to highlight a particular aspect of the classroom discourse, trying to relate it to the tension between univocal and dialogic functional dualism of texts: patterns in the process of changing referential perspectives in classroom discourse. This process generally can be documented in classrooms where students are invited to talk about everyday concepts and experiences in the process of learning scientific concepts. This kind of classroom discourse generally involves shifts from empirical to theoretical referentiality. In the first, the referent - a world or a referring expression - is directed linked to the thing or process it refers to in the world. In the second, referents are not directly linked with things and processes "out there". Reference is mediated by other meanings as part of a theoretical and conceptual system. In this paper we analyze the teacher's strategy to make this kind of shift.

Linguistic criteria can be used to distinguish whether an utterance belongs to the first or to the second type. If an utterance is set up in terms of referents visually present in a system or in terms of referents linked to objects brought from everyday experience - for example the pans referred to in the episode analyzed here - this utterance fall in the first type - empirical referentiality. If an utterance, on the contrary, uses referents that are not visually present in the system, but are entities created through intralinguistic relationships (sign/type - sign/type), which characterize the symbolic systems (Wertsch, 1991), this utterance falls in the second type - theoretical referentiality. It is important to note that, in science classrooms, entities such as electrons, atoms, molecules, etc., are created mainly through discourse, independently of their realistic status. In this sense we are going to refer to such entities as theoretical referents, although scientists, science students and science educators alike tend to consider these entities as real objects.

2. METHODS AND SAMPLES

The 45 minutes lesson analyzed in this paper is the eighth of 40 lessons recorded from two different classroom of one high school chemistry teacher in the year of 1998. Lower middle class students attend the school, localized in suburban area of Belo Horizonte, Brazil. Around 45 students attend each class. The teacher adopts a strategy that alternates group activities and whole class discussion, highly framed by the instructional material that she and the students are using. She tries to be interactive and dialogic in conducting the lesson.

The researcher, with a video camera, observed and recorded the lessons since the beginning of the school year, did field notes and interviewed students, the teacher and school staff. Immersing herself in these classrooms, the researcher adopts a ethnographically based perspective on constructing detailed register of the classroom life, which allows a multidimensional analysis, which includes verbal, non-verbal and emotional aspects of the process of meaning making (Santos, 2001).

For each recorded lesson we constructed a map of events to contextualize the episodes to be analyzed. The map of events is constituted of 5 columns. The first one shows the time spent in each event or action, which is specified in the second column. Each event constitutes a unit of meaning. In defining a unit of meaning we follow Bakhtin (1986) suggestion that the utterance is the unit of speech communication. In Bakhtin's view, the term utterance covers a broad range of phenomena, from a line in a dialogue to an entire piece of work, for example, an article. Based on Bakhtin's work, we define a unit of meaning as a segment of an ongoing discourse that has clear boundaries in terms of its thematic content. In this sense, an event of meaning, in the classroom discourse, is defined as a segment of this discourse that has clear boundaries in terms of its thematic content, in the sense that it has a clear beginning, when the theme is introduced, and a clear end, when the teachers, considering that everything about the theme into consideration has been said, changes to another theme. An event can be broke in sub-events, as a meaning is generally treated in classrooms in terms of other meanings that are used to define or specify it. Each sub-event corresponds to a sense or meaning that is brought into consideration in building the broader meaning that constitute the event. In differentiating between sense and meaning, Vygotsky states that:

> A word's sense is the aggregate of all psychological facts that arise in our consciousness as a result of the word. Sense is a dynamic, fluid, and complex formation which has several **zones** that vary in their stability. (...) In different contexts, a word's sense changes. In contrast, meaning is a comparatively fixed and stable point, one that remains constant with all the changes of the word's sense that are associated with its use in various contexts. (...) Isolated in the lexicon, the word has only one meaning. However, this meaning is nothing more than a potential that can only be realized in living speech, and in living speech meaning is only a cornerstone in the edifice of sense. (Vygotsky, 1934/1987, p. 275-276

In the case of the classroom analyzed here these units of meaning are highly constraint by the activities proposed in the instructional material. The time spent in each event allows for the evaluation of the importance given to each activity or theme under discussion. The third column in the map of events presents the themes and meanings discussed in each event approximately in the same sequence as they occurred in the lesson. The fourth column shows the activities that took place in the event and the fifth includes some comments about the students' engagement (or not) on these actions. The disposition of the texts alongside the columns follows a chronological order, exception given to the third column, where some times we opt for a logic sequence. The map of events allows us to establish the relationship between the different episodes to be analyzed. The map can tell us, for example, how the students' engagement with the activities changes during a lesson and how this aspect is related to the other aspects (theme, time spent, type of activity) coded in the map.

Some of the data presented here, that cannot be meaningful if we look only to this specific lesson, emerged from this multidimensional analysis. In constructing data from this particular lesson, we broke up it in units of actions and meanings and selected one unit of meaning as an episode to be analyzed. In this unit of meaning, different materials that can be used to build a pan are considered in terms of time of heating (and consequently consumption of energy) and time of cooling. The analysis will emphasize the shift in referential perspective and will relate this to the shift in the characteristic of the "text" being produced, from dialogical to univocal (Mortimer, 1998).

3. RESULTS

The teacher spend the first seven minutes of the lesson organizing the class in groups and setting the rules for the whole class discussion that will follow. At the seventh minute, one of the students was still out of any group and the teacher addressed her in a very emphatic way to insert her in a group. The action of this student is part of a repertory of tactics the students developed to resist the way of conducting the lesson adopted by the teacher, that is different from more traditional teacher centered expositive ways used by other teachers in the same school. This means that the teacher set up a different work contract for her lessons, which is new for the students. The action of this student only acquires meaning if we look at the sequence of lessons that preceded this one, as the establishment of a new work contract begun in the first lesson and are still in process in this lesson.

The episode selected constitutes one of the units of meaning in which the whole lesson can be broken. The units, in this particular lesson, are highly determined by the questions posed by the instructional material. The question that generated this episode asks the students to discuss the advantages and disadvantages of the different materials that can be used to make a pan in terms of consumption of energy for cooking and of conservation of the food's temperature after cooking. The students worked in groups answering the question. The episode is part of the whole class discussion that follows this group discussion, occurred in a previous lesson. This discussion is used by the teacher to introduce the concept of thermal conductivity. Thermal conductivity is, therefore, the theoretical referent being constructed, from the empirical referents "time of heating" and "time of cooling", which are highly established by the students from their everyday experiences with cooking.

The episode started 14 minutes after the beginning of the lesson and lasted 25 minutes. In these 25 minutes it happened 168 turns of speech, half of them from the teacher and half from the students. In this sense the episode can be considered as an example of interactive discourse, as the teacher is elaborating it through an interaction with the students.

Although the episode as a whole can be considered as a unit of meaning in which the theoretical referent "thermal conductivity" was elaborated from the empirical referents "time of heating" and "time of cooling", we can break it in smaller units of meaning and/or sense and construct a map of sub-events for the episode itself, which helps us to give a sense of how this central meaning was elaborated in the discussion between the teacher and her students. Although we based our analysis in this map, in figure 2 we present only a synthesis of results that can be highlighted from the use of it.

Duration of the sub-event (in minutes and in number of turns)	Themes, meanings and senses introduced in the sub-event	Sense or meaning	Prevailing referentiality
2' and 17 turns (1 to 17)	Which materials can be used to make a pan	Sense: materials from everyday experience - context of use of objects (pans)	empirical
9' and 56 turns (17 to 72)	What properties do the materials have that make them appropriate to make pans	Sense → meaning Properties of an object - pans made of different materials	empirical
8' and 70 turns (72 to 141)	What are the advantages and the disadvantages of each material	Meanings: pans, materials, thermal conductivity (which pans heat up fast; which pans cool down slowly)	empirical
6' and 27 turns (142 to 168	Thermal conductivity - good and bad conductors	Meaning: thermal conductivity; good and bad conductors	theoretical

Figure 2. Summary of the meanings and senses introduced in the 4 sub-events

The first part of the episode, constituted by the first three sub-events, is highly dialogical and the students contribute to the discourse with complete utterances in their own "voices", suggesting materials that can be used to make a pan (sub-event 1), discussing the properties of these materials that make them suitable to make pans (sub-event 2) and comparing the time of heating and the time of cooling of the different pans (sub-event 3). Most of the referents used are empirical, based on everyday experience and discourse. The teacher uses a set of pedagogical interventions (Scott, 1998, Mortimer and Scott, 2000) to maintain the flow of discourse, selecting some contributions, ignoring others, marking key meanings, rephrasing and sharing meaning produced by individual students with the whole class. It is worth noticing that the discussion begin with the use of two words - materials and their properties - that are general, unspecific, dynamic and fluid, characterizing what Vygotsky defines as word's senses. As the discussion progresses, different meanings related to this general senses are brought into consideration, contributing to fix and stabilize one meaning related to these two words: thermal conductivity, presented as a property of different materials.

The following transcripts, taken from the third event of the episode, exemplify the kind of students' and teacher's interventions when the empirical referentiality prevails in the discourse. It should be noticed that the students say complete sentences in answering the teacher's question, referring to the materials that are used to make pans. These referents are brought from their everyday experience and, in this sense, are "out there", what characterize empirical referentiality.

72. […] T.: Well! Now it asks to discuss the advantages and disadvantages of these materials you have mentioned, which can be used to make pans. Advantages and disadvantages related to the consumption of energy. Which one of these materials used to make pans will consume less energy when you are cooking, ok? And what are the disadvantages of these elements, these materials that consume less energy, ok? It has

advantages? Yes, ok, you have already mentioned. To consume less energy, but has it also disadvantages? Ok? Let's see what you have answered.

73. S1: The aluminum, it can be heated easily, but iron and stone are better to keep the food hot, stone more than iron. Because stone, it takes a long time to heat it, iron also takes a long time, although stone conserve the calorie better than iron.

74. T: Ok, then it is what she said. That she took aluminum, took iron and took stone. Then what she says, the aluminum cooks faster but the iron and the stone conserve more, and the stone conserve still more.

75. S(?): Better than iron.

76. T: Better than iron, ok. Let's see what this group (the teacher points to another group of students) answered.

77. S8: Iron heat faster than clay, aluminum faster than steel, steel faster than clay.

78. T: Then you said that aluminum and steel heat faster than iron, and what else? Than stone, than clay.

79. Several students: Than clay, than stone.

80. T: Then, what are the advantages you can see there? What is the advantage, it is that they get hot faster, isn't it? And what is the disadvantage of them?

81. S8: They cool down faster.

The strategy used by the teacher to shift from empirical to theoretical referentiality is to classify the pans in two groups: one for pans that get heated and cooled quickly and the other for pans that get heat and cooled slowly. This strategy begins at the third event (turns 88 to 103), at that time in terms of the empirical referents - the pans. This is used, in the fourth event, to distinguishing between good and bad thermal conductors, and the theoretical referents thermal conductivity, conductors and insulators are introduced. As this happens, the discourse become univocal and the students' participation is restricted to single words, completing slots in the teacher discourse normally by answering, in choir, evaluative question asked by the teacher. The following transcripts, taken from the fourth event of the episode, exemplify the kind of students' and teacher's interventions when the theoretical referentiality prevails in the discourse.

142. T: Then let's see here. Let's go back to here. Heat conductivity, isn't it? Thermal conductivity. When I talked about a property that is common to them. Thermal conductivity, we can talk in this way, good thermal conductors, can't we? Some are good thermal conductors and other are not so good thermal conductors. How do you classify these groups here (she points to the groups that she have written down in the blackboard)? First this one with aluminum, cooper and steel and then this other one, with clay, iron and stone. Which one you think are the best thermal conductors?

143. Some students: Clay.

144. S13: aluminum, copper and steel.

145. The best conductors?

146. S13: aluminum, copper and steel.

147. S(?): Hold on a minute....

148. T: Previously I had asked you to say that... when you put the pan on the fire it begins to get heated near the flame. Then someone said that the heat goes up, it goes up to the entire material. Where this is going to happen quickly, in the aluminum or, for example, in the stone, in the clay.

149. Several students: In the aluminum.

150. T: In the aluminum. Then what is the best thermal conductor?

151. Several students: The aluminum.

152. T: The aluminum, do you agree? When I talk about good conductors. The good conductors are those, which heat faster, ok? [...]

From these transcripts we can notice that when the teacher tries to introduce the theoretical referents "thermal conductivity" and "good conductors", the discourse becomes univocal, following an I-R-E pattern (**I**nitiation from the teacher, normally using a question, **R**esponse from the students and **E**valuation from the teacher) (Mehan, 1979). The students' participation is restricted to answering through single words, completing slots in the teacher discourse. The evaluation from the teacher can be stated through repeating the question, as showed in turn 145.

It also can be noticed that the teacher herself fails in making the shift from empirical to theoretical referentiality, as this would demand considering two variables when passing from the pans to the material from which the pan is made: the thermal conductivity of the material and the mass of the pan. The variable "mass" of the pans is referred to by the teacher in the turns 39 to 41, but is not recovered in this final discussion. As a consequence, the students failed in recognizing that the pan made of iron, although classified in the group of pans that get heated slowly, is made of a material that is a good thermal conductor - iron. In this case it is the mass of the pan and not the material, from which it is made, what makes the pan being heated slowly. The statement from the teacher in turn 152 ("The good conductors are those which heat faster, ok?") is true only when applied to the theoretical referent - the material. It is not completely true if the referent is the empirical one - the pans. As it is not clear from the discourse whether the teacher is referring to the material or to the pan, as she did not make this distinction clear, probably the students failed in recognizing this difference.

4. CONCLUSIONS AND IMPLICATIONS

The analysis of this particular episode confirms the pattern of discourse documented in other works (Mortimer, 1998, Scott, 1998, Mortimer and Machado, 2000, Scott and Mortimer, 2002), which alternates dialogical and univocal "texts" in constructing scientific meanings. In addition, we observed a high correlation between dialogical discourse and empirical referentiality on the one hand and between univocal discourse and theoretical referentiality on the other hand. The students' participation tends to decrease as the referential perspective shift from empirical to theoretical. Figure 2 shows that the first three sub-events, in which the empirical referentiality prevails, have a higher number of turns per minute when compared with sub-event 4, where the theoretical referentiality prevails. In addition, the quality of the student's talk decreases as the theoretical referents were introduced: from complete sentence uttered in the first three sub-events to single words to complete slots in the teacher's talk in the fourth sub-event The theoretical referents and scientific meanings constructed in this discourse are, therefore, highly controlled by the teacher and the students have very few opportunities to elaborate a theoretical discourse. If learning science is, in a sense, learning to "talk science" (Lemke, 1990), it is very important that the students learning to talk also when the referents are the theoretical entities of science.

By documenting and analyzing these patterns of discourse in science classroom we can generated knowledge about the ways teachers use language to make scientific meaning available in the classroom, which can be a useful tool in prepare reflective teachers in programs of professional development. Several curricular proposals for Science Education, from constructivism to S-T-S and S-T-L (science and technology literacy) have emphasized the importance of teacher-students and student-student interactions but little is known about how these interactions are enacted. The analysis carried out in this and in other papers in the same tradition shows that talk is just as important as the activities developed to address students' conceptions. Science teachers should be informed about the variety of ways in which talk can be done in science classrooms and about the ways to support students talk not only in relation to their everyday concepts and experiences but also in relation to the various theoretical ways of talk that populate science classrooms.

ACKNOWLEDGMENT

This work was supported by grants from CNPq, Brazil.

REFERENCES

Bakhtin, M.,M. (1981). *The dialogic imagination*, ed. by Michael Holquist. Austin: University of Texas Press.

Bakhtin, M.,M. (1986). *Speech Genres & Other Late Essays.* Austin: University of Texas Press.

Edwards D & Mercer N. (1987). *Common Knowledge*. London: Routledge

Lemke, J.,L. (1990). *Talking Science. Language, Learning and Values*. Norwood, NJ: Ablex Publishing Corporation.

Mehan, H. (1979). *Learning Lessons: Social organization in the classroom.* Cambridge, MA: Harvard University Press.

Mortimer, E., F. (1998). Multivoicedness and univocality in classroom discourse: an example from the theory of matter. *Int.J. of Sci. Ed.*, 20(1), 67-82.

Mortimer, E.,F. & Machado, A.,H. (2000). Anomalies and Conflicts in classroom discourse. *Science Education*, 84, 429-444.

Mortimer, E.,F. & Scott, P. (2000). Analysing discourse in science classroom. In: Leach, J., Millar, R. & Osborne, J. (Eds), *Improving Science Education: The contribution of research* (pp.126-142). Milton Keynes: Open University Press.

Mortimer, E., F. & Scott, P. (in press). *Making meaning in science classroom.* Milton Keynes: Open University Press.

Ogborn J, Kress G, Martins, I. & McGillicuddy, K. (1996). *Explaining science in the classroom.* Buckingham: Open University Press.

Santos, F.,M.,T. (2001). *Múltiplas dimensões das interações em sala de aula*. Unpublished PhD thesis. Belo Horizonte, Faculdade de Educação da UFMG.

Scott, P.,H. (1998). Teacher talk and meaning making in science classrooms: A Vygotskian analysis and review. *Studies in Science Education*, 32, 45-80.

Scott, P., H. & Mortimer, E., F. (2002). *Discourse activity on the social plane of high school science classes: a sociocultural tool to analysing and planning teaching interaction.* Paper presented at the annual meeting of AERA. New Orleans, April 2002.

Vygotsky, L.,S. (1934/1987). *Thinking and Speech*, in *The collected works of L.S. Vygotsky, Volume 1.* R.W. Rieber and A.S. Carton (Eds.), N. Minick (Trans.). New York: Plenum.

Wertsch, J.,V. (1991). *Voices of the mind: a sociocultural approach to mediated action.* Cambridge, MA: Harvard University Press.

Wertsch, J.,V. (1998). *Mind as action.* New York: Oxford University Press.

PART II

Learning and Understanding Science

STUDENTS' POSITIONS IN PHYSICS EDUCATION. A GENDERED PERSPECTIVE

Gertraud Benke, Helga Stadler
IFF, Wien, Universität Wien

ABSTRACT

In this paper we present first findings from a study on gender differences in physics education based on a questionnaire addressing four different areas: 1. Attitude and interest, 2. Epistemology, 3. Knowledge of basic physics, 4. Awareness of gender differences in physics. We discuss the interplay of interest, epistemological stances, ways of conceiving physics and gender. Contrary to our expectations, we did not find clear relations between gender and epistemological stances as well as between gender and conceiving physics. Instead we found class effects (self-selection in tracks), and answers on open-ended questions hint at an intricate relationship between understanding and (gendered) interest patterns.

1. PROBLEM STATEMENTS

Girls are only moderately interested in physics. This is repeatedly found not only in studies asking for students' interests (e.g. Hoffman et al. 1998) but is also reflected in students' choice of college program as well as their later profession. Studies show that girls and women lack self-esteem with respect to physics in comparison with their male peers. On top of that they show a poorer performance in physics assessments (TIMSS Studies, Stadler, 1999).

In our previous research (Stadler, Benke, Duit, 2001; Stadler, Duit, Benke, 2000), we developed the hypothesis that (most) girls and (most) boys approach physics differently, that they employ different comprehension strategies. Boys tend to believe that they understand something, if they can position the new "concept" within their overall understanding of physics, i.e. relate it to other (physical) concepts. In contrast, girls try to understand physical notions relating them back to elements and configurations of elements in everyday life.

In our present research, we include additional student properties, and try to map out the relation of these properties to gender and the overall contribution of these factors for engagement in and understanding of physics. Thus, we raise the following general question:

How does an individual's identity influence their participation in school physics? In this, we see the following aspects as crucial elements of an individuals' identity construction. (In attribution theory terms, these three aspects address stable intrinsic properties of a person which co-determine their actual behavior):

D. Psillos et al. (eds.), Science Education Research in the Knowledge-Based Society, 81–87.

- A person's relation to knowledge (the epistemological stance, as originally put forward by Belenky et al. 1986) – this describes the level of empowerment in dealing with given information – may I question it, do I have to accept it, etc. An important part of a person's relation to knowledge is also determined by the person's conception of knowledge (i.e. what constitutes "proper" knowledge in physics).
- A person's self-perception of interests. In our research, again and again we found students expressing a relation between "understanding" something and being interested in something. If they cannot "figure it out" (being active/empowered), they state they loose their interest. In other words, we see "interest" a comprehensive category capturing "motivation".
- A person's gender identity (and the relation to the perceived gender roles). In that, we are guided by concrete interactions and conversations with girls in addition to several accounts in the literature on gender differences (e.g. Boaler, 1997). Based on these accounts, we expect that girls and boys differ with respect to their consciously expressed description of what it means to understand something (in physics). Furthermore, we hypothesize that their notion of understanding will have implications for their engagement in and with physics.

Using this framework, we are addressing the following concrete research questions:

- What is the distribution of epistemological stances, ways of conceiving of physics, self-perception of interests, and the perception of gendered interests for the classes analyzed?
- Do we find gender differences in these areas?
- Is there a relation between interest in physics and epistemological stances?

Study design

Questionnaire, interview, and video-data were collected in three classes of high-school students (age 15/16) taking their first year of advanced physics. Our results are predominantly based on the analysis of (two sets of) 48 questionnaires administrated in the three classes (with a total of 25 boys and 23 girls). The classes differed in their natural science orientation: two classes were from a humanities tracked program, one from a natural science tracked program. The questionnaires were distributed before and after a unit on the earth's movement. The unit itself was videotaped, and will be analyzed in further research. Additionally, one month after the unit in-depth-interviews were done with a small number of the students.

The questionnaire poses questions in four different areas: 1. Students' interests and attitude towards physics, 2. students' knowledge of gravitation (which allows us to discern their use of argumentative patterns in physics), 3. students' epistemological position, and 4. students' perception of gender differences with respect to physics.

The majority of questions were open ended; a subsection of questions concerning students' interest and gender stereotypes contained rating scales. The questionnaires were handed out by the physics teacher as a homework assignment. Answers were overall surprisingly detailed.

For further analysis, answers were read carefully, clustered into categories derived from the data (Strauss, 1987), and subjected to descriptive statistics.

2. RESULTS

In the following, we will discuss our findings concerning students' interests, students' implicit self-positioning with respect to knowledge, students' ways of thinking about physics and interactions between these areas for both girls and boys.

Students' Interests

In the questionnaire, students were asked to name up to three of their favorite and of their least favored subjects. For each of the subjects, we calculated the overall frequency (of being liked and being disliked), and looked at affinities between subjects. We then clustered the natural sciences (physics, chemistry – but not biology, which seems to be a category of its own) and mathematics, and compared the frequency over all classes with interest in the language arts. We found the frequently reported gender breakdown with girl preferring language arts, and boys opting for natural sciences. However, when looking at each class individually, girls in the science track tended to be overall more interested in mathematics and science than the boys. Girls in the humanities track on the other hand were more interested in language arts than the boys of their classes and the boys overall.

Besides asking students to name subjects, we also asked them to rate their own as well as boys and girls general interest in particular topics in physics ("How to construct a car", "to understand how glasses work" etc.). In our analysis, we compared for each class the (average) interest girls/boys professed for themselves in the particular topic with the interest that was attributed to girls/boys by their own and the other (gender) group. Utilizing t-tests (based on the whole population, calculated separately for each topic), we found that both boys and girl exaggerate the actually existing differences: Boys believe girls overall to be less interested in the various topics than they (girls) actually are, and girls believe boys to be more interested than they (boys) are. Overall, the difference between the (actually expressed) interests of girls and boys is significantly smaller (frequently there is no statistically significant difference between the actual interests of boys and girls) than the difference attributed to the genders by either gender (that is the difference girls (boys) attribute to "the interest of girls in general" and "boys in general"). Frequency tabulations for each of the classes are congruent with the general picture.

Students' self-positioning with respect to knowing and knowledge

To determine students' epistemological attitudes, we asked the following questions (among others):

1. How do you know whether something is true or not [post test addition: "in physics"]?
2. Are all opinions [post test: "in physics"] equally good, or are some answers better than others?
3. Based on Belenky et al.'s (1996) typology, we categorized the answers into the following types:

- Appeal to authority. Answers in this category either appeal to characteristics of the person uttering the questionable statement ("it depends where s/he got their information from, or whether they deal with it professionally") or they refer to anonymous sources (like encyclopaedias, TV etc.) ("It is good, if it is to be found in a encyclopaedia".)
- Subjective position: e.g. "Every opinion is equally good", "I know by 'feeling'" Answers in this category appeal to ones feelings about something rather than arguments and logic.
- Argumentative position: Answers in this category require "proof", logic, an argument for something to be believable and "good". ("It depends whether something sounds logical, then I'll believe it." "Arguments need to be put forward (which I can follow)".)
- Additionally, we categorized the "mixed" forms of "authority and subjective position", and "subjective position and argument", etc. ("If it has been proven, I believe it. If not, it depends on who is telling me and whether it sounds believable." "Either I've heard the same somewhere else, or I simply believe, that it could be true".)

On a qualitative level, when trying to apply the categories derived from the literature, we found ourselves repeatedly unsatisfied with our original conception of "authority". Students frequently displayed an awareness of the complexity of the subject matter, and the overall social organization of knowledge, which puts people into expert positions, constituting a social system for granting believability to particular individuals. Yet, on the other hand, deference to (such) authorities does frequently imply a passive position with respect to the subject matter, as the student does no longer submit something to his or her own reason. Other students, when considering physics, appeal to encyclopaedias etc. Again, formally, they are deferring judgments to some "outside authority". On the other hand, we see an active involvement in doing so, as students need to go out, discern what kind of information they need, and search for an entry in the encyclopedia, etc. The distinction of level of involvement with "the expert" and "the encyclopaedia" is also reflected linguistically, in that students state "I look it up", but never "I go see an expert". Instead "the expert says something", i.e. they incidentally (or in their physics classes) hear something attributed or said by an expert. Thus, we ended up classifying "encyclopaedia" statements as belonging to the argumentative discourse, as it afforded an active awareness of a lack of knowledge and an action to overcome it. (In our first look at the data, we introduced a specific category for these answers, however, it turned out that in almost all cases, students answering this way, were answering either additionally or in the following questionnaire item from an argumentative position.) In some sense, that is what the argumentative position is all about anyways: to have an individual as an active participant in the construction of his or her knowledge.

Interestingly, we did not find a general pattern of epistemological positioning for all classes with respect to gender. Instead, we found that the girls of one of the humanities classes asked more for argumentative answers than the boys of the same class, who reasoned from a subjective point of view. In the other two classes, this was just the opposite with boys asking more for arguments than girls, who were appealing to authority or taking a subjective point of view.

Overall, we find that girls as a group are more heterogeneous; it is more often the case that girls answer a question which contains elements from more than one position (see example below), and in almost every class at each test, the number of categories girls are in is larger than the number of categories which were needed to classify the boys answers. Furthermore, it was mostly girls (11 of 13 cases) who combined an argumentative position with a subjective position as in "[I believe something] if it sounds logical, or if I trust the person."

Concerning the epistemological question, we did not expect to find a short-term physics instruction about the movement of the earth to make any real impact, and varied the question. (See above). Looking at the differences between the two questions, addressing beliefs about what it takes for something to be believable, and what it takes for something to be believable in physics, we found that students were less inclined to answer from a subjective point of view when thinking about physics. Thus, "general" "subjective thinkers" moved to either an argumentative position dealing with physics, or they deferred to experts (authorities).

In our analysis, we also looked at the relation between epistemological position and affinity for a particular subject. Would students interested in language arts differ from students expressing an interest in physics? We found the following tendencies:

- Students appealing to authority are mostly quoting physics as a subject they dislike, and generally state an interest in language arts.
- Students who express a dislike of physics are more likely to appeal to authority (6 of 11) than answering from a subjective or argumentative position.
- Students who favor physics (and say so) tend to request arguments (argumentative position). Yet, the converse is NOT valid: students who request arguments are just as likely to favor physics as they favor the language arts.

<u>Way of conceiving physics</u>

In the questionnaire, we asked students to respond to the following question:

- Imagine you are in a school of the future and you just learnt "everything" about the sun. What questions are you able to answer now? What do you think do you need to be able to answer, in order to be able to claim that you know "everything" about the sun?

Answers were classified according to whether the answers were

1. questions about quantities (how big is the sun),
2. a phenomenon (ozone, what is it made of) or
3. qualitative (why does the sun not move).

(Additionally, we categorized responses that contained questions of more than one type).

In other words, qualitative questions asked for causal explanations, answering some deeper – and sometimes philosophical – question. In contrast "phenomena" were concerned with the description of the functioning of some phenomena, the inner workings without any overt concern for anything beyond the phenomena itself.

In our analysis, we found a tendency for students to replace questions about quantities as well as qualitative questions by questions about phenomena after the instruction. Overall we did not find any significant gender pattering although girls tended to use slightly more qualitative questions, and they retained more quantitative questions after the instruction than the boys. Interestingly, we did not find any class effects (i.e. difference between the tracks) in this case either. Additionally, we did not find any tendency of an interaction between ways of conceiving physics and students' epistemological stances.

We see this as an indication that the image of physics is quite homogeneous in our population with respect to the analysed variables. Girls and boys, students choosing a natural science track and a humanities track, students favoring different epistemological positions, they all show a similar (and similarly varied) approach to raise questions about some object of (physical) interest. There is always roughly the same breakdown of students (boys, girls, humanities students etc.) who want to know about quantities, others who ask about qualities and still others who raise questions about physical phenomena.

3. CONCLUSIONS

Overall, we find students of the different classes to maintain a similar image of physics, which was probably formed by their previous schooling experiences. The perspectives of what constitutes a physical statement/question are quite varied, yet not linked to any observed individual variables. The unit itself – something very different from their up-to-then mostly frontally taught physics experience – changed the ways students formulated physical questions. Likewise, and contrary to our expectations, the epistemological stances show less a difference between different gender groups, than a difference between positions taken "in general" and with regards to physics. Addressing physics, students took more often the more "objective" argumentative position.

Epistemological stances, however, interacted in our sample with interest, with students being less interested in physics making more appeals to authority, students appealing to authority being generally interested in language arts and students showing an interest in physics asking for arguments. Due to the small sample size, we could not discern a further gender effect in this interaction, however, it is clear that interest itself is highly gendered, as discussed above.

We thus find it is mostly interest patterns and attributions of interest, which show a clear gender breakdown in our data. The lack of a gender pattern with respect to the other variables seems surprising considering the different approaches girls and boys choose to approach a task in physics, and their differing ways of relating to physical phenomena, as discussed in our previous research. We explain this partly by a "class effect". Students self-select themselves into the natural science and the humanities tracked programs (in Austria usually in the 8th grade), which are generally displaying a skewed gender distribution (with the humanities programs chosen predominantly by girls, the natural science programs by boys).

Correspondingly, we found great differences between the individual classes, although they were all taught the same unit(s) by the very same teacher.

But the story does not end here. The answers in the questionnaire point to an intricate relationship between interest and understanding, with students expressing: "[I'm interested in] physics (because I understand everything)", and [I don't want to become a physicist because] I don't understand it completely, it is complex and therefore boring". Again and again we find students setting up this relationship, sometimes also in the other direction: "I'm not interested, therefore I don't pay enough attention". Looking at these answers (which we have encountered numerous times also in student interviews in other schools), one gets the impression that students are caught in a loop of des/interest and non/understanding, with one fuelling the other. Yet, students tend to blame themselves, as not being interested enough, as not paying enough attention, to be able to comprehend. Additionally, girls face the negative perception of the overall interest in physics of girls in general, making it easier for themselves, as well as the people around them, to give up (on them), to explain failure to understand in terms of gendered interest patterns.

We conclude that in order to archive greater gender equity in physics classes, we need to pay particular attention to girls' modes of understanding and relating to physics, in order to retain and re-install girls' interest in physics. To do so, we need a better understanding of what it means for girls (and boys) to "prove" something, to have something "which is logical". In our study, girls display a greater in-group variation concerning their requirements to believe in something (than boys). Addressing girls' needs thus implies a classroom culture, which allows multiple pathways – for girls and boys alike – to approach physics.

REFERENCES

Belenky, M. F., Clinchy, B. M., Goldberger, N. R., & Tarule, J. M., (1997, 1986[1]): *Women's Ways of Knowing*. New York: Basic Books.

Boaler, J., (1997): *Experiencing School Mathematics. Teaching styles, sex and setting*. Buckingham: Open University Press.

Hoffmann, L, Häußler, P., & Lehrke, M., (1998): *The IPN-Study of Interests in Physics*. Kiel, Germany: IPN - Institute for Science Education at the University of Kiel.

Stadler, H., (1999): TIMSS 3 in Österreich – geschlechtsspezifische Aspekte. In: *Tagungsband der 62. Physikertagung der Deutschen Physikalischen Gesellschaft*. Didaktik der Physik, Ludwigsburg.

Stadler, H., Duit, R., & Benke, G. (2000): Do boys and girls understand physics differently? In: *Physics Education*, 35(6), November 2000, 417-422.

Stadler, H., Benke, G., & Duit, R. (2001): How do boys and girls use language in physics classes. In: Behrendt, H., Dahncke, H., Duit, R., Gräber, W., Komorek, M., Kross, A., & Reiska, P. (Hrsg.), *Research in Science Education Past, Present, and Future*. Dordrecht: Kluwer.

Strauss, A. L.(1987): *Qualitative Analysis for Social Scientists*. Cambridge: Cambridge University Press.

SITUATED CONCEPTIONS AND OBSTACLES. THE EXAMPLE OF DIGESTION / EXCRETION

Pierre Clément
Didactique Biologie et Environnement, LIRDHIST, Université Claude Bernard Lyon-1, France

ABSTRACT

The concept of epistemological obstacle, as defined by Bachelard (1938), is often used in French science education research and teaching. New data are presented on the classical example of conceptions on digestion / excretion. Differences between biology and other sciences students' samples appear significant, except in the first year of University. The work reported in this paper illustrates the notion of "situated conceptions", which differs when a student answers different questions on digestion / excretion. We discuss how the concept of obstacle is useful to interpret the conceptions. We propose three complementary categories of obstacles: epistemological, didactical and psychological.

1. BACKGROUND

The notion of conceptions is to-day central in the European research in Science Education; "conceptual change" is now used for "learning science" (Duit 1999). Nevertheless, the terminology "obstacles" to learning (obstacles to change conceptions) is only used in France and in Southern Europe countries (e.g. Astolfi et al 1997a). The Science philosopher Gaston Bachelard (1938) introduced the concept of epistemological obstacle when the common thinking is an obstacle to the (historical or individual) building of a scientific knowledge.

The "common thinking" of Bachelard is included in the notion of "conceptions" (or, for some authors, "preconceptions" or "misconceptions"[1]). The obstacle is identified after a deeper analysis of the conceptions: the obstacle is more stable, more transversal than the conception. The identification of the obstacle is useful to explain the conception and to imagine appropriate pedagogical strategies.

[1] The notion of " misconceptions " is dangerous, because we have to analyse the conceptions, and not to judge them. Every conception is useful until it is falsified in precise situations. A scientific conceptualisation is also a not yet falsified conception. No cortical areas are devoted for "scientific conceptions", and other for "misconceptions" ! (Clément 1994)

D. Psillos et al. (eds.), Science Education Research in the Knowledge-Based Society, 89–97.

2. AIMS AND FRAMEWORK

I have been working on digestion / excretion since 1979 (Clément et al 1981, 1983, Clément 1991, 1994)[2]. I ask different samples of students to draw and label the pathway that a litre of beer tapes through their body. Drawings are then classified in 3 categories of conceptions (figure 1). From my first results (Clément 1991), the proportion of the 3 categories is characteristic of each kind of samples.

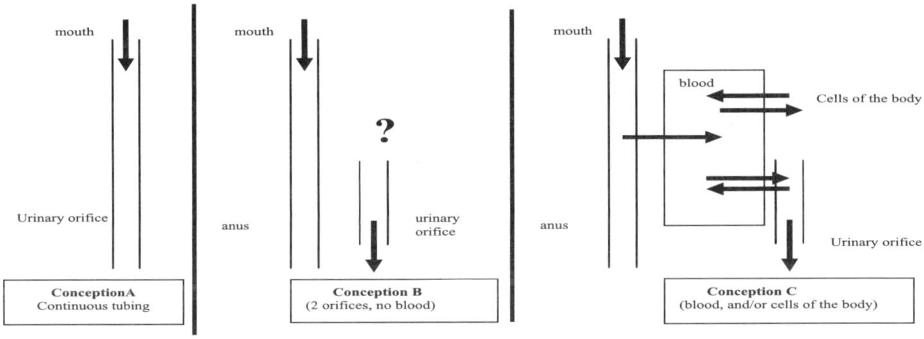

Figure 1. The 3 conceptions on digestion / excretion

Using the classical example of digestion / excretion to teach in the University the notion of "conceptions", I obtained a lot of new data. Their analyse shows:

- The same amount of the 3 conceptions since 1991, for each type of sample;
- A difference between Biology and other students after the 1st year in University;
- Variations of the same student's conceptions according to the situations ("situated conceptions": Clément 1994, 1999);
- Different obstacles: epistemological, didactical, and psychological.

3. METHODS AND SAMPLES

Samples: A total of 752 persons (students, teachers and researchers: table 1) wrote answers to one or two successive questions.

[2] Other teams also worked on this example: e.g. in France: Giordan & de Vecchi 1987, Sauvageot-Skibine 1993, Astolfi et al 1997b, … ; in other countries : Carey 1985, Nunez & Banet 1997, Teixeira 1998, Psarros & Stavidrou 1999, …

Table 1. Main characteristics of the samples

Number	Size	Scientific level	Scientific area	Localisation	Date(s)	Question(s)
A	81	Researchers	Cognitive sc.	Univ. Compiègne	1997	(I)
B	16	Students 4th year	Museology	Univ. Mulhouse	1997	(I), (II)
C	28	Students 5th year	Cultural Managem.	ENESAD Dijon	1995, 1997	(I), (II)
D	80	Students 2nd year	Info. & Com.	Univ. Lyon 1	1993, 4, 5, 6	(I)
E1 E2	55 54	Students 1st year Students 2nd year	Medicine Medicine	Univ. Nantes Univ. Nantes	1994 1994	(I) (I)
F1 F2	43 71	Students 2nd year Students 2nd year	Biology Other Sciences	Univ. Lyon 1 idem	1996, 7, 8, 9 idem	(I), (II / III) (I), (II / III)
G	76	Students 2nd year	Biology & other Sc.	Univ. Lyon 1	2000, 2001	(I)
H	58	Students 1st year	Biology & other Sc.	Univ. Lyon 1	1992	(I), (II)
J	43	Students 1st year	Biology & other Sc.	Univ. Lyon 1	2001	(I), (II)
K1 K2	47 62	Students 1st year Students 1st year	Biology Other Sciences	Univ. Montpellier 2 idem	2000, 2001 idem	(I), (IV) (I), (IV)
L1 L2	9 14	Teachers (sec.) Teachers (sec.)	Biology Other Sciences	Univ. Clermont Fd idem	1994 idem	(I) (I)
M	15	Doctorants (begin to teach in the Univ.)	Biology	CIES (Univ. Lyon 1, Dijon, Clermont, St Etienne)	2001	(I)

Questions

Question (I). The first question is always introduced by this oral text : "You are very thirsty. You go in a pub and you drink a litre of beer. You know that, some minutes after, you will have to go to urinate ! Draw and label quickly the way of this litre of beer in your body, after you drank it and before you go to urinate. Where is the litre of beer going in your body?"

Question (II). For some samples (table 1), a second question is asked, introduced by this oral text : "You visit a diamond mine. You see a fine little diamond, you swallow it to keep it inside you and recover it at the exit, the day after, at home. Then, you drink a litre of beer because you are very thirsty. Draw now (with labels) the way of this diamond and of the litre of beer in your body."

Question (III). In some other samples (table 1), the second question was shorter and more academic: "Just draw now, and legend, the anatomy of your digestive tract".

Question (IV). In one case, there were two successive questions after the question (I): "The intestine opens outside by an orifice. What is the name of this orifice?" "You drink water. The water is going into your stomach, then into your intestine: and then? The water is not living your body by the orifice of the intestine! Where is the water just after the intestine?"

Context. I always asked the question(s) in the beginning of a conference or a lecture. The answers are anonymous. In some cases, the students write down if they are or not biologists (table 1: samples E, F, K, L). The answers to question (I) were gathered before they answered the second question. To compare their 2 answers, students had to choose a code.

Analysis of data.

Question (I): Answers are classified in 3 categories of conceptions (figure 1).

Other questions (II, III, IV): the goal of these questions is to verify if the students know or not the name of the exit orifice of the intestine (anus); and if they think or not that the liquid we drink is going into the blood (circulatory system).

4. RESULTS FROM QUESTION (I)

Various samples of Human and Social Sciences: The conception A is the most frequent (60 to 90 % of the answers: table 2). The same results had been obtained for comparable samples by Clément (1991).

Table 2. Answers to question (I) by various samples of Human and Social Sciences

Sample	**Conception A** (continuous tubing)	**Conception B** (only dig. + excretion)	**Conception C** (dig. + circ. + excretion)	Total
A -Researchers in **Cognitive Sc.**	57 **70 %**	12 **15 %**	12 **15 %**	81 100 %
D – Students 2nd year, **Info. Com.**	57 **71 %**	14 **18 %**	9 **11 %**	80 100 %
B – Students 4th year, **Social Sc.**	14 **88 %**	0 **0 %**	2 **13 %**	16 100 %
C – Students 5th year, **Human Sc.**	29 **60 %**	11 **23 %**	8 **17 %**	48 100 %

Science students of the <u>first</u> year of the University (Table 3) : There is no difference before and after 1991 (K1 and before 1991), nor between Biology and other Sciences students (K1 and K2) : the amount of conception A stays around 70 % (table 3). The amount of conceptions A is less important (55 %) for the samples E1 (medicine) and J (teachers to be) : Chi2 = 19.65 ($\alpha < 0.006$ %).

Table 3. Answers to question (I) by Science students of the 1st year of the University.

	Conception A (Continuous tubing)	**Conception B** (Only dig. + excretion)	**Conception C** (Dig. + circ. + excretion)	Total
Before 1991 **Biologists** (Clément 1991)	53 **72 %**	11 **15 %**	10 **13 %**	74 100 %
Sample **K1** **Biologists**	36 **73 %**	4 **8 %**	9 **18 %**	49 100 %
Sample **K2** Other sciences	43 **69 %**	7 **11 %**	12 **19 %**	62 100 %
Sample H **Bio + Other Sc.**	41 **71 %**	7 **12 %**	10 **17 %**	58 100 %
Sample **J** **Bio + Other Sc.**	*24* ***56 %***	*4* ***9 %***	*15* ***35 %***	*43* *100 %*
*Sample **E1*** **Medicine**	*30* ***55 %***	*2* *4 %*	*23* *42 %*	*55* *100 %*

Science students of the <u>second</u> year of the University (Table 4) : The conception A is very important in the sample F2 (other science students : 83 %, table 4). It remains present in biology student samples (Table 4: 40 to 55 %; there is no significant difference before 1991 and F1).

The conceptions differ between Biologists and other Sciences students (samples F1 and F2 : Chi2 = 22.87, α =0.001%). The sample G, with Biology and other Sciences students, is intermediate between F1 and F2.

Significant decreasing of conceptions A (28 %) and increasing of conceptions C (65 %) are observed in the sample E2 : compared to Biologists before 1991, Chi2 = 19.37, α =0.006%. The explanation is the severe selection of the students E2 after the very competitive examination at the end of the first year of Medicine.

Table 4. Answers to question (I) by Science students 2nd year of the University

	Conception A (Continuous tubing)	**Conception B** (Only dig. + excretion)	**Conception C** (Dig. + circ. + excretion)	Total
Before1991 ***Biologists***	*71* ***55 %***	*19* ***15 %***	*38* ***30 %***	*128* *100 %*
Sample **F1** **Biologists**	17 **40 %**	6 **14 %**	20 **46 %**	43 100 %
Sample **F2** Other sciences	59 **83 %**	4 (1 irony) **6 %**	8 **11 %**	71 100 %
Sample **G** **Bio + other sc.**	46 **61 %**	15 **20 %**	15 **20 %**	76 100 %
*Sample **E2*** **MEDICINE** (*)	15 **28 %**	4 (4 irony) **7 %**	35 **65 %**	54 100 %

Biology doctorants and Science teachers (Table 5) : 0 % of conceptions A & B, and 100 % of conceptions C, are only observed for Biology teachers (secondary school). The Biology doctorants (who begin to teach Biology at the University first cycle) are very near 100 % : the 2 doctorants with conceptions A & B work in biochemistry. The other Sciences teachers do not significantly differ before and after 1991 (e.g. for conceptions A : 48 % in Clément 1991 ; 36 % sample L2).

Table 5. Answers to question (I) by Biology doctorants and Science teachers

	Conception A (Continuous tubing)	Conception B (Only dig. + excretion)	Conception C (Dig. + circ. + excretion)	Total
Sample **M** Biology doctorants	1 **7 %**	1 **7 %**	13 **87 %**	15 100 %
Sample **L1** Biology teachers	0 **0 %**	0 **0 %**	9 **100 %**	9 100 %
Sample **L2** Other Sc. teachers	5 **36 %**	1 **7 %**	8 **57 %**	14 100 %

5. RESULTS FROM QUESTIONS (II), (III) & (IV)

The goal of questions (II) or (III) or (IV) was to strongly induce other conceptions than continuous tubing. I have no place enough to present all the results announced in table 1. I just take an example for each question (the other samples present the same type of results).

Question (II) (ingestion of a diamond before the litre of beer: cf above: methods). The question (II) was asked just after the question (I), to science students (2nd year university, 1996: part of the sample F).

Table 6. Answers to the question (I) and then to the question (II) by science students 2nd year of the University (n = 19)

QUESTION (I)	QUESTION (II)		
	Conception A	**Conception B**	**Conception C**
Conception A **9 (47 %)**	Single tube: 1 Bifurcated tube: 6 Single tube + blood: 1	1	0
Conception B **7 (37 %)**	Bifurcated tube: 4	2	1
Conception C **3 (16 %)**	0	0	3
Total ***19 (100 %)***	***12 (63 %)***	***3 (16 %)***	***4 (21 %)***

- The students who expressed conceptions C to answer to the question (I) mobilised this conception C to answer to the question (II) (table 6).
- The majority of conceptions A after question (I) stayed conceptions A after question (II), with an innovation: the bifurcated tube: a tube for the diamond (to the anus) and another for the liquid (to the excretory orifice). Only one student A moved to conception B ("?" between the digestive and excretory systems) after the question (II).
- More unexpected was the evolution of initial conceptions B: 4/7 students drew a bifurcated tube (conception A) !

Finally, the amount of conceptions A is more important after the question (II) !

Question (III). (*"Draw schematically your digestive tract"*). The question (III) was asked just after the question (I), to science students 2nd year university (part of the sample F). The results are presented in tables 7 & 8

Table 7. Answers of students 2nd year of the university to question (I)

	Conception A	Conception B	Conception C	Total
Biologists	3 **27 %**	0 **0 %**	8 **73 %**	11 100 %
Maths, Physics, Chemistry	23 **77 %**	2 **7 %**	5 **17 %**	30 100 %

Table 8. Answers of students 2nd year (the same as in table 7) to question (III)

	Correct schema	Incorrect schema	Total
Biologists	-->anus : 10 →rectum: 1 **100 %**	0 **0 %**	11 100 %
Maths, Physics, Chemistry	anus:22, colon, rectum:4, exit: 1 **90 %**	Stomach: 1, bladder: 2 **10 %**	30 100 %

These results are a good illustration of the notion of "situated conceptions" (Clément 1994, 1999): the same students displayed different conceptions for the exit of intestine. In an academic situation (question II : *"Draw the digestive tract"*), the exit was the anus (table 8). In the new situation (question I : the way of the litre of beer), 27 % and 77 % of them solved the problem by the invention of a new conception (A : continuous tubing digestion – excretion : table 7).

Questions (IV). After the question (I), some students (part of the sample K) had to answer on a new sheet to 2 new questions: questions IV (cf. above).

Table 9. Answers of students 1st year University to question (I)

	Conception A	**Conception B**	**Conception C**	Total
Biologists	6 **60 %**	0 **0 %**	4 **40 %**	10 100 %
Maths, Physics, Chemistry	13 **65 %**	2 **10 %**	5 **25 %**	20 100 %
Total	**19 (63 %)**	**2 (7 %)**	**9 (30 %)**	30 (100 %)

Table 10. Answers of the same students as in table 9, to question (IV-1)

	Correct answer	Incorrect answer	No answer	Total
Biologists	Anus : 9 **90 %**	Rectum : 1 **10 %**	**0** **0 %**	10 100 %
Maths, Physics, Chemistry	Anus : 15 **75 %**	Duodenum 1; Rectum 1 **10 %**	3 **15 %**	20 100 %
Total	**24 (80 %)**	**6**	**(20 %)**	30 (100 %)

Table 11. Answers of the same students as in tables 9 & 10, to question (IV-2)

	An urinary organ **(conception A)**	Blood (Conception C)	Total
Biologists	Bladder 4, Kidney 1, (male sex) 1 **60 %**	4 **40 %**	10 100 %
Maths, Physics, Chemistry	Bladder 8, Kidney 7, Urinary tract 1 **80 %**	4 **20 %**	20 100 %
Total	**22 (73 %)**	**8 (27 %)**	30 (100 %)

Table 10 confirms that the great majority of science students know well the anatomy of the digestive tract, and the outside opening of the intestine at the anus. Nevertheless, in the same time, more than 60 % of them end the intestine in the bladder (or in the kidney) when they answer to question (I) (table 9). That is a new confirmation of the notion of "situated conceptions".

More unexpected are the results concerning the question IV-2 (table 11). Saying that the ingested liquid is not going to the anus, this question wanted to suggest that, after the intestine, the water goes into the blood. Nevertheless, only 8/30 students answered that ! Less than the 9 conceptions C after the question (I) (table 9). So, around 3/4 of these students had apparently forgotten, even in this situation, that the ingested water is going into the blood.

6. GENERAL DISCUSSION: THE DIFFERENT OBSTACLES.

In a socioconstructivist perspective, any cognition is related to a precise context. Brown, Collins & Duguid (1989) argued that knowledge is situated, and proposed the notion of "situated cognition". The present work showed that, according to different situations, the same student can build or mobilise different **situated conceptions** (Clément 1994, 1999) that the intestine: ends at the anus, or it ends into the bladder, or it bifurcates in two tubes (digestive and excretory). These situated conceptions can be explained by the identification of some complementary kinds of stable obstacles.

(1) Epistemological obstacles:

Epistemological obstacles come from contradictions between the everyday life and the scientific knowledge. The easiest solution to explain the necessity to urinate after drinking is to imagine continuous tubing between the mouth and the excretory orifice (conception A). The obstacle comes from the contradiction between the good pipes of our everyday life (their walls are never permeable) and several tubes in our body. Body tubes walls are very permeable (intestine, capillaries, tubes of the nephron in the kidney, ...) ! The epistemological obstacle here is clearly **the permeability of the biological walls** of a lot of tubes in our body.

There are other epistemological obstacles to understand the anatomy and physiology of the 3 concerned systems (digestive, circulatory, excretory): chemical digestion, closed circulation, ... Nevertheless, in the context used in this work (questions and samples), these other obstacles did not play a decisive role.

(2) Didactical obstacles:

Didactical obstacles come from contradictions between previous teaching and the scientific knowledge. Our results show that most of the students had forgotten that the ingested liquid is going from intestine into the blood and into the whole body via the circulatory system. Nevertheless, this message had been given to them at several times in Primary and Secondary School. This lapse of memory can be the result of several didactical obstacles :

- The earlier French primary school textbooks often draw "the way of food" from mouth to anus, with precise times: a boy is eating an apple at midday; 1 minute after, the apple is in the stomach, and 14 hours it reaches the anus! There is a big mistake: that is not the "way of food", but the way of what is not digested! In these drawings, the blood is absent.
- In secondary school, the chapters digestion, circulation, excretion are separate, without any cross referencing question. Each chapter have several interesting messages: the permeability of the tube walls is only mentioned, without much relevance.
- The teaching and evaluation of biology is often too academic, and would need more connections with everyday life situations, as this one proposed by the question (I).

(3) Psychological obstacles :

The question of the permeability of biological walls is also a deeper problem. We analysed its presence after a work on conceptions of students on the living cells: only the external cyto-membrane was generally drawn as a "double membrane", and the exchanges through it were often forgotten (Clément et al 1983). We did this work with a psychologist who identified, besides epistemological obstacles (related to the process of the permeability), a psychological obstacle, called "le moi peau" ("the self skin"): our skin is also a permeable wall, supporting our communication with the others ...

More classical psychological obstacles are the steps of development of Piaget, and the relation with the matter "Biology", or with the school and the scholar learning. Coming from a recent sociological approach (Charlot 1997), these last dimensions must be taken into account by science education research.

7. CONCLUSIONS

This work shows the interest of the "obstacles" concept to analyse deeper the conceptions of students for research in didactics of Science. In this example, the methodology of the analysis used successive layers of conceptualisation: (1) Situated conceptions from the productions of the students; (2) Categories of conceptions; (3) Identification and analyse of the epistemological, didactical and psychological obstacles underlying the main conceptions. Then more effective pedagogical strategies can be proposed, based upon the identified obstacles.

REFERENCES

Astolfi, J.P. et al., (1997a). *Mots clés de la didactique des sciences.* Bruxelles: De Boeck

Astolfi, J.P. et al. (1997b). *Pratiques de formation en didactique des sciences.* Bruxelles : De Boeck.

Bachelard, G. (1938). *La formation de l'esprit scientifique.* Paris : Vrin.

Brown, J.S., Collins, A., Duguid, P. (1989). Situated cognition and the culture of learning. *Educational Researcher*, 18 (1): 32-42.

Carey, S. (1985). *Conceptual change in childhood.* Cambridge (Ma): MIT Press.

Charlot, B. (1997). *Du rapport au savoir. Eléments pour une théorie.* Paris :Anthropos.

Clément, P., Serverin, J.L., Luciani, A. (1981). Quelle digestion des représentations initiales dans la pratique pédagogique ?. *Pédagogiques*, 1(3), 20-22.

Clément, P., Serverin, J.L., Luciani A. (1983). Les représentations en biologie et les objectifs de la pédagogie : digérer ou régurgiter ? *Actes J.I.E.S.*, 5, 453-460.

Clément, P. (1991). Sur la persistance d'une conception : la tuyauterie continue digestion-excrétion. *Aster (Paris : INRP),* 13, 133-155.

Clément, P. (1994). Représentations, conceptions, connaissances. In A.Giordan et al, *Conceptions et connaissances* (pp.15-45). Berne : Peter Lang.

Clément, P. (1999). Situated conceptions. Theory and methodology. From the collection of data (on the brain) to the analyse of conceptions. in *Fourth ESERA* Summerschool (pp.298-315). U.Paris 7.

Duit, R. (1999). Towards inclusive views of conceptual change. *Proceedings ESERA-Kiel* (pp. 64-66).

Giordan A., de Vecchi G., (1987). *Les origines du savoir.* Genève : Delachaux Niestlé.

Nunez, F. & Banet, E. (1997). Students'conceptual patterns of human nutrition. *Int. J. Sci. Educ,* 19(5), 509-526.

Psarros, N. & Stavridou, H. (1999). The adventure of food inside the human body : Primary students'conceptions about the structure and the function of the human digestive system. In: Psillos, D. et al. (eds), *Science Education Research in the Knowledge Based Society* (pp. 745-747). ESERA / Aristotle University of Thessaloniki.

Sauvageot-Skibine, M. (1993). De la représentation en tuyaux au concept de milieu intérieur. *Aster* (Paris : INRP), 17, 189-204.

Teixeira, F.M.(1998). What happens to the food we eat ? Children's conceptions of the structure and function of the digestive system. In: Andersson, B. et al. (eds), *Research in Didakik of Biology,.,* Göteborg Universitet : IPD-rapporter, Na-Spektrum, 22, 97-110.

About Some Of The Difficulties In Learning Thermodynamics At The University Level

[1]Chantal Duprez, [2]Martine Méheut

[1]*University Of Lille*, [2]*IUFM Of Créteil, L D S P, University Denis Diderot, Paris, France*

ABSTRACT

The research work, which is set out here, constitutes one stage of a project whose goal is to develop situations for the teaching and learning of thermodynamics at university level. In a constructivist perspective, the knowledge of the way of reasoning among students is an essential element in the didactical reconstruction work of teaching objectives and contents. This research work has allowed us to define the important conceptual difficulties that students came across when studying thermodynamics and more precisely the second law of thermodynamics: Difficulty in recognising the changes in the state of the thermal reservoir; Difficulty in reasoning on the causes of transformation. These results will be used to elaborate contents and situations for teaching and learning thermodynamics at university level.

1. INTRODUCTION

The research work, which is set out here, constitutes one stage of a project whose goal is to develop situations for the teaching and learning of thermodynamics at university level.

Thermodynamics is regarded by students as a field of physics which is particularly abstract and difficult. There are different reasons for this: The relative lack of perceptive aspects (especially visual) in this field, a poor use of iconic representations and diagrams, teaching contents which are often very formal and poorly related to everyday life problems.

To overcome these difficulties, different resources can be mobilized: Problem situations rooted in everyday life situations (Welzel & Pospiech 2000) and visual representations (Boohan & Ogborn 1996).

Even then the conceptual difficulties met by students must be clearly recognized. In a constructivist perspective, the knowledge of the way of reasoning among students is an essential element in the didactical reconstruction work of teaching objectives and contents (Artigue 1988, Kattman et al. 1995). In this field, many studies have underscored some of the difficulties encountered by young pupils (Erickson 1985, Tiberghien 1985) or by teenagers (Kesidou & Duit 1993) in order to separate the concepts of heat and temperature. Other research works show that students have a tendency to focus on only one object (Rozier 1988), to reason only on one variable at a time (Rozier & Viennot 1991). They favour the variables, which

D. Psillos et al. (eds.), Science Education Research in the Knowledge-Based Society, 99–106.

are the most accessible to common sense. For instance, students only remember from entropy everything that is related to spatial variables (Grea & Viard 1995).

2. THE THEME OF THE RESEARCH WORK

The analysis of students' mistakes and first preliminary interviews brought us to study the way of reasoning linked to the notion of reversibility more precisely (How students define the reversibility of a transformation? What signs make them decide whether a transformation is reversible or not ?). When the situation studied deals with one or more thermal reservoirs, we have brought to light special difficulties linked to the notion of thermal reservoirs as these difficulties bring about mistakes in recognizing if a process is reversible or not. We will present in this paper these difficulties linked to the concept of thermal reservoirs.

A thermal reservoir is a system of very large dimensions and therefore of very large thermal capacity. When it interacts with a system (noted σ) of far lower dimensions (therefore of much lower thermal capacity), it exchanges energy by thermal transfer, while submitting an infinitesimal variation of its temperature, which is thus considered as constant. The notion of thermal reservoir is interesting for the physicist, because only its temperature needs to be known. It is not necessary to know the values of other variables such as pressure, volume, etc.

The questions, which are used to test the students' representations about the concept of thermal reservoirs, involve reservoirs, which are brought into contact with a system σ of different temperature. A thermal transfer takes place between the system with the highest temperature and the system with the lowest temperature. Such a transformation can be regarded as reversible when the difference, which is the cause of the transformation, in this case the difference of temperatures, trend towards zero.

3. METHODS AND SAMPLES

The population studied is made up of students attending the first year of the IUFM (Post-Graduate Teacher Training Institute). They have a university degree in Physics (minimum three years studies after secondary education). A preliminary study involved semi-directive interviews among students studying in one of the IUFM as well as questionnaires that second and third year of university students were asked to answer. From the results obtained, we evolved questions and divided them into three questionnaires. These questionnaires were then handed out to students from four different IUFM. They were distributed randomly among the students of a same IUFM. We can therefore conclude that the groups answering the different questions were of the same level. The total number of students was 130, that is to say than about 40 students answered each question. The questionnaires preceded the teaching of thermodynamics.

4. RESULTS

These preliminary interviews led us to consider the hypothesis that in situations where a system σ of finite dimensions is brought into contact with a thermal reservoir, students have clearly understood that there is no change in the temperature of the thermal reservoir but they have not taken into account the variation of energy in the thermal reservoir. We then produced several questions to confirm this hypothesis.

The first question (A) asks explicitly if the internal energy of the thermal reservoir changes. Two other questions are based on an identical situation where a system σ (a cube of small dimensions) is brought into contact with a hot reservoir and then restored to its initial state by contact with a cold reservoir. In these transformations, there is an irreversible transfer from the hot reservoir to the cold reservoir through the intermediary of the system σ. Question (B) asks explicitly if the thermal reservoirs have changed. The terms of question (C) include some obvious mistakes from the physicist's point of view with a localised type of reasoning, focusing attention on the system σ without taking the thermal reservoirs into account.

4.1 An explicit question about the variation of energy in the thermal reservoir

A cube (m=10g) is initially at a temperature t_1= 20°C. It is put in a vessel which contains 1000L of water at a temperature t_2= 80°C. As the quantity of water is important, we can consider it as being a thermal reservoir. The cube reaches a temperature t_2= 80°C.

Indicate below if the variables which characterize the state of the thermal reservoir have changed during the process. Give the details answers.

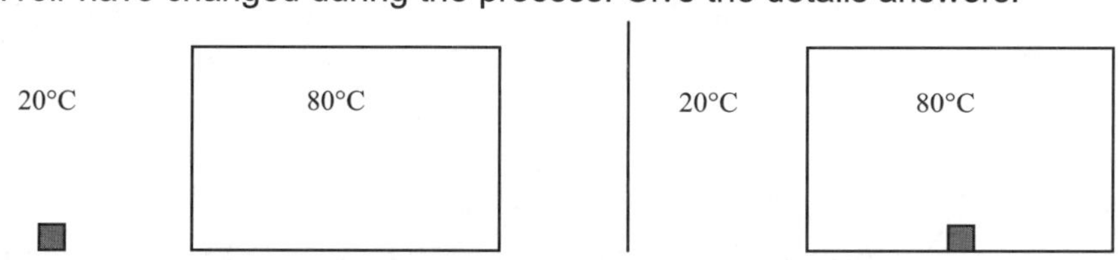

The temperature of the thermal reservoir changes
yes ❑ no ❑ I don't know ❑ Justify.
The internal energy of the thermal reservoir changes
yes ❑ no ❑ I don't know ❑ Justify.

Figure 1. Question A

* 79% answer that the temperature does not change,

– Either by definition (38%)

- Or because of the important difference of the dimensions between the thermal reservoir and the system of finite dimensions interacting with the thermal reservoir (17%).

* 17% answer that the temperature has changed, in contradiction with the definition of a thermal reservoir.

* 5 % give no answer.

Results are less satisfactory concerning energy.

* 48% answer correctly that there is a change in the internal energy,

- 19% specify that there is a heat flow.

* 29% answer that the energy of the thermal reservoir has not changed,

- 10% specify that the temperature is constant.

* 24% give no answer.

4.2 A question about a thermal reservoir change

This question (Question B, see figure 2) asks if the source has changed without specifying the nature of the change. It is explicitly written that the process is irreversible. This piece of information should prompt the students to find a change in the final situation because the universe changes when a process is irreversible.

Consider the following irreversible process: a cube (m=100g) is in contact with a thermal reservoir at 20°C (situation I). The cube is taken and put in contact with a thermal reservoir at 80°C. Situation II is when the temperature of the cube reaches 80°C. The cube is put in contact with the thermal reservoir at 20°C. Situation III is when the temperature of the cube is 20°C again.

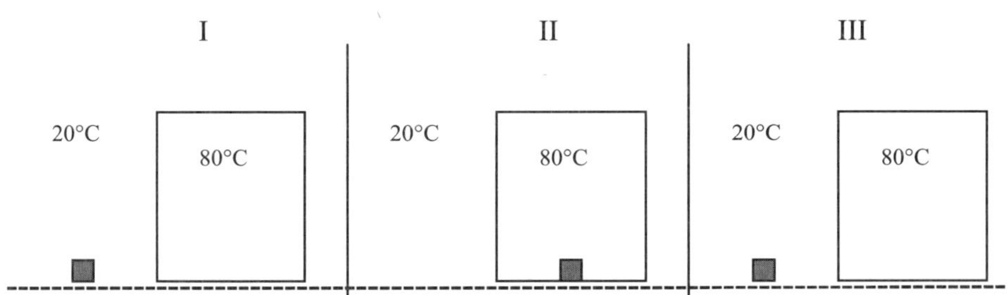

The state of the hot reservoir changes between situations I and III

yes ❑ no ❑ I don't know ❑

If you answer yes, specify the change or the changes of the reservoir. If you answer no, justify your answer.

The state of the cold reservoir changes between situations I and III

yes ❑ no ❑ I don't know ❑

If you answer yes, specify the change or the changes of the reservoir. If you answer no, justify your answer.

Figure 2. Question B

* Half of the students (48%) assert that the state of the thermal reservoir does not change,

- The majority argument is that temperature remains constant (36%) « A thermal reservoir is such that its temperature change is nil, so there is no change in the thermal reservoir »
- Some students (36%) specify that there is a heat flow between the cube and the thermal reservoirs, but they do not deduce the change of energy in the thermal reservoirs: « It is not modified by the heat supply from the cube ». Changes are negligible: « The quantity of heat given out by the thermal reservoir is considered as negligible and therefore its temperature does not change ».

* One out of four students (23%) answers correctly that the state of the thermal reservoir changes,

- Three students (7%) justify it by a change of temperature in the thermal reservoir, without specifying that the variation of temperature in the thermal reservoir was slight
- One student justifies the change by a slight variation of temperature, without referring to a change of energy in the thermal reservoir
- Only three students (7%) associate the change of the thermal reservoir with a variation of its energy and only one student gives a complete answer: « The hot reservoir released some energy to the cube by thermal transfer but its temperature has remained constant », the two other students specify that the thermal reservoir supplies some energy. Two other students (5%) mention a heat flow but it is difficult for us to be sure from this answer that they associate this heat flow with a variation of energy, especially as some of the students who answer that the thermal reservoir has not changed, also mention the heat flows.
- To conclude, the percentage of students who associate the change of the thermal reservoir with a variation of its energy is very low (one student out of ten).

* 20 % give no answer.

* The replies of 9% of the students mention that one of the thermal reservoirs has changed even though they consider that the other thermal reservoir has not changed.

4.3 An indirect question on the change in a thermal reservoir

Answers to the previous question bring to the foreground the fact that no attention has been paid to the modification of the energy in the thermal reservoir. It seems that it would be wise to check if the students are capable of using this knowledge in situations in which the fact of not being aware of the variation of energy would lead them to make obvious mistakes. With this aim in mind, we produced a question (C) (see figure 3) referring to the irreversible situation of question (B) where the system σ (the cube) is restored to its initial state. A transformation is irreversible if it is impossible to find one or several transformations, which restore the universe to its initial state. In question (C), a reasoning based solely on the restoration of the cube to its initial state is proposed without taking the thermal reservoirs into account. The incorrect affirmation here is that the process is reversible. In order to come to an accurate conclusion, the modification of the thermal reservoirs must be taken into account.

A cube (m=100g) is in contact with a thermal reservoir at 20°C (situation I). The cube is taken and put in contact with a thermal reservoir at 80°C. When the temperature of the cube is 80°C (situation II), the cube is replaced in contact with the thermal reservoir at 20°C. The temperature of the cube returns to 20°C (situation III). Is the transformation reversible?

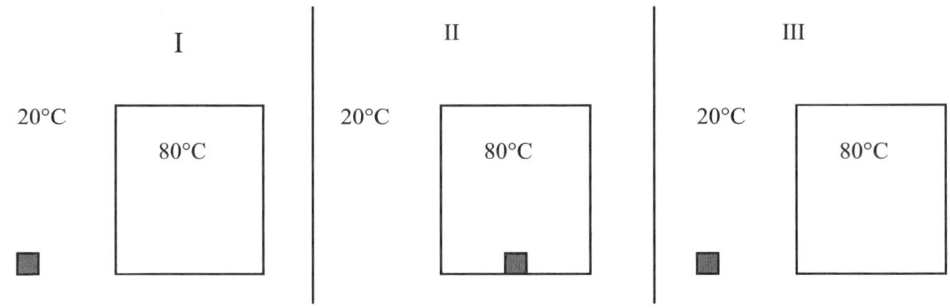

A student proposes the following answer: "The cube is restored to its initial state, so the process is reversible". Do you agree with this answer?

yes ❑ no ❑ I don't know ❑

If you answer yes, give a detail reply.
If you answer no, give a detail reply and answer the following question: Are there any differences between situation I and III? If so, what are they?

Figure 3. Question C

* 48% answer accurately that the process is irreversible but do not pay any attention to the changes of energy in the thermal reservoirs. Indeed, for the question about the differences between situation I and situation III,

- 16% of the answers are that the situations one and three are the same, in contradiction with the definition of an irreversible process
- 13% give diverse arguments which have nothing to do with the changes in the thermal reservoirs
- Only 9% of students point out that there is an irreversible transfer from the hot system to the cold system. « The cube which has a temperature of 80°C gets cooler by giving some thermal energy to the thermal reservoir at 20°C. This energy cannot come back spontaneously to the cube »
- 7% imagine that changes occurred in the cube. « If it has liquefied between situations I and situations III, it will have changed shape».

* 36% answer wrongly that the process is reversible,

– 23% of these specify that the cube goes back to its initial state or that situation I is identical to III.

* 16% give no answer.

4.4 Synthesis of the results

The results obtained from this set of questions confirm our hypothesis: the students do not take into account the fact that the internal energy of a thermal reservoir changes when it is brought into contact with another system of a different temperature. Indeed, when the question explicitly mentions the variation of energy of a thermal reservoir during an interaction, less than half of the students consider that the energy of the thermal reservoir is modified; when the question is asked in terms of change without any explicit reference to the energy, 10% of the students at the most mention the variation of energy; as for the indirect question, only 9% of the arguments are correct for question (C), the variation of energy does not appear in students' argumentation whereas they should realise that it is this variation of energy which accounts for the changes that occurred between the initial situation and the final situation and consequently justifies the irreversibility of the transformation.

Different hypotheses concerning students' difficulties in this domain can be suggested:

– In the training courses, the model of thermal reservoirs is presented as a system, which has a constant temperature, and not as a system whose variation of temperature is extremely low. The change of the energy in the thermal reservoir is often expressed by a relation of the following type $\Delta U = mC_P\ \Delta T$. If the temperature is constant, ΔT is nil and the student deduces from this that there is no variation of energy.

– When the reasoning is correct, considering ΔT as quasi nil and the mass extremely important, the lifting of the uncertainty in the expression of ΔU leads to a finite value. The thermal reservoir is a extremely important reservoir of energy, that is to say that the relative variation of its energy is extremely low. To focus on the absolute variation ΔU corresponds to a method of reasoning which is rarely used as in most reasoning processes in physics, it is necessary to take the relative variation into account.

– The energy variable is not a natural variable, so the student only focuses on the variable whose meaning is familiar to him: the temperature but, as the temperature is constant for a thermal reservoir, the student deduces that the thermal reservoir does not change when brought into contact with another system. The focus on a variable such as the temperature is probably emphasized by the terms found in exercises where the systems are always described by their volume, their temperature and their pressure.

– In situations of interactions between a system σ and a thermal reservoir, there is a dissymmetry between the presentation of the system σ whose states and/or transformations are described in detail whereas Σ is only characterized by the constancy of its temperature, that is to say more succinctly. This situation strengthens the trend noted by S. Rozier (1988) that of focusing one's reasoning on a sole object and not taking the modification of the thermal reservoir into account.

5. CONCLUSION

This study has allowed us to characterize an important conceptual difficulty facing students concerning the notion of thermal reservoir. They find it difficult to conceive that the thermal reservoir can have a variation of energy while having a constant temperature. This involves difficulties in the study of irreversibility of transformations caused by the interactions of a finite dimension system with a thermal reservoir.

We have, moreover, studied other difficulties based on different transformations: a transformation due to a difference of pressure with restoration of the system to the initial state, without the environment returning to its initial state, transformations due to differences of temperature or of pressure where the interactions are slowed down, transformations due to slight differences of pressure and temperature.

These results will be used to elaborate contents and teaching/learning situations for university level studies in thermodynamics. These situations will provide students with the opportunity to become aware of the contradictions they express when asserting that the energy of the source does not vary. Thus students will be led to regard the transformation of a system as a part of a transformation involving the whole universe.

BIBLIOGRAPHY

Artigue M. (1988). Ingéniérie didactique. *Recherches en Didactique des Mathématiques*, 9(3), 281-308.

Boohan R. & Ogborn J.(1996). *Energy and Change*, Project sponsored by the Nuffield Foundation

Erickson G. (1985). Heat and Temperature, part A. In: R. Driver, E. Guesne & A. Tiberghien (Eds), *Children Ideas in Science* (PP.52-84). Milton Keynes: Open University Press.

Grea J. & Viard J. (1995). From Language to Concept Appropriation in Physics. Two Cases Studies. In: C. Bernadini, C. Tarsitani & M. Vincentini (Eds), *Proceedings of the International Conference: Thinking Physics for Teaching, Roma 1994* (pp. 97-106). New York: Plenum Press.

Kattmann U., Duit R., Gropengieber H. & Komorek M. (1995).A Model of Educational Reconstruction. Paper presented at The NARST Annual Meeting, San Francisco.

Kesidou S. & Duit R. (1993). Students' Conceptions of the Second Law of Thermodynamics, An Interpretative Study. *Journal of Research in Science Teaching*, 30(1), 85-106.

Rozier S. (1988). *Le raisonnement linéaire causal en thermodynamique classique élémentaire.* Thèse de doctorat, Université Denis Diderot (Paris 7).

Rozier S. & Viennot L. (1991). Students'reasonings in thermodynamics. *International Journal of Science Education*, 13(2), 159-170.

Tiberghien A. (1985). Heat and Temperature, part B. In: R. Driver, E. Guesne & A. Tiberghien (Eds), *Children Ideas in Science* (pp.52-84). Milton Keynes, Open University Press.

Welzel M. & Pospiech G. (2000). A Teaching-learning Sequence on Thermodynamics: Starting with Experiments and Reaching Theoretical Explanations. Paper presented at The International Symposium Designing and Validating Teaching-learning Sequences in a Research Perspective, Paris.

METACOGNITIVE EXPERIENCES IN THE DOMAIN OF PHYSICS: DEVELOPMENTAL AND EDUCATIONAL ASPECTS

Eleftheria Gonida, Grigoris Kiosseoglou, Dimitris Psillos

Aristotle University of Thessaloniki, Greece

ABSTRACT

The study aimed to investigate metacognitive experiences in the domain of physics. Specifically, the experiences of task difficulty, certainty about the solution provided to a task, and personal satisfaction with the solution were examined. A total of 138 fifth and sixth grade pupils were asked to (a) solve a battery of tasks measuring thermal phenomena, and (b) evaluate their experienced task difficulty, certainty and personal satisfaction with their solutions on 4-point scales. Data analyses indicated that fifth and sixth graders exhibit relatively limited metacognitive awareness; their metacognitive estimations were not accurate in regard to their cognitive performance. However, the results suggest that pupils begin to associate their metacognitive estimations with their cognitive performance during the examined age period, although they usually overestimate it. The results are discussed in regard to the importance of metacognition to science education and implications for instructional practices are pointed out.

The importance of metacognitive awareness in educational practice has been widely acknowledged by a number of developmental and educational psychologists (Boekaerts, 1997; Brown, 1995; Davidson, Deuser, & Sternberg, 1996; Glaser, 1991; Vosniadou, 1996a). In general, metacognition presupposes self-reflection and implies monitoring, control and regulation during problem solving, which result in facilitation and promotion of learning. In science education, metaconceptual awareness is considered as a facilitatory condition for conceptual change and reorganization of the existing cognitive structures into new more scientifically plausible ones; moreover, new learning environments that involve metacognitive instructional practices or practices that would result in metaconceptual awareness are suggested (Mason, 1994a; Vosniadou, 1996b; Vosniadou, Ioannides, Dimitrakopoulou & Papademetriou, 2001; White, 1994; Wiser & Amin, 2001). For example, Greeno (1992) advocates the adoption of a 'thinking curriculum' for science teaching. Shayer and Adey (1993), having applied a 2-year intervention programme in science learning to 11 year old children, reported increased achievement in science, mathematics and English; this improvement was attributed to those aspects of the intervention which were designed to improve children's metacognition. Mason (1994a, 1994b) discusses how analogy facilitates conceptual change in fifth grade children by promoting their metaconceptual awareness. Wiser and Amin (2001) argue that metaconceptual teaching based on the integration of the two languages/conceptualizations of thermal phenomena (the lay and the scientific one) facilitate young adolescents to achieve a reorganization of the relevant

D. Psillos et al. (eds.), Science Education Research in the Knowledge-Based Society, 107–115.

concepts. Vosniadou et al. (2001) constructed a learning environment for teaching mechanics to fifth and sixth graders which -among other characteristics- also promoted metaco-nceptual awareness.

However, to design such a learning environment, researchers have first to analyze the different aspects of metacognitive awareness (e.g., metacognitive knowledge, metacognitive experiences, see Flavell, 1979) in terms of their structure, development and interrelated factors. Besides, if the development of scientific thinking is also metacognitive (Kuhn, 1989), they have to point out the metacognitive gains as well as the deficiencies in different age periods, in order to be able to build powerful learning environments that would promote and sustain metacognitive skills, and/or organize training programs focused on students' metacognitive awareness.

Although, most relevant studies in the physical sciences have emphasized the role of metacognition in science understanding and suggest alternative learning environments that improve metaconceptual awareness, the evidence available regarding the developmental level of metacognitive awareness related to physics in different age periods is very limited. In a study by Craig and Yore (1995) fourth to eighth grade students' metacognitive knowledge about science reading, science text and science reading strategies was found to be at a surface level.

The present study aims to shed some light towards the above direction. It focuses on metacognitive experiences within the domain of physics and for the ages of 10-11 years. Metacognitive experiences consist a dynamic aspect of metacognition which reflects the level and the accuracy of the learner's awareness about the task and her/his cognitive performance in the form of feelings of task difficulty, required time for a task to end, certainty about the solution provided, etc. They are feelings or sensations that may arise before, during or after a cognitive enterprise, interact with the cognitive processes themselves and affect the learning outcome (Flavell, 1979; see also Efklides & Vauras, 1999). Our aim was to study pupils' metacognitive awareness, as it is indicated by their metacognitive experiences, in relation to thermal phenomena. More specifically, we aimed to investigate the accuracy of pupils' metacognitive experiences in regard to their actual cognitive performance. Three metacognitive experiences were examined in this study: feelings of task difficulty, certainty about the solution provided to the task, and personal satisfaction with the solution. The science concepts examined were on thermal phenomena. Thermal phenomena, although commonly used in both educational curricula as well as in everyday life, are resistant to conceptual change and raise significant difficulties to pupils (Wiser & Carey, 1983; Wiser & Amin, 2001). The age of 10-11 years was selected because physics as a school subject is taught during the fifth and sixth grade in the Greek primary school.

1. METHOD

Participants A total of 138 pupils (78 girls and 60 boys) attending the fifth (N= 67) and sixth grade (N=71) of primary school participated in the study. Their mean age was 10,3 and 11 years, respectively for each grade. All pupils had been taught the phenomenon of heat in the beginning of their school year; however, pupils of each grade had worked mostly on different thermal phenomena according to their curriculum.

Tasks / Measures. Two protocols consisting of cognitive tasks and metacognitive measures, one for each age group, was assigned to the participants.

Cognitive tasks. A battery of 5 cognitive tasks examining thermal phenomena was given to each group. The tasks were different for the two groups according to the curriculum of each grade. In particular, the pupils of the fifth grade were tested on tasks measuring aspects of (a) the difference between heat and temperature, (b) air expansion, (c) liquid expansion, and (d) thermal equilibrium. The pupils of the sixth grade were tested on tasks measuring (a) vaporization, (b) transfer of heat, (c) liquefication, and (d) thermal equilibrium. For example, in one task concerning thermal equilibrium, pupils were asked whether ice cubes, butter, vegetables, meat and ice cream which have been in the freezer for many days had the same or different temperature and to justify their answer; concerning air expansion, they were asked to explain why a balloon which was put on a radiator got inflated after some hours; concerning transfer of heat, pupils were asked how children get warm from the radiator, since they sit far away from it.

The difficulty level of the tasks was evaluated, first, by a group of teachers, who were teaching physics in the fifth and sixth grade of primary school, and, second, by an expert in physics (university professor). Task difficulty ranged from low to middle to high level. There were only two cases of disagreement in their difficulty evaluations, one for each grade.

Metacognitive measures. As it has been noted earlier, the metacognitive experiences examined in the study included difficulty estimations about the task to be solved, certainty estimations about the solution provided to each task, and personal satisfaction estimations with the solution provided. After they had solved each task, pupils were asked to state their experiences on three 4-point Likert–type scales, one for each experience. Specifically, the points of the scale were as follows: 1: not at all, 2: slightly, 3: quite, 4: very (difficult / certain / satisfied).

Procedure and Scoring Criteria. All participants were group tested in their class during a regular school hour. Preliminary instructions about the protocol were given to the pupils by the experimenters. The order of the tasks was randomized among the pupils' protocols.

Pupils' replies to each task were classified according to their content in the following four categories:

a) Fully phenomenological answers (usually incomplete and/or wrong);
b) Phenomenological answers towards the correct direction;
c) Incomplete scientific replies;
d) Correct and complete scientific replies.

A scoring scale from 0 to 3 was assigned to the above categories, respectively. We present an example of scoring for the air expansion task described above:

Reply 1: 'A large quantity of air entered into the balloon' (score=0).
Reply 2: 'Hotness entered into the balloon and the balloon became big' (score=1).

Reply 3: 'Because it was expanded' (score=2).
Reply 4: 'During expansion the balloon gets heat and its volume increases' (score=3).

2. RESULTS AND DISCUSSION

Descriptive Data. Table 1 presents mean cognitive performance and standard deviations in each task both for fifth and sixth graders. The difficulty level of each task is also indicated on the table. According to Table 1, both groups of pupils exhibited very low scores and there was a great variability of scores among the tasks in both groups. It is clear that different thermal phenomena raise different cognitive demands. For example, both fifth and sixth graders scored extremely low in all three tasks examining thermal equilibrium independently of their difficulty level. What also seems interesting is the level of the cognitive scores in relation to the task difficulty indicated by the pupils' teachers and the expert. Obviously, participants scored differently from their teachers' expectations in most of the cases. Pupils' scores were very poor or relatively poor, even in tasks that had been characterized as posing difficulty of middle or low level to them.

Table 1. (a) 5th Grade: Mean scores in the five cognitive tasks, (b) 6th Grade: Mean scores in the five cognitive tasks

(a)

Tasks	Mean	SD	Task Difficulty
Task 1 (Air Expansion)	1.075	1.235	HD/MD*
Task 2 (Thermal Equilibrium)	.224	.794	MD
Task 3 (Heat - Temperature)	1.343	1.503	LD
Task 4 (Thermal Equilibrium)	.642	1.177	LD
Task 5 (Liquid Expansion)	.881	1.237	HD
Total Mean	.833		

(b)

Tasks	Mean	SD	Task Difficulty
Task 1 (Transfer of Heat)	1.352	.927	LD
Task 2 (Vaporization)	1.972	1.362	LD
Task 3 (Transfer of Heat)	.535	.790	MD
Task 4 (Liquification)	.380	.663	MD
Task 5 (Thermal Equilibrium)	.085	.405	MD/HD*
Total Mean	.865		

Note: The symbols HD, MD, and LD stand for the words High Difficulty, Middle-level Difficulty and Low Difficulty, respectively.

- Cases of disagreement between the evaluators.

Furthermore, the question that arises is how pupils' metacognitive estimations were formulated since their scores have been so low. Table 2 presents the frequencies and percentages of fifth graders' estimations about task difficulty, certainty and personal satisfaction for the solution provided to each task. Similarly, Table 3 presents the above data for sixth graders.

Comparing Table 2 with Table 3 we observe almost the same tendency in regard to pupils' metacognitive estimations. In most cases, both fifth and sixth graders tend to estimate low task difficulty (values 1 and 2 on the relevant 4-point scale), high certainty about their solutions to each task and high personal satisfaction with them (values 3 and 4). Even for the tasks in which the pupils' performance was extremely low, such as Task 2 for the fifth grade (mean=.224) and Task 5 for the sixth grade (mean=.085), the percentages of pupils who experienced the above feelings were very high (see Tables 2 and 3). Thus, their metacognitive experiences seem to be not accurate in regard to their actual cognitive performance. However, in the next session we will proceed to a further level of analysis in order to be able to pinpoint any possible relations between cognitive performance and metacognitive estimations.

Table 2. 5th Grade: Frequencies (and percentages) of pupils' metacognitive estimations about task difficulty, certainty and personal satisfaction for the solution

Difficulty Level	**Task 1**	**Task 2**	**Task 3**	**Task 4**	**Task 5**
1 & 2 *not at all / slightly difficult*	46 (69.7)	32 (52.5)	50 (76.9)	28 (52.8)	35 (59.3)
3 & 4 *quite / very difficult*	20 (30.3)	29 (47.5)	15 (23.0)	25 (47.2)	24 (40.6)
Certainty Level	**Task 1**	**Task 2**	**Task 3**	**Task 4**	**Task 5**
1 & 2 *not at all / slightly certain*	20 (30.3)	32 (52.5)	16 (23.9)	26 (50.0)	28 (47.4)
3 & 4 *quite / very certain*	46 (69.7)	29 (47.5)	51 (79.2)	26 (50.0)	31 (52.5)
Satisfaction Level	**Task 1**	**Task 2**	**Task 3**	**Task 4**	**Task 5**
1 & 2 *not at all / slightly satisfied*	19 (28.8)	33 (49.2)	10 (15.4)	23 (44.2)	23 (39.0)
3 & 4 *quite / very satisfied*	47 (71.2)	28 (41.8)	55 (84.6)	29 (55.7)	36 (61.0)

Table 3. 6th Grade: Frequencies (and percentages) of pupils' metacognitive estimations about task difficulty, certainty and personall satisfaction for the solution

Difficulty Level	**Task 1**	**Task 2**	**Task 3**	**Task 4**	**Task 5**
1 & 2 *not at all / slightly difficult*	61 (87.2)	57 (86.4)	54 (79.5)	51 (77.3)	39 (60.0)
3 & 4 *quite / very difficult*	9 (12.8)	9 (13.6)	14 (20.6)	15 (22.7)	26 (40.0)
Certainty Level	**Task 1**	**Task 2**	**Task 3**	**Task 4**	**Task 5**
1 & 2 *not at all / slightly certain*	18 (25.7)	14 (21.2)	22 (32.4)	22 (33.3)	33 (50.7)
3 & 4 *quite / very certain*	52 (74.3)	52 (78.8)	46 (67.6)	44 (66.6)	32 (49.2)
Satisfaction Level	**Task 1**	**Task 2**	**Task 3**	**Task 4**	**Task 5**
1 & 2 *not at all / slightly satisfied*	20 (28.9)	21 (16.9)	18 (26.5)	14 (21.2)	25 (39.1)
3 & 4 *quite / very satisfied*	49 (71.0)	74 (83.1)	50 (73.6)	52 (78.8)	39 (61.0)

Correlations and Chi-square Test. Beyond the descriptive level, Pearson-r correlations between cognitive performance and metacognitive experiences were performed. Table 4, which presents the statistically significant correlations for both grades, shows the following: (a) Significant correlations between cognitive scores and metacognitive estimations were found in both grades; however, more significant correlations were found in the sixth than in the fifth grade. In particular, significant negative correlations were found between cognitive scores and difficulty estimations and positive correlations were found between cognitive scores and certainty as well as satisfaction estimations.

(b) Among the three metacognitive experiences examined, certainty estimations were mostly significantly correlated to respective cognitive scores. Actually, significant positive correlations between certainty experiences and cognitive scores were found in all but thermal equilibrium tasks, which may raise particularly high cognitive and metacognitive demands. The formation and the development of different metacognitive experiences in regard to tasks examining different physical phenomena merits further study.

Table 4. Significant correlations between cognitive scores and metacognitive estimations for5th (a) and 6th Grade (b).

(a)

Tasks	Task Difficulty	Certainty	Personal Satisfaction
Task 1 (Air Expansion)	-----	.25*	-----
Task 2 (Thermal Equilibrium)	-----	-----	-----
Task 3 (Heat - Temperature)	-----	.29*	.26*
Task 4 (Thermal Equilibrium)	-----	-----	-----
Task 5 (Liquid Expansion)	-.30*	.44**	-----

(b)

Tasks	Task Difficulty	Certainty	Personal Satisfaction
Task 1 (Transfer of Heat)	-.28*	.27*	-----
Task 2 (Vaporization)	-.35*	.51**	.25*
Task 3 (Transfer of Heat)	-----	.26*	.25*
Task 4 (Liquefaction)	-----	.28*	-----
Task 5 (Thermal Equilibrium)	-----	-----	-----

Note: The symbols (*) and (**) stand for p< .05 and p<.001, respectively.

Furthermore, in order to test whether cognitive performance and metacognitive estimations are dependent on each other, we applied the chi-square test for independence, using also the adjusted standardized residuals (d_{ij})* (Everitt, 1977). Some statistically significant associations between the two variables were found in both grades. In particular, in the fifth grade significant associations were found between certainty estimations and Tasks 1, 3 and 5 [$x^2(2)= 6.34$, $p<.05$; $x^2(1)= 4.46$, $p<.05$; $x^2(2)= 8.32$, $p<.05$, respectively for the three tasks]. In the sixth grade significant associations were found between certainty estimations and Tasks 1, 2 and

* Values of dij>1.96 denote significant association between i (tasks) and j (metacognitive estimations) of the contingency table.

4 [$x^2(2)$= 6.92, $p<.05$; $x^2(2)$= 11.60, $p<.01$; $x^2(1)$= 4.34, $p<.05$, respectively for the three tasks], and between Task 2 and difficulty [$x^2(2)$= 9.78, $p<.005$], as well as satisfaction estimations [$x^2(2)$= 9.25, $p<.005$]. For example, the significant association between certainty estimations and Task 5 in the fifth grade means that fifth graders who had scored 0 in Task 5 stated low certainty (value 1 or 2, d_{ij}=2.6), whereas fifth graders who had scored 3 in the same task stated high certainty (value 3 or 4, d_{ij}=2.6). Similarly, sixth graders who had scored 1 or 2 in Task 2 stated high difficulty (value 3 or 4, d_{ij}=2.4), whereas those pupils who had scored 3 stated low difficulty (value 1 or 2, d_{ij}=2.9).

We consider that the above results indicate that cognitive and metacognitive performance begin to be dependent on each other during this developmental period; that is, pupils of this age have just started to associate their metacognitive estimations to their cognitive scores, although unsuccessfully in most of the cases, especially in the fifth grade.

3. CONCLUSIONS AND IMPLICATIONS FOR EDUCATION

The results of the present study in agreement with those of the study by Craig and Yore (1995) suggest that metacognitive awareness related to physics tasks -at least to those tasks examining aspects of thermal phenomena- is relatively limited during the examined age period. Pupils' metacognitive deficiencies have been detected at the level of all three metacognitive experiences. Fifth and sixth graders are not aware of their actual cognitive performance as this has been measured in the present study (academic or taught knowledge performance); they usually underestimate task difficulty and overestimate their performance when they are called to report on experiences such as certainty and personal satisfaction in regard to their own solutions. The results of this study suggest that metacognitive experiences have just begun to be associated with cognitive performance during the examined developmental period. Although in most of the cases examined pupils' experiences were not accurate in regard to real cognitive performance, in some cases they were found not only to be correlated to the respective cognitive scores but also to depend on them.

However, one could possibly argue that pupils of this age do not lack metacognitive awareness in regard to their performance to the tasks, but rather the knowledge and skills to solve such tasks. Since they are still bound to their alternative conceptions, they may think that they provide the correct answers to the tasks at hand, and consequently exhibit positive metacognitive experiences. In other words, pupils' metacognitive estimations could be accurate in regard to their alternative conceptions, but inaccurate in regard to their academic/taught knowledge performance. We think that this may be a fair criticism to many metacognitive studies including ours, a criticism that raises issues meriting further investigation at a micro-level through the use of a different methodology. It should be mentioned, however, that apart from the inaccurate metacognitive experiences, which are considered in our study as indices of limited metacognitive awareness, the explanations pupils provided for their answers were also poor. Most pupils did not provide any explanation for their solutions to the tasks, while those explanations, which they did provide, were either poor or wrong. It seems, thus, that pupils were not aware of their everyday experientially conceptions either; if they had been aware, they would have been able to provide complete and accurate explanations of these alternative conceptions.

We believe that the development of metacognitive awareness can facilitate science understanding. In order to re-organize their pre-existing conceptions, pupils have to (a) be aware of the two conceptualizations, the everyday -which is the one they possess- and the scientific (see Wiser & Amin, 2001), and (b) realize their limitations, weaknesses and shortages both at a very specific level (e.g., task at hand) and at a more general level (e.g., as problem solvers in particular subject-matters or domains).

Researchers claim that metacognition develops along age and through thinking experiences (e.g., Flavell, 1988; Wellman, 1985). New instructional practices focusing on such thinking experiences would contribute a lot towards the development of metacognitive awareness. It is important for teachers to realize the above necessity and to emphasize metacognitive instruction (see Manning & Payne, 1996). In the present study, teachers had underestimated the difficulty level of the tasks for the pupils and expected much higher scores than the ones pupils obtained. We consider that teachers should be educated in order to facilitate the enhancement of pupils' metacognitive knowledge of tasks and goals and the formation of accurate metacognitive estimations regarding their actual performance. Such advanced instructional practices would also potentially enable the teachers themselves to form a more accurate picture of their pupils' understanding, to have realistic expectations about their cognitive performance, and to organize their own teaching more effectively.

In conclusion, we suggest that, in science teaching, learning environments that involve metacognitive instructional practices could contribute to a deeper understanding of physical phenomena and the adoption of scientific explanatory frameworks through two paths: (a) the improvement of pupils' awareness of both task demands and their performance, and (b) the improvement of teachers' awareness about their pupils' knowns and unknowns, strengths and weaknesses.

REFERENCES

Boekaerts, M. (1997). Self-regulated learning: A new concept embraced by researchers, policy makers, educators, teachers, and students. *Learning and Instruction*, 7 (2), 161-186.

Brown, A., L. (1995). Advances in learning and instruction. *Educational Researcher*, 23 (8), 4-12.

Craig, M., T., & Yore, L., D. (1995). Middle school students' metacognitive knowledge about science reading and science text: An interview study. *Reading Psychology*, 16 (2), 169-213.

Davidson, J., E., Deuser, R., & Sternberg, R., J. (1996). The role of metacognition in problem solving. In Metcalfe, J. & Shimamura, A., P. (Eds.), *Metacognition: Knowing about knowing* (pp. 208-226). Cambridge, MA: The MIT Press.

Efklides, A., & Vauras, M., (Eds.), (1999). Metacognitive experiences and their role in cognition. *European Journal of Psychology of Education*, XIV (4) (Special Issue).

Everitt, B., S. (1977). *The analysis of contingency tables*. London: Chapman & Hall.

Flavell, J., H. (1979). Metacognition and cognitive monitoring: A new area of cognitive-developmental inquiry. *American Psychologist*, 34, 906-911.

Flavell, J., H. (1988). The development of children's knowledge about the mind. In: Astington, J., W., Harris, P., L. & Olson, R., O. (Eds.). *Developing theories of mind* (pp. 244-267). Cambridge: Cambridge University Press.

Glaser, R. (1991). The maturing of the relationship between the science of learning and cognition and educational practice. *Learning and Instruction*, 1, 129-144.

Greeno, J., G. (1992). Mathematical and scientific thinking in classrooms and other situations. In: Halpern, D., F. (Ed.), *Enhancing thinking skills in the sciences and mathematics* (pp. 39-61). Hillsdale, NJ: Erlbaum.

Kuhn, D. (1989). Children and adults as intuitive scientists. *Psychological Review*, 96, 674-689.

Manning, B., H., & Payne, B.,D. (1996). *Self-talk for teachers and students: Metacognitive strategies for personal and classroom use*. Boston, MA: Allyn & Bacon.

Mason, L. (1994a). Cognitive and metacognitive aspects in conceptual change by analogy. *Instructional Science*, 22 (3), 157-187.

Mason, L. (1994b). Analogy, metaconceptual awareness and conceptual change: A classroom study. *Educational Studies*, 20 (2), 267-291.

Shayer, M., & Adey, Ph., S. (1993). Accelerating the development of formal thinking in middle and high school students: Three years after a two-year intervention. *Journal of Research in Science Teaching*, 30 (4), 351-366.

Vosniadou, S. (1996a). Towards a revised cognitive psychology for new advances in learning and instruction. *Learning and Instruction*, 6 (2), 95-109.

Vosniadou, S. (1996b). Learning environments for representational growth and cognitive flexibility. In: Vosniadou, S., De Corte, E., Glaser, R. & Mandl, H. (Eds.). *International perspectives on the design of technology-supported learning environments* (pp. 13-23). Mahwah, NJ: Lawrence Erlbaum.

Vosniadou, S., Ioannides, C., Dimitrakopoulou, A., & Papademetriou, E. (2001). Designing learning environments to promote conceptual change in science. *Learning and Instruction*, 11 (4-5), 381-419.

Wellman, H., M. (1985). The child's theory of mind: The development of conceptions of cognition. In: Yussen, S., R. (Ed.). *The growth of reflection in children* (pp. 169-206). New York: Academic Press.

White, R., T. (1994). Conceptual and conceptional change. *Learning and Instruction*, (1), 117-121.

Wiser, M., & Amin, T. (2001). "Is heat hot?" Inducing conceptual change by integrating everyday and scientific perspectives on thermal phenomena. *Learning and Instruction*, 11 (4-5), 331-355.

Wiser, M., & Carey, S. (1983). When heat and temperature were one. In: Gentner,D. & Stevens, A., L. (Eds.), Mental models (pp. 267-297). Hillsdale, NJ: Lawrence Erlbaum.

HOW CHILDREN REASON FROM DATA TO CONCLUSIONS IN PRACTICAL SCIENCE INVESTIGATIONS

Zoe Kanari, Robin Millar
University of York

ABSTRACT

This study examines the understandings that children bring to a practical science investigation, focussing on their reasoning as they collect and evaluate experimental data and draw conclusions about relationships between variables. Two practical investigations were used in the study, each carried out by 10 children aged 9, 11 and 13 (total n=60) from 6 schools in the North of England. Both investigations involved one independent variable, which has an effect on the outcome, and one, which does not. Children worked individually and their activities were filmed. After the practical investigation, each child was interviewed about a video recording, which showed two other children carrying out the other practical investigation of the pair. The child's responses were recorded on interview sheets and were also audio-taped. The analysis shows that, in situations of co-variation, most children use the presence of a trend to conclude that a variable has an effect. In situations of non-covariation, their reasoning is much more varied and fewer reach a well-argued conclusion. Such tasks are therefore useful to teachers and researchers for eliciting children's reasoning in greater depth, and for helping to develop it further.

1. BACKGROUND, AIMS AND FRAMEWORK

Learning about scientific enquiry constitutes a core element of science education and children's investigative performance has been the focus of several major studies. Research in this area has been undertaken in the fields of science education and cognitive psychology. In general, it is suggested that whilst pupils can design and carry out investigations, they often have difficulties in reasoning from their data.

One of the most influential studies within science education was undertaken in the UK by the Assessment of Performance Unit (APU). Children at different ages were observed carrying out multivariable tasks, that is, explorations of the effect of one or more independent variables on a dependent variable (e.g. which material is best for keeping you warm on a cold day?). Findings from these surveys suggest that when the conclusion to be drawn from the data is unclear, children may impose patterns on data (Austin et al., 1991). A similar conclusion is drawn by the Procedural and Conceptual Knowledge in Science (PACKS) project, which built on the work of the APU. The findings from this project suggest that in situations of unambiguous data, which conflicted with their prediction, many children appeared to cling on to their prediction if this was based on an underlying model of the process involved (Millar & Lubben, 1996).

In their attempts to describe aspects of scientific reasoning, cognitive psychologists have also examined how individuals see the relation between

D. Psillos et al. (eds.), Science Education Research in the Knowledge-Based Society, 117–125.

experimental data and the hypothesis or theory under consideration, and how prior ideas affect the conclusions drawn from data. Kuhn et al. (1988) introduced children to a problem and elicited their "theories" (i.e. prior ideas) about it. They then asked children to say whether given pieces of data supported or conflicted with a particular theory and to explain their reasoning. Kuhn et al. (1988) postulate that the process of theory and evidence coordination undergoes developmental change and that many children do not consider the possibility that their theory might be false or that alternative theories might exist. On the other hand, Chinn and Brewer (1998) suggest that scientists regularly discount anomalous data in favour of their theories and argue that children are like scientists in their fundamental abilities to think about evidence and theories and to use evidence to choose between theories. Furthermore, Koslowski (1996) criticizes the work of Kuhn et al., and argues that, when they have an underlying mechanism in mind, subjects treat their ideas as working hypotheses and modify them when confronted with anomalies in a theoretically motivated way, rather than an ad hoc one. However, in Koslowski's work subjects were presented with conclusions from investigations rather than with primary data, and asked to discuss their implications from theories. So the step of data interpretation in the process of going from data to conclusions is left unexplored.

The study reported in this paper builds on the previous work summarized above and looks at children's investigative performance and their reasoning from data to conclusions. The study addresses the following questions:

1. What understandings of scientific inquiry do children bring to their science investigations?
2. Are there any common patterns in children's actions and reasoning as they collect experimental data and draw conclusions from it?
3. Given that measurement error is always present, how do children's actions and reasoning differ when investigating variables that do not covary, as compared with variables that do?
4. How do children's actions and reasoning change with age and school experience?

2. METHODS AND SAMPLES

A practical science investigation was given to each of 60 (3x20) children aged 9, 11 and 13 of a representative range of abilities, in 6 schools in the North of England. Each was asked to undertake one of the following tasks:

1. A Pendulums investigation, to explore the relation between the length and the weight of a pendulum, and its time of swing (the time needed for 10 swings)
2. A Boxes investigation, to explore the relation between the mass of an object and the area of its bottom surface, and the force needed to pull the object along a horizontal surface.

Thirty children were given the Pendulums investigation as a practical task and thirty undertook the Boxes investigation.

The apparatus provided for the Pendulums task consisted of:

- Five pieces of string of different lengths (20cm, 40cm, 60cm, 80cm and 100cm) each with a hook on both ends.
- Five objects of the same shape and size but of different weights (20g, 40g, 60g, 80g and 100g) each with a hook on one end so that they could be hung on the strings to make pendulums of different lengths and weights.
- A support frame with a hook in the middle from which a pendulum could be hung.
- A stop clock for measuring the time for ten swings. This was chosen to give readings to a precision of 0.1s.

The equipment for the Boxes task consisted of:

- Five small wooden boxes of the same depth and all made to weight exactly 100g, with bottom surface areas ranging from $40cm^2$ to $200cm^2$. Each box had a piece of string attached to one side that could be used to pull it along a horizontal surface.
- Five 100g weights that could be put into the boxes to make them heavier. Velcro pads on the box and the weights prevented these slipping around.
- A carpet strip 0.4x2m along which the boxes would be pulled.
- A force meter, which could be used to measure the force, needed to pull the boxes along the carpet strip.

It was decided to use multivariable tasks as these have the advantage of being relatively familiar to pupils and teachers, and enable us to relate our findings to those of previous research, much of which also used investigation tasks of this type. Both investigations were specifically designed so as to involve one independent variable which co-varies with the dependent variable (we will refer to this as the 'covariation case') and one independent variable, which does not (we will refer to this as the 'non-covariation case'). Both, however, were variables that the pupil might expect to have an effect on the outcome. The intention behind this design was to test the hypothesis that pupils would find the non-covariation case much more demanding than the covariation case, because measurement error would mask the absence of covariation. We used two tasks with the same structure (a covariation and a non-covariation case) but different contexts (pendulums and pulling boxes) to check on the extent to which pupils' actions differ with context.

In each task, pupils were asked to investigate the covariation case first followed by the non-covariation case. This enabled us to compare pupils' actions and reasoning in situations where they are likely to obtain data showing a clear trend and in situations where there is no obvious pattern. In both investigations, pupils were given three hypotheses to choose between. These were written on three cards, the chosen card being placed on the table as the investigation proceeded. For the *covariation* case in the Pendulums task, for example, the cards read:

a) when you increase the *length* of a pendulum, its time of swing gets greater
b) when you increase the *length* of a pendulum, its time of swing does not change
c) when you increase the *length* of a pendulum, its time of swing gets smaller

For the *non-covariation* case, the cards read:

a) when you increase the *weight* of a pendulum, its time of swing gets greater
b) when you increase the *weight* of a pendulum, its time of swing does not change
c) when you increase the *weight* of a pendulum, its time of swing gets smaller

Pupils could then change the card they chose initially for another card, if they wanted to alter their hypothesis at any stage as the investigation proceeded. They were also asked to indicate their final conclusion by choosing one of the cards. This approach is modelled on that used by Koslowski (1996) in her series of studies of scientific reasoning.

Children worked individually and their activities were filmed. Each practical investigation was followed by a clinical interview, during which the children who undertook the Pendulums investigation as a practical task watched a video, which showed two other children carrying out the Boxes investigation. Conversely, the children who undertook the Boxes investigation as a practical task watched a video, which showed two other children carrying out the Pendulums investigation. The video consisted of five short extracts, in each of which the two children collected some data, which was shown, clearly on the screen. The reason for using these video extracts was to ensure that all pupils told us their interpretation of the same data, which had particular characteristics that we were interested in, since in the practical investigations pupils' data, and hence their reasoning about it, might vary widely. At the beginning of the interview the same sets of hypotheses used in the practical investigations were given to pupils to choose from. After each video-extract the child was asked to say, based on this information, how sure he/she was about his chosen hypothesis. The children could review their chosen hypothesis and the interview ended when they were sure that their current hypothesis was correct. The children's responses during the interviews were recorded on interview sheets and on audio-tape.

3. RESULTS

The data for analysis consisted of video recordings of sixty pupils doing two practical science tasks (30 of each), and audio recordings of interviews with them about the video-extracts described above. For each task, the pupil's exploration of the covariation case and of the non-covariation case were considered separately. The video recordings of pupil's investigations were transcribed and used to develop records like that displayed in Figure 1.

HYPOTHESIS	MEASUREMENTS				VERBAL BEHAVIOUR
longer – less time	1	20cm	100 g	8.1s	(1) Picks 20cm and 40g for the next one R- Remember you are trying to find out what happens when you increase the length of a pendulum - Yeah! Picks 100cm and 100g
	2	100cm	100 g	20.0s	(2) Changes card R- How sure are you about that? - I'm pretty sure because I used the same weight all the time so you know that wouldn't change it at all and then I used the smallest one and the biggest one so I'm pretty sure because they are different. R- Do you think it's necessary to try anything else or... - I might try one more just [?...?]
longer – more time	3	60cm	100 g	15.5s	(3) - Yeah, I'm sure about that R- [introduces second investigation]

Figure 1. Extract form the record of Sarah (age 11) carrying out the Pendulums task (covariation case)

The first column in Figure 1 shows the pupil's current hypothesis at each stage of the investigation, in an abbreviated form (e.g. 'longer-less time' means the pupil has selected the card: 'when you increase the length of a pendulum its time of swing decreases'). The middle column lists the measurements in the order in which they were taken by the child. Finally, the third column shows descriptions of the child's actions and verbal interactions with the researcher (indicated by an R). The number in brackets alongside the text in this column indicates the measurement to which the text relates.

The audio-recordings of the interviews were used to develop records like that shown in Figure 2. This has one row for each video-extract (VE) being discussed. The middle column indicates the current hypothesis, at the start and after each video-extract. The words 'change' and 'keep' in the first column indicate whether the hypothesis was changed or retained. The third column summarises the verbal response of the pupil, with direct quotations where these are useful to illuminate the pupil's reasoning.

Video extract	**Current hypothesis**	**Verbal response**
	heavier - less time	
VE1 change	heavier - more time	Its definitely wrong 'because it took longer time with the 80g weight'
VE2 change	heavier - less time	She is not absolutely sure about her new hypothesis. She needs to see more: 'Maybe on the longest string it might take a shorter time with the 20g one'
VE3 keep		She thinks she might be wrong because: 'Because the two of the weights took shorter time than the two heavier weights on the pendulum and this says it needs less time, but it actually took more time' Needs to see more: 'Maybe on the longest string the heavier weight might take a shorter time'
VE4		she is absolutely sure now: 'I thought that if the string is longer, with the heavier weight it might take a shorter time because of the force pulling it up and down [?...?]ing backwards and forwards but if the force is lighter it takes longer cause there isn't so much force coming down'
VE5		Can't decide because 'sometimes it like took a shorter time in the lighter weight and sometimes it took a longer time in that weight so it's kind of in the middle...' R- asks her if she can choose between one of the cards 'It could be all of them really...'

Figure 2. Extract form the record of Isabel's (age 9) response to the interview about the Pendulums video-investigation (non-covariation case)

A set of four tables was constructed this way to summarise the performance and reasoning of each pupil on the covariation and non-covariation cases in one practical and one task observed on video (in total 240 tables for 60 pupils). Using these as the starting point, a set of descriptors was used to characterise each distinct aspect of each child's response to each case investigated. Aspects recorded included things like the pupil's initial hypothesis and final conclusion, the number of measurements taken, the number of these repeated more than once, whether the other independent variable was held constant, and so on. By entering all these descriptors on to a large data base, it was then possible to look for patterns in pupils' responses, and ways of reasoning. There is only space here to report a small number of these.

Table 1 shows the frequency of correct and incorrect conclusions in the covariation and non-covariation cases in the practical task.

Table 1. Children's conclusions in the practical investigations

Conclusion	Covariation case	Non-covariation case	Total
Incorrect	0	30	30
Correct	60	30	90
Total	60	60	120

All the children got the correct answer in the covariation case, probably because the existence of a trend was very obvious in their data. On the other hand, only 30 children (50% of the sample) got the correct answer in the non-covariation case. A chi-square test indicates that the differences in Table 1 are statistically significant at the 1% level (χ^2=40, df=1).

Table 2 shows the corresponding data for the investigation observed on video. Again a significantly higher proportion of children reached a correct conclusion in the covariation case than the non-covariation case (χ^2=36.9, df=1, $p<0.01$).

Table 2. Children's conclusions in the investigations observed on video

Conclusion	Covariation case	Non-covariation case	Total
Incorrect	2	32	34
Correct	58	28	86
Total	60	60	120

This strongly corroborates our initial hypothesis that pupils would find the non-covariation case more difficult. Indeed many had great difficulty in interpreting their data. This difference between the covariation case and the non-covariation case contrasts with the absence of any clear differences between the two investigation contexts: pendulums and pulling boxes. For most indicators used, there was no significant difference between the contexts.

Returning to the differences between covariation and non-covariation cases, repeat measurements were more common in the non-covariation case (Table 3), as pupils checked their initial measurements and tried to find a pattern. An analysis of the transcribed pupils' responses shows that a measurement was usually repeated when the first one gave an unexpected result, or to check and confirm a result. Most pupils used the results of a repeat measurement to replace some or all of their previous ones, and many did not record all the individual measurements they took.

Table 3. Taking repeat measurements in the practical investigations

Took repeat measurements	Covariation case	Non-covariation case	Total
No	50	32	82
Yes	10	28	38
Total	60	60	120

It appeared that children, when exploring the covariation case, were able quickly to see a trend in the data, and based their conclusion on this. In exploring the non-covariation case, however, reasoning involved a variety of interacting ideas, including ideas about the physical situation involved, the measuring equipment being used, and the possible variability of repeat measurements.

Since the pupils' actions during the practical investigations varied widely, it is useful to examine their responses to the video-investigations where all pupils were asked to comment on the same data. The last video-extract (VE5) of the non-covariation case presented a set of repeated measurements. In the children's responses to these, the ideas similar to those discussed above were evident. In approaching this kind of data, only four children (all aged 13) referred to the averages of the repeat measurements to reach the correct conclusion. One, for example, said, "The averages are the same; they do change a little bit but the eighty square centimetre and the hundred and sixty square centimetre still have the same average" (Mandy, age 13). Ten children did not refer to the averages of repeated measurements, but drew conclusions directly from the individual measurements, arguing that the variations they observed were due to error. So, although they seemed to have an awareness of measurement error, they did not use the average as the 'best value'. For example, Les (age 9) said: "I think it might be a tiny little bit right... I'm not that sure because they are all different [...] They all keep changing [...] they haven't like pulled it... They haven't been pulling it too hard or too low".

Sixteen children appeared puzzled by the data in VE5 and thought that these were inconclusive, e.g. "The first two are right and then the next one it's coming all different so I don't really know what to believe. And then the last one I'm wrong [referring to his hypothesis]. So, I don't know..." (Alex, age 11). Alex seemed to consider that each measurement had to be treated as 'correct' and did not refer to the idea of error, or consider taking an average value. A further nine children judged differences within the sets of repeat measurements to be significant and decided that the data did not provide firm support for their current hypothesis. However, they still thought that their hypothesis was correct. For example, Tina (age 9), after watching VE5, was not sure that her current (incorrect) hypothesis was correct, but decided to keep it: "I'm not sure because those two there have got eight point six and those two there got eight point six and it's going up and down all the time so I think it's right but I'm not completely sure". Finally, in their attempt to make sense of the data in VE5, twelve children chose to focus on those parts of the data that seemed to support their hypothesis, and largely disregarded the rest.

4. CONCLUSIONS AND IMPLICATIONS

Several of the findings above corroborate those reported by previous studies. We found that children, in general retained their current hypothesis unless the data was clearly at odds with it. If the data were ambiguous, children were likely to hold on to their hypothesis, often claiming that the data supported it. Our more interesting and important findings, however, are those, which go beyond the scope of previous studies. Although Kuhn et al. and Koslowski term the object of their studies 'scientific reasoning', in fact they are exploring 'logical reasoning', i.e. reasoning from given propositions as opposed to reasoning from primary empirical data. The children in our study had greater difficulty in interpreting measurements to decide if a variable had increased, decreased or stayed the same, than with the logical reasoning involved in relating such decisions to their current hypothesis, or in controlling variables in the first place. The need for an understanding of data and measurement became much more clearly apparent in the non-covariation cases, as we had hypothesised. For these, some understanding of measurement error is critically important. In fact children's explorations of the covariation case did not raise very interesting issues in data interpretation since data almost always showed a clear trend, in contrast to their explorations of the non-covariation case.

We would suggest that investigations designed to include exploration of both covariation and non-covariation cases are of particular value in both teaching and research. In school science, children are often asked to investigate situations where two variables covary. Often the existence of a trend is clear from the data they collect. They are seldom asked to investigate the effect of a variable, which does not (in fact) covary with the outcome. However, in everyday life we often have to decide if we believe an outcome covaries or does not covary with a specific factor (for example: Does this food increase my risk of heart disease? Do mobile phones cause cancer?). If our aim in school science is to promote scientific literacy, we need to include a mixture of investigations of covariation and non-covariation cases in order to develop pupils' ability to reason about the implications of empirical data. The non-covariation case is also very valuable for research, highlighting features of children's thinking, which are hidden if only covariation cases are considered.

REFERENCES

Austin, R., Holding, B., Bell, J. & Daniels, S. (1991). *Assessment Matters No. 7: Patterns and Relationships in School Science*. London: School Examinations & Assessment Council (SEAC).

Chinn, C.,.A. & Brewer, W.,F. (1998). An empirical test of a taxonomy of responses to anomalous data in science. *Journal of Research in science teaching*, 35, (6), 623-654.

Koslowski, B. (1996). *Theory and Evidence: The Development of Scientific Reasoning*. Cambridge, MA: MIT Press.

Kuhn, D., Amsel, E. & O'Loughlin, M. (1988). *The Development of Scientific Thinking Skills*. London: Academic Press.

Millar, R. & Lubben, F. (1996). Investigative work in science: the role of prior expectations and evidence in shaping conclusions. *Education 3 to 13*, March, 28-34.

Mechanistic Reasoning On The Concept Of Wave Surface, And On The Huygens' Principle

Laurence Maurines

I.U.F.M. of Créteil, L.D.S.P., University of Paris 7, France

ABSTRACT

This study is the unpublished part of a research analysing the general ways of reasoning used by students in understanding the propagation of waves in a three-dimensional medium in the presence or the absence of obstacles. It focuses on the concept of wave surface and on the Huygens' principle. Sets of questionnaires were designed and about 300 French physics students (aged 20-23) were questioned after lessons on waves. The same tendencies as those pointed out in our other research on waves are revealed: reasoning on a travelling shape considered as a material object and as a whole, reducing the number of variables. Certain pedagogical implications are discussed.

1. BACKGROUND AND AIMS

This particular study is part of a research programme exploring the difficulties encountered by students when learning wave phenomena in a three-dimensional medium. Having previously analysed how students use the concepts of ray and wave surface when describing and explaining wave phenomena by means of drawing (Maurines, 1997), we are interested here in the way they consider the physical quantities, which the wave surface depends on, and in the way they handle the Huygens' principle. The main question is to examine whether the tendencies towards a mechanistic and single-notion based reasoning, first brought out for the propagation of a transverse signal on a string (Maurines, 1992), also appear for the propagation of waves in a three-dimensional medium in the presence or the absence of obstacles. This study aims to support discussions on teaching goals and to help to elaborate pedagogical means adapted to the difficulties identified.

As far as we know, these questions about the scalar three-dimensional wave model, which need to be used in wave situations such as diffraction and interference and which can also account for geometrical situations such as propagation and reflection, have hardly been explored by science education researchers. Indeed, most investigations about students' conceptions in the domain of waves concern light and situations, which geometrical optics can account for (e.g. Galili, 1996). There are also some studies on the propagation of visible mechanical signals and of sound along one direction (Linder and Erickson, 1989 ; Wittmann and al., 1999) and on light diffraction (Ambrose and al.,1999; Colin, 1999).

D. Psillos et al. (eds.), Science Education Research in the Knowledge-Based Society, 127–135.

First, the problems explored in this research and the methods used will be presented. Secondly, we shall give the results and a first level of analysis. Then, a model, that is to say, a synthetic and predictive description of the students' main trends of reasoning will be proposed. Finally, we shall discuss some of the pedagogical implications.

2. FRAMEWORK

2.1 Some remarks on the concept of wave surface and on the Huygens' principle

In the scalar three-dimensional wave model, a wave is described by a function F of two variables : one concerns space (M), and the other time (t). In the case of progressive waves, the function F(M, t) depends on a function Φ(M, t), which connects the variables of space and time, and which is called the phase of the wave. For example, for a spherical wave propagating in a non dispersive three-dimensional medium, we have $F(M,t)=A(M)G(r-ct)$ and $\Phi(M,t)=r-ct$.

A wave surface of a wave travelling in a three-dimensional medium is a phase surface, that is to say, the locus of the points of the medium reached at the same instant by the wave. They verify $\Phi(M,t)=\text{constant}$. On a wave surface, the amplitude of the wave, A(M), is not necessarily constant. For example, the amplitude of the sound emitted by a loud-speaker is not constant on a wave surface. It is given by the radiating diagram. The shape of a wave surface depends on the propagation velocity (consequently, on the medium), and also on the shape of the source when the point of observation is "near" to the source.

The principle known as "the Huygens' principle" or more generally as "the wave- envelope principle" says that the wave surface of the wave resulting from the superposition of waves emitted by coherent adjacent point sources is the envelope of the wave surfaces of the elementary spherical waves. The point sources can be fictitious Huygens' sources of a wave surface or material sources, as for example the points of a reflecting surface.

In order to draw a wave surface in a specific situation, two kinds of methods can be used. One is based on the Huygens' principle and links two levels which we have called the elementary waves level and the macroscopic level. The other deals with the wave resulting from the superposition of the elementary waves and concerns only the macroscopic level. In the first method, a wave surface is considered as the envelope of the wave surfaces of elementary spherical waves. In the second method, a wave surface is considered as innumerable points moving at the wave velocity along perpendicular rays.

2.2 Questions explored

Our previous research about waves propagating along one direction (Maurines, 1995) showed that students have difficulties in differentiating the concept of phase from the concept of amplitude. So, we first examined whether, for students, the wave amplitude was necessarily constant on a wave surface.

Second, as we pointed out that students consider the shape of the signal propagating on a string as if it was the shape of a material object (Maurines, 1992), we examined whether, for students, the shape of a wave surface depended on the propagation velocity and on the shape of the source.

Third, as a reasoning based on a travelling shape considered as a material object is inconsistent with a reasoning based on the Huygens' principle, we focused on situations where this principle needed to be used. So, we examined whether students used the Huygens' principle in order to draw the wave surface of a reflected wave, and also of a wave resulting of the superposition of two coherent spherical waves.

3. METHODS AND SAMPLES

In order to validate our hypotheses on the way students reason, a set of nine paper and pencil questionnaires consisting of one or two questions were designed. Half of the questionnaires do not specify the type of waves whereas the others do (light, sound, ripples on water). All the questions are qualitative. Some ask students to compare and draw the shape of the wave surfaces of two different waves; others ask them to draw the wave surface of a specific wave or to choose a method in order to obtain a sound wave surface by experiment. Justifications were required.

The questionnaire was distributed to different groups of about 50 students. A total of about 300 French students (aged 20 to 23) were questioned. Most of them were in classes training them to teach physics (fourth or fifth university year). The others were in the second and third years of scientific universities. All had received lessons on waves at university. As the answers and justifications given by the different subgroups described above were similar, no distinction has been made.

The students' answers and comments were processed in three steps according to a method close to the one proposed by Minstrell (1992). They were first analysed question-by-question and questionnaire-by-questionnaire. Representative categories of explanatory patterns or strategies employed by the students in addressing particular situations were identified. These categories were then grouped following the problematic framework presented above (the schemes or clusters of facets of knowledge as they are called by Minstrell). Finally, these groups of categories were accounted for with a few trends of reasoning.

4. RESULTS

For the sake of brevity, only the main results are given here. For each problem explored, the text of one questionnaire is first presented briefly. Then, the main types of answer are illustrated with examples and their percentages of occurrence are given. These percentages have to be considered as tendencies of reasoning and not for themselves. Indeed, the number of answers obtained for a particular question is not always high (from 22 to 44) because many students did not complete all the questions. We consider that a percentage of incorrect answers higher than 25% are the sign of a difficulty. This is the same for the percentages where no answer is given.

4.1 Students and the wave surface

For students, a wave surface is not a phase surface.

One of the questionnaires exploring this problem concerns the way of obtaining a wave surface of a sound emitted by a loudspeaker by means of experiment. The experimental device is described and two methods are proposed. Students are asked to choose one method, and if necessarily, to give another one. In the first method proposed, which is incorrect, the amplitude of the sine observed on the screen of the oscillograph is kept constant. In the second one, which is correct, the sine observed on the screen of the oscillograph does not move horizontally.

Only 12% of the students (N=49) choose the correct method. 27 % do not answer and 61% choose a wrong method. Consequently, for a majority of students, a wave surface is not a phase surface. The comments reveal that, for half of them, a wave surface is an iso-amplitude surface : *"wave surface : the same amplitude"*. For the others, a wave surface is, at the same time, an iso-amplitude surface and a wave surface, that is to say, an iso-displacement surface : *"we have to combine the two methods because the two conditions on the amplitude and on the phase have to be verified"*.

For students, the shape of a wave surface does not depend on the medium.

Two questionnaires dealing with the propagation of light and sound in air have been designed in exactly the same way. In the first question, students are asked to draw a wave surface of the wave emitted by a source, considered as a point, when the temperature of the air is constant. In the second question, they are asked to compare this wave surface with the wave surface of the wave emitted by the same source when the temperature of the air changes with altitude. The text stated that the wave velocity depends on the temperature and how this velocity changes with the temperature. In the first situation, the wave surface is spherical. In the second situation, it is non-symmetrical.

For the first question dealing with the propagation in an homogeneous medium, a majority of students (89%, N=61) give a correct answer.

For the second question dealing with the propagation in an inhomogeneous medium, only 31% of the students (N=61) give a correct answer and 28% do not answer. 41% of the students give a wrong answer by drawing a symmetrical wave surface. Consequently, for many students giving an answer, the shape of a wave surface does not depend on the medium. The justifications point out that, for these students, the shape of a wave surface does not change when the wave velocity changes. The wave surface just seems to be moving faster : *"the wave surface 2 is the same as the wave surface 1. It does not depend on the light velocity"*, *"yes, changing the temperature has an effect on the velocity but not on the shape of the wave surface"*, *"yes, the wave surface is the same but it reaches the point M faster"*.

For students, the shape of a wave surface does not depend on the source.

The questionnaire exploring this problem concerns the propagation of ripples on water. Students are asked to compare and draw the wave surfaces of the ripples propagating on the surface of a lake whose depth is constant, just after a ball and a cube have been thrown in. In the first case, circular ripples propagate on the lake whereas, in the second case, "squared" ripples propagate.

Only 24% of the students (N=46) give a correct answer and 35% do not answer. 41% of the students give a wrong answer. For them, the shapes of the ripples propagating on water, just after that a cube and a spherical ball have been thrown in, are the same : *"the wave surfaces are always circles. They do not depend on the shape. Only, the amplitude will change in relation to the mass of the objects", "because it is not the shape of the object which is important. It is the contact with the surface of the water".*

Even if these answers can be induced by the familiar situation dealing with stones thrown in water, we find the justifications very significant. Indeed, they reveal the same tendencies of reasoning as in our research on the propagation of a signal on a string. First, the source seems only to put the wave surface into motion and not also give it a shape; second, on the drawings given by students, the shape of the wave surface is considered as a whole. It is not decomposed into innumerable points propagating at wave velocity as soon as created by the source. Before going on, we have to be aware that, in this situation where a wave surface can be obtained by two different methods, students never chose the method based on the Huygens'principle.

4.2 Students and the Huygens' principle

In a wave situation such as interference.

One of the questionnaires exploring this problem concerns the superposition of two coherent spherical waves. The text asks firstly that a single point source emitting a sine wave be considered and then two point sources, similar to the first one, emitting simultaneously. The drawing of a wave surface of the wave emitted by the single point source is given. At the first question, students are asked to compare the wave surface of the wave propagating when the two point sources are emitting to the wave surface of the wave propagating when the single point is emitting. The two point sources are said to be in phase. At the second question, students are asked whether it is possible to obtain a non-symmetrical wave surface when the two point sources are emitting at the same time and how this can be done.

In this interference situation, a single resulting wave can be defined. When the point sources are emitting in phase, the shape of the resulting wave surface is symmetrical (it is elliptical). It coincides with the envelope of the wave surfaces of the waves emitted by the two point sources when the point of observation is far off, because, in this case, the point sources can be considered very close to each other. The wave amplitude is not constant on the wave surface: there are hyperbolic bright and dark interference fringes. When the two point sources are not emitting in phase, the wave surface of the resulting wave is not symmetrical.

In the first question concerning two point sources emitting in phase, 84% of the students (N=32) give a diagram. Only 41% of these diagrams (N=27) are correct: a resulting wave exists and its wave surface is the envelope of the wave surfaces of the waves emitted by the two point sources (fig.1 in insert 1). The other diagrams (59%) are wrong. On half of these diagrams, interference fringes are drawn instead of wave surfaces (fig.2 in insert 1). Here we come across the same difficulty in differentiating the wave phase from the wave amplitude as the one encountered in the question on the method to use in order to obtain by experiment a sound wave surface. On half of these diagrams, the wave surfaces of the waves emitted by the two point sources are drawn but not their envelope (fig.3 in insert 1). For these students, the two circular wave surfaces seem to have their own identity, not being able to disappear when the waves superpose one another.

Fig. 1 correct answer: an envelope is drawn

"These dotted lines have interfered and no longer belong to the wave surface"

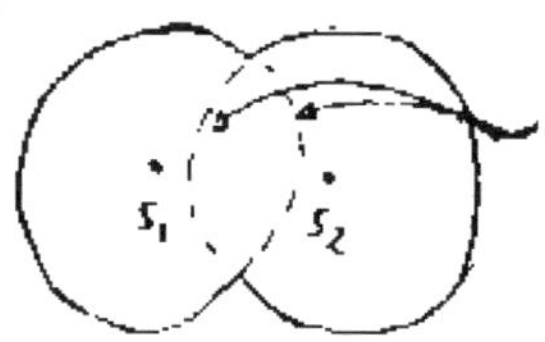

Fig. 2 incorrect answer : the phase surfaces are confused with the iso-amplitude surfaces

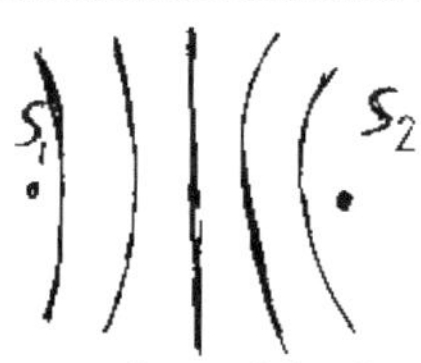

Fig.3 incorrect answer: only the wave surfaces of the elementary waves are drawn

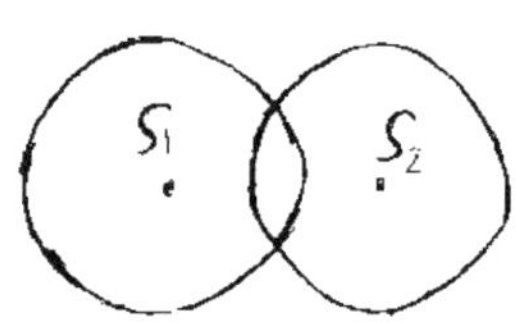

Insert 1. The shape given by students to the wave surface of the wave propagating when two coherent point sources are emitting in phase

In the second question asking students whether it is possible to obtain a non-symmetrical wave surface, 31% of the students do not answer. Only 36% of the students giving an answer (N=22) say that it is possible to have a non-symmetrical wave surface. As students tend to merge the wave amplitude with the wave phase, it is not surprising that very few of them give a correct method (the two point sources have not to be in phase) and that most of them (27%, N=22) give an incorrect method: *"the point source S_2 has to be moved", " the intensity of the point source S_2*

has to be higher". This is certainly also why many students answer that it is not possible to obtain a non-symmetrical wave surface (64%, N=22).

In a geometrical situation such as reflection.

One of the questionnaires dealing with this problem concerns the reflection of an incident plane wave on a "parabolic" mirror. The shapes of the wave surfaces of the incident plane wave and of the mirror are given. Students are asked to draw a wave surface of the reflected wave using the Huygens' principle.

Fig. 1 correct answer: students link the macroscopic level and the elementary waves level correctly.
"Wave surface of the reflected wave"

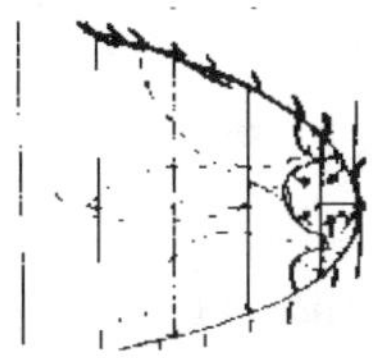

Fig.2 incorrect answer: students reason only on the elementary waves level
"According to the Huygens' principle, a spherical wave propagates from a point source, therefore the wave surfaces on the reflecting surface"

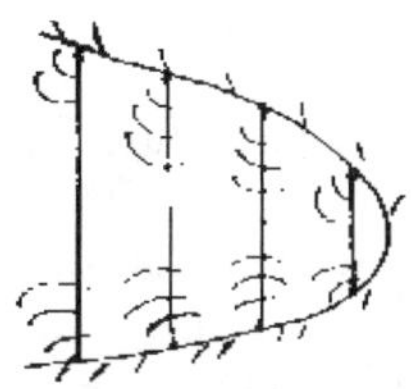

Fig. 3 incorrect answer: students' reason only on the macroscopic level. The method used gives a correct shape. *"For a parabolic surface, geometrical optics says that the reflected rays converge at the focus. The wave surfaces are perpendicular to the* rays"

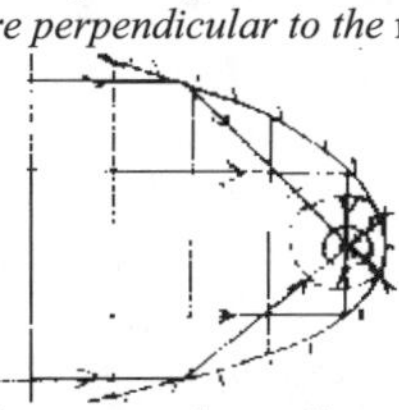

Fig. 4 and 5 incorrect answers: students' reason only on the macroscopic level. The method used gives a wrong shape.

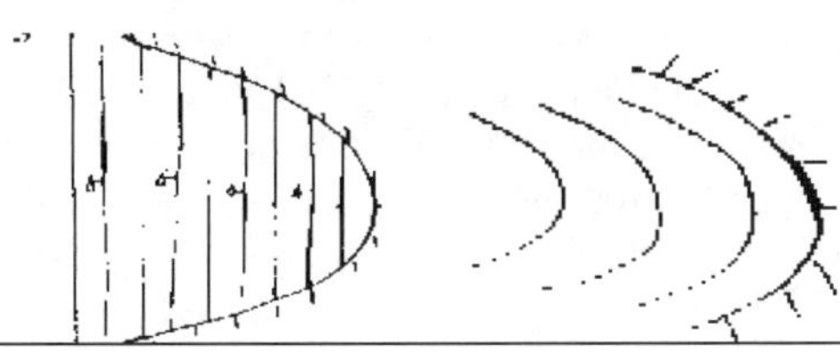

Insert 2. The shape given by students to the wave surface of the reflected wave of an incident plane wave reflecting on a "parabolic" mirror

37% of the students (N=49) do not answer the question. Only 33% of the students giving a drawing (N=27) take into account the elementary waves level. Only half of them also consider the macroscopic level and few of them (13%, N=27) link these two levels correctly (fig.1 in insert 2). Half of them seem to focus only on the elementary waves level (fig.2).

Most of the students giving a drawing (67%, N=27) use another method than the one imposed and focus only on the macroscopic level. Only one third of these students decompose the incident wave surface into an infinity of points and give a correct shape to the reflected wave surface (fig.3 in insert 2). The others consider the incident wave surface as a whole and give to the reflected wave surface the shape of the incident wave (fig.4 in insert 2) or of the mirror (fig.5 in insert 2).

5. CONCLUSION AND IMPLICATIONS

The previous results validate our hypotheses on students' reasoning. Indeed, they reveal the same connected trends of reasoning as those we already put forward in our research on the propagation of waves along one direction.

On the one hand, students tend to reason on a single and more concrete notion. Indeed, they do not differentiate the concepts of phase and amplitude: for them, the amplitude of a wave has to be constant on a wave surface; in an interference situation, the wave surfaces of the resulting wave are confused with interference fringes.

On the other hand, students tend to follow a travelling shape considered as a whole and as a material object. Indeed, they consider the shapes of a wave surface and of a solid in the same way. For them:

- The shape of a wave surface does not depend on the wave velocity and on the source creating the wave as the shape of a rigid ball does not depend on its velocity and on the hand that throws it up.
- The shape of a wave surface does not change when reflecting on a mirror, as if the wave surface were a rigid ball bouncing on a wall, or keeps the print of the mirror as if the wave surface were a ball made of clay thrown on a wall.
- In an interference situation, two wave surfaces keep their own identity and cannot disappear as if they were two key rings hooked together.

Besides these trends of reasoning, another one, specific to the three-dimensional wave phenomena is revealed. Students tend to reason in a geometrical manner not using the Huygens' principle. This tendency is encountered not only in geometrical situations, such as the reflection situation, but also in wave situations such as the interference situation presented here or such the diffraction situation analysed previously (Maurines, 1997, 1999).

If these trends to follow a travelling whole and to reason on the basis of a single and more concrete notion can be noticed in the results of many investigations about students' conceptions in the domain of waves, they are not specific of waves. Indeed, a lot of research has pointed out that students reason this way in various physical fields (Di Sessa, 1983 ; Minstrell, 1992 ; Resnick, 1989 ; Viennot, 1996). They are certainly the characteristics of the common reasoning.

The way, in which waves are taught today, at least in France, hardly confronts this kind of reasoning with the theory. Indeed, the difficulty of differentiating the concepts of phase and amplitude can be reinforced when reading some university books : " *wavefronts : surfaces on which Ψ=constant at each instant*" (Soum, 1995). Moreover, French books do not help students to overcome the trend to consider a wave surface as a whole and as an object : the wave surfaces properties are rarely emphasised nor is the Huygens' principle used.

In order to help students not to confuse a wave surface with an iso-amplitude surface and not to consider a wave surface as a material object, it is worthwhile to discuss the parameters, on which a wave surface depends and does not depend. In order to put into question the trend to reason on a travelling shape, it seems fruitful to propose exercises, based on the Huygens' principle, in which a wave has to be considered as the superposition of elementary waves. Qualitative questions, such as those presented here, can be used for these purposes. The graphical work they induce must not be neglected even at university level because it permits qualitative questioning of the way students reason. Moreover, the visual impact of drawings must not be underestimated. Computer simulations can be helpful.

REFERENCES

Ambrose B. S. & Shaffer P. S. & Steinberg R. N. & Mc Dermott L. C. (1999). An investigation of students' understanding of single-slit diffraction and double-slit interference. *Am.J. Phys*, vol.67, n°2, pp. 146-155.

Colin P. (1999). *Deux modèles dans une situation de physique : le cas de l'optique. Difficultés des étudiants, points de vue des enseignants et propositions pour structurer des séquences d'enseignement.* Thèse, Université Denis Diderot Paris 7.

Di Sessa A. (1983). Pnenomenology and the Evolution of Intuition. In D. Gentener and A.L. Stevens, *Mental Models*, Lawrence Erlbaum Associates Inc., Hillsdale, New Jersey, pp.15-34.

Galili I. (1996). Students' conceptual change in geometrical optics. *Int. J. Sci. Educ.*, vol. 18, n°7, pp. 847-868.

Linder C. & Erickson G. (1989). A study of tertiary physics students'conceptualizations of sound. *Int. J. Sc. Edu.*, vol. 11, special issue, pp. 491-501.

Maurines L. (1992). Spontaneous reasoning on the propagation of visible mechanical signals. *Int.J.Sc.Edu.*, vol.14, n°3, pp. 279-293

Maurines L. (1995). Les étudiants et la phase d'une onde progressive : résultats d'une enquête exploratoire. In *actes du cinquième séminaire national de recherche en didactique des sciences physiques*. Reims, I.U.F.M. de Reims, pp. 107-123.

Maurines L. (1997). Students and the wave geometrical model of the wave propagation in a three-dimensional medium. *Selected papers of the First Int. Conf. of ESERA*. In *Research in science education in Europe*. Kluwer Ed., 1999, pp.103-112.

Maurines L. (1999). Spontaneous reasoning on light diffraction and coherent illumination optical imaging. In *synopsis book of the Sec. Int. Conf. of ESERA*. Kiel. IPN-Institute for Science Education/University of Kiel (Ed.), pp.92-94.

Minstrell J. (1992). Facets of students' knowledge and relevant instruction. In R. Duit, F. Goldberg and H. Niederer (eds) *Research in Physics learning : Theoretical Issues and Empirical Studies*. Kiel, IPN, pp.110-128.

Resnick L.B. (1989). Convictions ontologiques dans l'apprentissage de la physique. In N. Bednarz et C.Garnier, *Construction des savoirs, Obstacles et Conflits*. CIRADE, Agence d'Arc inc, Ottawa,pp.103-109.

Soum G. & Denizart M. Jagut R. (1995). *Optique,* tome 1. Hachette, Paris.

Viennot L. (1996). *Raisonner en physique*. Paris, De Boeck.

Wittmann M. & Steinberg R.N. & Redish E.F. (1999). Making sense of how students make sense of mechanical waves. *The physics teacher*, n°37, pp.-21.

ATOMIC PHYSICS IN UPPER SECONDARY SCHOOL: LAYERS OF CONCEPTIONS IN INDIVIDUAL COGNITIVE STRUCTURE

Jürgen Petri, Hans Niedderer

Institute of Physics Education, University of Bremen, Germany

ABSTRACT

Several learning studies in the field of science education describe the final state of students' concepts of, e.g., air pressure or chemical bonding as a multi-layered cognitive conceptual structure (manifold conceptions, multiple frameworks). The hypothesis of a multi-layered cognitive structure assumes that students - e.g., when asked to explain a specific problem - are able to activate significantly differing and competing understandings of a single concept and that therefore *no* essential shift of context is necessary. Concerning atomic physics, our hypothesis of cognitive layers is based on an explorative case study (Petri & Niedderer, 1998). In a new analysis of our interview data, we investigated the cognitive structure of six students with respect to their individual layers at the end of the course (Petri & Niedderer, 2001). Our findings, which confirm the important role of cognitive layers during conceptual change, are summarized and discussed in this paper.

1. BACKGROUND, AIMS AND FRAMEWORK

Some research literature on conceptual change, and conceptual structure respectively, describes individual students' final cognitive states as multi-layered or containing manifold conceptions. By multi-layered or manifold conceptions we mean that students, even in a single particular context, may consciously apply significantly differing and competing conceptions or explanatory schemes.

Maloney & Siegler (1993), referring to tasks on momentum and energy in mechanics, formulate an individual cognitive structural relation of different conceptions, which is independent of the context and non-situated. To determine which of the competing conceptions will be applied in a particular case, they use the term strength to quantitatively characterize a particular understanding.

Tytler (1998) distinguishes three levels of conceptions of air pressure, i.e. "pre-explanatory", "intermediate" and "advanced". He speaks of "a multi-layered understanding" or "multiple perspectives". The meaning of these terms includes the application of different conceptions to a single task.

Taber (2000) explicitly deals with a model of cognitive layers of qualitative conceptions. He proves *"that an individual learner can simultaneously hold in cognitive structure several alternative stable and coherent explanatory schemes that are applied to the same concept area"* (Taber, 2000, 399). Taber describes a pre-university A-level student's ways of answering when reasoning the generation of

D. Psillos et al. (eds.), Science Education Research in the Knowledge-Based Society, 137–144.

chemical bonds using three different concepts independent in their content and persistent in time. The student applies these concepts partly within different contexts but also partly to the same problem. Taber emphasizes that he speaks of different concepts used in a single scientific content area.

These findings match with some results of our study on atomic physics (Petri & Niedderer, 1998): The development of one particular student's (Carl) conceptions of atoms was analysed as a learning pathway (Scott, 1992) in terms of the model of the cognitive system proposed by Niedderer & Schecker (1992). The pathway comprises of initial, final, and several meta-stable intermediate states of Carl's “cognitive element atom”. The final state of Carl’s "cognitive element atom" is characterized by the coexistence of three clearly distinguished conceptions. During his conceptual development, Carl is solving a conflict on the level of his meta-cognitive beliefs related to physics which leads to a considerable change in his "framework of reasoning" (Schecker, 1985) or his "framework theory" (Vosniadou, 1994), respectively. The term conception should be understood here in the sense of a “specific theory” (Vosniadou, 1994) or a “scheme of knowledge” (Galili & Hazan, 2000). Both constructs consist of a certain number of “beliefs” (Vosniadou, 1994) or “facets” (Minstrell, 1992), which at least from a student's point of view are closely linked with regard to their physical contents.

The notion of cognitive layers implies that the conceptions defining the different layers *may* be generated in a certain succession and under certain conditions, and that - as in Carl’s atomic conceptions – they may be based on each other in content and structure. The single components of the individual multi-layered structure also differ with respect to the strength of their cognitive grounding and regarding their status in the sense of Hewson & Hewson (1992). If individual students are able to attribute a certain status to their individual, qualitative conceptions relating to the same context, this may hint at a highly developed consciousness of the corresponding cognitive layer structure. One of the most significant examples is given by Scott (1992): Referring to the structure of the student Sharron's final conceptual state, Scott states that Sharron is able to use different models of gas the way she deems them appropriate in various given situations.

In addition to "status" (the value assigned by the student) and "strength" (the frequency of usage depending on cognitive grounding), we introduce the term "level of application" as a third variable to characterize the different layers. The level of application describes the students' quality and complexity of reasoning with regard to a particular layer.

One of our basic hypotheses is that there is a strong connection between the evidence of cognitive layer structures and students' severe learning difficulties with several science concepts labelled the conceptual change problem (cf e.g. Taber, 2000). From a constructivist perspective, conceptual change means graded modifications in cognitive structures marking a transition from intuitive to scientific conceptions. It is *not* supposed to mean an exchange or a simple addition of conceptions as misleadingly suggested by the term (cf Duit, 1996). Tyson, Venville, Harrison & Treagust (1997) distinguish between three different qualities of conceptual change: "addition", "weak revision" and "strong revision". According to this distinction, Newtonian mechanics and quantum physics, e.g., require a strong revision of crucial concepts. Tyson et al. state: *"Conceptual change does not imply that initial conceptions are "extinguished". Initial conceptions, especially those that hold explanatory power in non-scientific contexts, may be held currently with new conceptions. Successful students learn to utilize different conceptions in appropriate contexts"* (Tyson et al., 1997, *402)*. As stated above, the capability to use different conceptions in various ways may be an expression of a meta-conceptual awareness of the multi-layered structure. Numerous authors therefore request - as remedy to counter the "situatedness of conceptions" - that teaching should develop a meta-conceptual awareness or train an adequate and thoughtful use of models and conceptions (e.g. Vosniadou, 1994; Taber, 2000).

Vosniadou (1994) emphasizes that the transformation of students' "framework theory" comprising of their epistemological and ontological opinions is necessary for a strong conceptual revision. In her findings, she does not speak of cognitive layers, but stresses the consistency of primary school students' mental models of in her case the shape of the earth. However, she classifies 11 out of the 60 mental models she found with her students as "mixed". Therefore, the existence of cognitive layer structures as transitory states in difficult and long-term learning processes seems to be a plausible hypothesis concerning Vosniadou's theory as well.

The main aim of our data analysis for individual cognitive layers is to contribute to a more detailed picture of the role of cognitive layers during conceptual change processes. The question of the relevance of cognitive layer structures to the appropriate description and assessment of successful teaching and learning is an issue already raised by Niedderer & Goldberg (1995).

2. METHODS AND SAMPLES

In 1993/94, ten students attending an A-level course were taught according to the "Bremen Approach on Atomic Physics" developed by H. Niedderer (e.g. Niedderer, Bethge & Cassens 1990; Niedderer & Deylitz, 1998). This approach favours an atomic model based on the Schrödinger equation and prefers to visualize bound electrons as continuous and stationary charge density distributions, to be described by their state functions (ψ-functions) and correlating orbitals. The Schrödinger equation and the interpretation of the state functions as density distributions are taught from multiple perspectives. They are applied to various cases (Hydrogen atom, several higher atoms, the H_2^+-molecule and solid states) in order to facilitate a non-situated conceptual development and a meta-conceptual consciousness. Bohr's "planetary model of the atom", already known to our students, was only referred to when necessary. Furthermore, the statistical interpretation of the state function (Born) was mainly discussed as the "probability of localization".

The results discussed below are based on a new analysis of the post-instruction interviews carried out with six out of the course's ten students, including Carl, three months after the end of the teaching unit. The semi-structured interview took about 45 minutes. It covered the whole unit but emphasized the qualitative and intuitive elements of the teaching approach in accordance with the aims of the study. The order of the questions was dealt with flexibly, only the initial questions (figure 1) for the different subjects were compulsory. More specific questions were posed only if the students did not mention the corresponding subjects themselves.

The data were interpreted in an iterative, hermeneutical process as described by Roth (1995, 333).

- What do you remember of Mr. Niedderer's instruction?
- What do the single components of the Schrödinger equation mean and what do you think the formula is about?
- What do you have to bear in mind when calculating atoms with more than one electron by means of the Schrödinger equation?
- How do you imagine a Lithium atom? Sketch and explain!
- What do you understand by localization? Please comment on the statement of a student: "Something materializes out of nothing."
- What do you understand by a "charge cloud"?
- Please comment: "In the double slit experiment the electron seems to pass through both slits."
- What is the meaning of the Uncertainty Principle and how can it be applied?

Figure 1. Some compulsory interview questions

3. RESULTS

In view of the content of teaching, a cognitive layer structure is bound to be found to a certain degree with any student. Even those students who are still using spontaneously the planetary atomic model, are nevertheless aware of alternative models and are able to mention, discuss and apply them to a limited extent - at least when asked directly. In these cases of weak conceptual restructuring and low-level application, the layers remain partly dependent on a specific context: for example these students reconstruct only two of the three layers they basically know.

Compared to Carl, there are significant differences characterizing the final state of the other students' "cognitive element atom" (figure 2). Such differences result from individual suppositions and attitudes; they reflect individual learning pathways and the quality of the conceptual change achieved.

Carl's cognitive layers were found to be very distinct. The following interview section gives an example of how his three cognitive layers manifest themselves in his language at a high level of reflection and within a single context.

> I: Please comment on the following statement: According to the orbital model the electrons don't move in circles and ellipses (Bohr) but inside their orbitals.
>
> C: These are - the conceptions (I had of it) before and after (the course)! The trajectories are no longer determinable circles or ellipses. You don't always know exactly where it (the electron) is, you only know where it can be and what the probability of this is. And in such an orbital, if you think of a p-orbital, you can see that it actually can't always move in circles, if you accept a movement at all. In the end I have stopped presupposing a movement of the electron but just assumed it would be there. And it is no longer that electron, that ball, but something blurred, a kind of energy, for example. It's very difficult to explain to myself what an electron actually is. Sometimes, when localized, it behaves like a tiny particle, but when not being localized it's indeed no particle!

Carl stresses his disagreement with the planetary model. Step by step he reconstructs the alternative quantum conceptions. His going beyond the context set by the interview demonstrates the marked degree of his multi-layered cognitive structure and the high quality of how he tackled "the problem electron".

Tim's cognitive layer structure is very similar to Carl's, however, it is more related to the context and articulated quite stereotypically. For him, the newly acquired atomic models are of high status, comparable to Carl, but he explains the charge cloud model less often. Tim does not show any inner conflict. Instead, his support of the quantum atomic models is based more on secondary motives (such as social appreciation, feelings of superiority over "non-insiders", etc.) which would suggest a less strong conceptual restructuring.

cognitive layer / student	"planetary model" 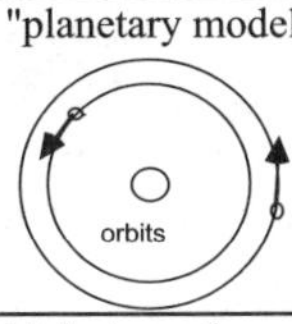	"probability model" 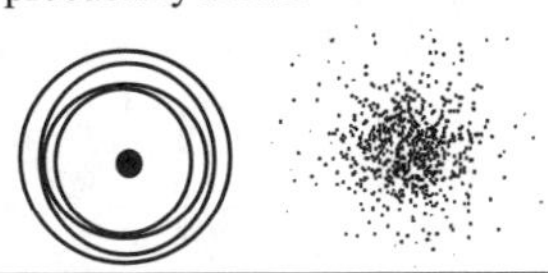	"charge cloud model" 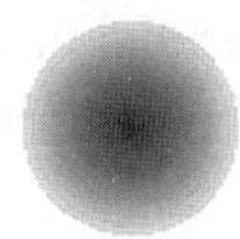
Carl	high strength low status	medium strength medium status	medium strength very high status
	authentic conflict solving, clear distinction and reflective use of models strong conceptual restructuring, advanced conceptual change		
Tim	high strength low status	medium strength medium status	low strength very high status
	secondary motivation, missing authenticity, lower application level weak conceptual restructuring, limited conceptual change		
Rolf	high strength high status	low strength medium status	very low strength low status
	fixation to formulas, scepticism towards probability calculation no conceptual restructuring, some quantitative conceptual gain		
Paul	medium strength low status	medium strength medium status	medium strength high status
	professional use of models including context transfer, very high application level, exceptional preconditions, no restructuring necessary		
Karin	very high strength very high status	low strength low status	very low strength very low status
	affective reservations about charge cloud and instruction no conceptual restructuring, some quantitative conceptual gain		
Stefan	very high strength very high status	medium strength low status	very low strength low status
	protest against instruction no conceptual restructuring, some quantitative conceptual gain		

Figure 2. Layer structures of six students doing atomic physics

For Paul, a cognitive layer structure is less characteristic as he does not have the problem to abandon a planetary concept of the atom, which - according to his own statement - has not been internalised by him before. He has a much more extended framework of reasoning, in an ontological as well as an epistemological respect, and is able to handle the three different models of the atom in a quite professional way.

Rolf, Karin and Stefan show a very rudimentary layer structure, which is mostly limited to an expected degree if someone seriously takes part in discussions of the subject. All of them are not able or willing to accept some of the crucial aspects of quantum physics. For Rolf, e.g., the statements of probability and measurement processes, which cannot be expressed in formulas, are highly dissatisfactory. For these students, quantum atomic models remain at a lower status than the planetary model, and new layers and their limits are often only to be observed if a "specific pressure" from outside is exerted. In such a case Stefan distinguishes between a non-plausible knowledge about quantum atomic models and an illustrative and plausible planetary model.

Karin's comment on the impulse already discussed in connection with Carl demonstrates that she is aware of the "ban imposed on orbits" by the uncertainty relation. But Karin is unable to develop a plausible alternative visualisation and thus cannot solve her problem in a satisfactory way. Stating that the electron "is whirling somehow" and "it's just there", she expresses features of the cognitive layer "probability model". When the interviewer asked her to give a "fundamental alternative" he aimed at the "charge cloud model". But Karin doesn't (want to) make use of this model on her own initiative during the whole interview. In contrast to Carl, the status, strength and level of application of her related layer are found to be very low:

> I: Please comment on the following statement: According to the orbital model the electrons don't move in circles and ellipses (Bohr) but inside their orbitals.
>
> K: ... If you imagine that as in Bohr's model, there are only two electrons on the first shell, I just have this picture: There is this circle, and the shell is always in the same distance to the nucleus. So there are really only these two permanently on the shell and the other ones are moving over there. Whereas in this model, this orbital model, these electrons move in these areas and not on determined orbits. You can't say that electrons move in circles, they just whirl around – they are whirling somehow. Otherwise I could immediately determine the location and speed of an electron: In that moment, at that time it will be exactly at that point. And that's not true, I can't say that!
>
> I: … Would you say it moves somehow?
>
> K: Yes, I don't know whether it accelerates when moving or whether it carries out a steady movement, or whether it flies straight ahead or in waves. I don't even know whether it moves at all!
>
> (…)
>
> I: ... Is there an alternative view that would avoid all these problems with movement?
>
> K: No, I just think: okay, it doesn't move, it's just there. How it has got there - I don't know, but it's there! I mean it could be in a completely different place in the next moment! But I don't know how it gets from this point to that point! Thus, it has to be moving - or, God knows!

4. CONCLUSIONS AND IMPLICATIONS

Our analysis based on a single interview has a primarily explorative character. For a more precise determination of individual layer structures, broader and especially designed investigations would have to be carried out. Nevertheless, some preliminary statements and thoughts can be deduced from the present results:

The organisation of individual knowledge after physics classes aiming at conceptual change can be appropriately described by cognitive layer structures. Differences in quality and content of the individual layer structures permit qualitative and half-quantitative comparisons. From this perspective, the question of an adequate description, assessment and explanation of successful learning and teaching is posed anew: Specific and profound interviews which activate cognitive layers of lower strength are expected to deliver a more differentiated and appropriate picture than written tests, particularly multiple-choice tests, can do.

Based on these results and ideas, Susann Hartmann of our research group started to investigate cognitive layers in the domain of Newtonian mechanics, where, compared to atomic physics, the students' alternative and probably non-Newtonian layers are scientifically less recognized. Nevertheless, these layers deserve and require a high level of attention and should be discussed extensively during teaching. Ms Hartmann developed a specific interview technique based on three already existing tasks referring to qualitative conceptual understanding. The main feature of the interviews is a newly developed system of instructions and questions without giving content information prompting the students to formulate, explain, rethink and possibly rephrase their answers. First results are expected in 2002.

Our results confirm the important role of a cognitive layer structure for conceptual change postulated, among others, by Taber (2000). As a hypothesis, we expect cognitive layer structures to appear more often and more differentiated in classes stressing qualitative conceptual understanding than in classes almost exclusively aiming at quantitative aspects. For example, central to the "Bremen Approach on Atomic Physics" are computers and especially a symbolic modelling software which enables extensive and widely autonomous modelling activities of the students fostering a conceptual understanding of the stationary Schrödinger equation and its basic solutions.

With regard to the students, it seems that a cognitive layer structure is most characteristic for a teaching process that will eventually achieve a deep conceptual change. From the significant differences of status, strength and level of application of the different layers, distinctions should be possible between a rudimentary, a basically successful, and a complete conceptual change. A complete conceptual change is characterized by the highest ratings relating to the status, strength and level of application of a student's final conceptions. This does not mean that the previous conceptions cannot be found anymore in the student's cognitive system. The level of application could be scaled using terms such as denomination, description, explanation, transfer and reflection of the domain of appropriateness and validity.

Further perspectives should arise by comparing different domains of physics. These domains differ among others in their relations to everyday life, in their degree of abstractness and in the scientific status of those alternative conceptions representing different cognitive layers. These differences should have distinct effects on the development of cognitive layer structures. Studies in mechanics by Ms Hartmann and in electric circuits by Andreas Redfors (Kristianstad University, Sweden) are still in progress. The results reported above together with these new studies are expected to allow comparisons of layer structures from different content domains.

REFERENCES

Duit, R., (1996). Lernen als Konzeptwechsel im naturwissenschaftlichen Unterricht. In: Duit, R., Rhoeneck, C. von (eds.), *Lernen in den Naturwissenschaften* (pp. 145-162). Kiel: IPN, Kiel.

Galili, I.; Hazan, A., (2000): Learners' knowledge in optics: interpretation, structure and analysis. *Int. J. Sci. Educ.*, 22, 57-88.

Hewson, P.; Hewson, M., (1992). The status of students' conceptions. In: Duit, R.; Goldberg, F.; Niedderer, H. (eds.): *Research in Physics Learning: Theoretical Issues and Empirical Studies* (pp.59-74). Kiel: IPN.

Maloney, D., P.; Siegler, R., S. (1993). Conceptual competition in physics learning. *Int. J. Sci. Educ.*, 15, 283-295.

Minstrell, J., (1992). Facets of students' knowledge and relevant instruction. In: Duit, R., Goldberg, F., Niedderer, H. (eds.): *Research in Physics Learning - Theoretical Issues and Empirical Studies* (pp. 110-128). Kiel: IPN.

Niedderer, H.; Bethge, T.; Cassens, H., (1990): A Simplified Quantum Model - a Teaching Approach and Evaluation of Understanding. In: Lijnse, P. L.; Licht, P.; De Vos, W.; Waarlo, A. J. (eds.) *Relating Macroscopic Phenomena to Microscopic Particles - A Central Problem in Secondary Science Education* (pp. 67-80). Utrecht: CD-ß Press.

Niedderer, H.; Deylitz, S., (1998). Introduction to Atomic Physics - A concept based on the Schrödinger equation. Textbook for students." Perth/Bremen. (http://didaktik.physik.uni-bremen.de/niedderer/projects/quanten/).

Niedderer, H.; Goldberg, F., (1995). Lernprozesse beim elektrischen Stromkreis. *Zeitschrift für Didaktik der Naturwissenschaften* ZfDN 1, Heft 1, 73-86.

Niedderer, H.; Schecker, H., (1992). Towards an explicit description of cognitive systems for research in physics learning. In: Duit, R.; Goldberg, F.; Niedderer, H. (eds.): *Research in Physics Learning - Theoretical Issues and Empirical Studies* (pp. 74-98). Kiel: IPN.

Petri, J.; Niedderer, H., (1998). A learning pathway in high-school level quantum atomic physics. *Int. J. Sci. Educ.*, 20, 1075-1088.

Petri, J.; Niedderer, H., (2001). Kognitive Schichtenstrukturen nach einer UE Atomphysik (Sek II). *Zeitschrift für Didaktik der Naturwissenschaften*, ZfDN, 7, 53-68.

Roth, W., M., (1995). Affordances of computers in teacher-student interactions: the case of Interactive Physics™. *Journal of Research in Science Teaching*, 32, 329-347.

Schecker, H., (1985). Das Schülervorverständnis zur Mechanik. Eine Untersuchung in der Sekundarstufe II unter Einbeziehung historischer und wissenschaftstheoretischer Aspekte. Bremen: Universität Bremen.

Scott, P., H., (1992). Pathways in learning science: A case study of the development of one student's ideas relating to the structure of matter. In: Duit, R.; Goldberg, F.; Niedderer, H. (eds.): *Research in Physics Learning - Theoretical Issues and Empirical Studies* (pp. 203-224). Kiel: IPN.

Taber, K. S., (2000). Multiple frameworks?: Evidence of manifold conceptions in individual cognitive structure. *Int. J. Sci. Educ.*, 22, 399-417.

Tyson, L., M.; Venville, G., J.; Harrison, A., G.; Treagust, D., F., (1997). A Multidimensional Framework for Interpreting Conceptual Change Events in the Classroom. *Science Education*, 4/1997, 387-404.

Tytler, R., (1998). The nature of students' informal science conceptions. *Int. J. Sci. Educ.*, 20, 901-927.

Vosniadou, S., (1994). Capturing and modeling the process of conceptual change. *Learning and Instruction*, 4, 45-69.

THE ELECTRIC CURRENT ON ITS WAY TO OUR HOUSE AND THE PARALLEL CONNECTION OF THE ELECTRIC APPLIANCES: PRIMARY STUDENTS' (11-12) REPRESENTATIONS

Vassiliki Pilatou, Heleni Stavridou

University of Thessaly, Department of Primary Education, Volos, Greece

ABSTRACT

The aim of this study was to detect primary students' conceptions about the origin of the electric current, the way it is conveyed and the connection of the household electric appliances. In total, 383 students (11-12), from the town of Volos-Greece, participated in this research. 213 took part in the experimental groups and 170 comprised the traditional classes. The results presented here derived from the elaboration of one writing/drawing activity included in a written questionnaire, which was answered by the students before and after teaching. The analysis showed that primary-grade children draw, before instruction, the electric current "coming" with one cable from the electricity pylons to our house and they represent each electric device working independently and having its own wire connected with a socket in the wall. After a nine-hour constructivist didactical intervention the students' representations of the experimental classes have been improved substantially. Children seem to have understood the notion of circuit, they draw two cables from the electricity factory up to our house and they represent the parallel connection of three household electrical appliances. The drawings of the traditional classes have not evolved after teaching.

1. BACKGROUND, AIMS AND FRAMEWORK

The ideas that children bring to science classes have been the subject of extensive research. Much work has been done in the field of preconceptions, which abound on the subject of children's understanding of electricity and, in particular, electric circuits and electric current (Shipstone, 1984; Psillos & Koumaras, 1988; Newton & Newton, 1996; Summers *et al.* 1998; Kibble, 1999).

However, little research attention has been given to students' ideas as for the electric current they "use" in their everyday life and the functioning of the household electrical devices. One of these researches showed the power grid's pylon to be a potent and powerful symbol of electricity and danger for the students, aged 11-12, while it was astonishing the percentage of them who believed that electricity exists in a disconnected plug of a table lamp (Solomon *et al.*1985). Bullock (1992) came to the conclusion that by experimenting with electricity, primary-grade children (6 years old) can begin to understand some basic concepts about it (e.g. where electricity "lives"). Also, the investigation showed that children (aged 9-12) are not clear about where electricity comes from, they hold many different theories about its origin (Qualter, 1994) and they are not aware of the parallel connection in the

D. Psillos et al. (eds.), Science Education Research in the Knowledge-Based Society, 145–153.

house's electric installation. Moreover, younger students think that electricity is included into the electric appliances people buy and they locate the appliances' functioning mainly to the switch, to the plug, to the wire or to the socket in the wall (Solomonidou & Kakana, 2000).

So, the aim of the present study was:

- To investigate students' ideas about: a) the origin of electricity, b) the way the electric current is conveyed from its origin to our house and c) the connection of the household electrical devices,
- To detect the evolution of students' initial ideas in two different learning situations: a) a traditional teaching environment and b) a co-operative learning environment inspired by a social constructivist view of learning (Duit & Treagust, 1998).

The study was part of a larger program aiming at the introduction of innovative ways of teaching and learning science in primary school.

2. METHOD AND SAMPLE

Seven primary schools from the region of Volos-Greece participated in this program. In total, 213 students (aged 11-12) took part in the 11 experimental classes (116 attended the 5th grade and 97 attended the 6th grade) and 170 comprised the 8 traditional classes (100 attended the 5th grade and 70 attended the 6th grade). All the teachers who participated in this programme were volunteers, 11 of them taught to the experimental groups and 8 to the traditional classes.

At the beginning of the program, a written questionnaire about electricity was administered and answered by both the students of the experimental and the traditional classes. The questionnaire comprised both multiple choice and open-ended questions. The students were also invited to make drawings in order to explain their ideas, and were encouraged to justify their answers.

At the following, nine innovative units about electricity had been developed (1 hour each) according the constructivist view of learning, which also used in teaching children of the *experimental classes*. These *units* were the following: 1) Bulbs, wires and batteries: the simple circuit, 2) The multimeter (Part I), 3) The multimeter (Part II), 4) The "train analogy"-The simple circuit (Dupin & Joshua, 1989), 5) Circuits with one battery and two bulbs connected in series, 6) Circuits with one battery and two bulbs connected in parallel, 7) The house's electric installation (Part I), 8) The house's electric installation (Part II), 9) The electric current from its origin to our house.

Four of these units (6, 7, 8 and 9) concerned the parallel connection of simple circuits with batteries and bulbs, the parallel connection of the household electric appliances and the house's electric installation.

Children worked in small groups of 4-5 persons by using worksheets and proper materials and had a lot of discussion about their personal ideas. Each group had a small suitcase on their desk, which included bulbs, wires, batteries, a switch and a multimeter for the experiments. As for the mains electricity, we constructed a 3D model in order to represent the origin of the electric current, the way it is conveyed, the parallel connection of two electric appliances inside a house and the position of simple electric components in the house's electric installation.

The students of the *traditional classes* had been taught electricity in the traditional/conventional way and as it is presented in the current Primary Curriculum (the teacher is the dispenser of knowledge, he/she is lecturing and the students pay attention to him/her, the experiments done and explained, usually, by the instructor, the children do not co-operate and do not interact with each other etc.).

The children of the traditional groups who attended the 5th grade had been taught 6 units (1 hour each) about electricity. These units concerned the electric charge, the electric sources, the electric current, the electric circuits, the transfer and "consumption" of the electric energy (house's electric installation, electricity board) and the results of the electric current.

The students of the traditional groups who attended the 6th grade had been taught 8 units about electricity (1 hour each). Some of these units concerned the ways of the electric current production (factories where the electric energy is produced – electric generators) and the function of the household electrical devices/children toys inside the house (e.g. the vacuum cleaner, the electric fan, the electric train).

In order to detect children's final ideas, both the students of the experimental and the traditional classes answered the post-test questionnaire (the same as pre-test).

Four of the questions which were included in the questionnaire (pre and post-test) referred to the connection of the household electrical appliances. More specifically, in the first three questions students were asked to express their ideas about whether we can bring, at the same time, three electric devices (an electric cooker, a washing machine and a vacuum cleaner) into operation or not and to justify, in each case, the answer they gave.

The results presented here derived from the elaboration of the fourth question -a writing/drawing activity- where students were asked to represent the way that the electric current "follows" from its origin to our house and according to what they had answered to the three previous questions, the connection of three electric appliances that are functioning, at the same time, inside the house or the connection of one electric appliance that is operating.

The results presented below, are referred to the students' drawings of the experimental and the control group classes before and after teaching, whereas the comparison between the drawings of the two levels involved: 5th and 6th grade did not detain the present study.

As about the schemes, we analysed children's drawings by taking into consideration 4 basic criteria and more specifically, whether students depicted in their schemes: a) where the electric current "comes from", b) how many external cables are connected to the building of the house, c) at which place the external, outside the house, cables are connected to the wires inside of it and d) how the electric appliances are connected to each other while they are functioning.

3. RESULTS

From the students' drawings, in the pre-test questionnaire, it is obvious that children hold before instruction a variety of initial conceptions about the origin of the electric current (table 1). The majority of them (53,3% of the experimental groups and 58,2% of the traditional classes) show that the electric current "comes" to the house, from the power grid's pylons (figure 1), the socket in the wall or from the electricity board.

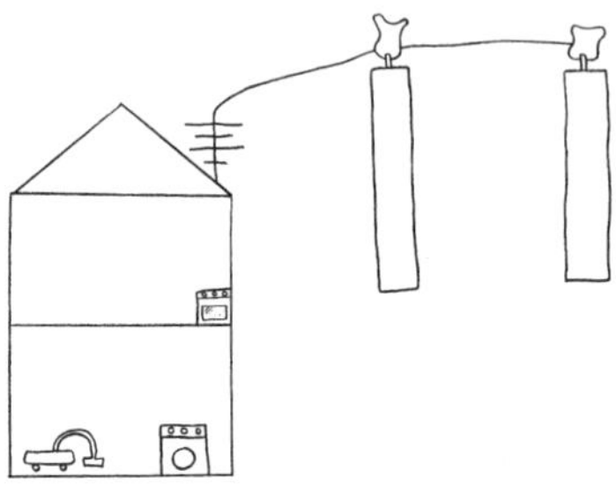

Figure 1. Students' drawing about the origin of electricity

Some other students (12% of the experimental groups and 12,3% of the traditional classes) give an insufficient answer to this question, since they draw the electric current "coming" to the house, through wires, from some place out of it but without showing which and where this place is. Only 33,5% of the experimental classes and 24,7% of the traditional groups, locate electricity's origin in DEH (the Greek 'Public Electricity Services') or another factory, before the instruction.

Table 1. Children's drawings-criterion a

a/a	Category responses Where does the electric current "come from"?	Experimental class		Traditional class	
		Pre-test (N=191) %	Post-test (N=191) %	Pre-test (N=146) %	Post-test (N=142) %
1	Power grid's pylon	46	6,3	53,4	38
2	DEH/ factory	33,5	67,5	24,7	20,4
3	Socket	5,2	12	4,1	14,1
4	Electricity board	2,1	5,8	0,7	11,3
5	Other	1	0,5	4,8	2,1
6	Insufficient answer	12	7,8	12,3	14,1

Regarding the way electricity is conveyed (table 2), children draw cables, while from their schemes it is obvious that even if many cables (3 or 4) are situated to the power grids, for the electricity's transmission, usually one is finally connected to the building of the house (figure 2). In more details, many of the students (34% of the experimental groups and 30,1% of the traditional groups) present, before teaching, only one cable that "helps" the electricity's transfer to our house and a few of them (16,8% of the experimental groups and 19,9% of the traditional groups) draw three cables for the electric current's transmission. Moreover, a lot of students (30,9% of the experimental groups and 41,5% of the traditional groups) give an insufficient answer to this question and they don't represent in their schemes how the electric current is conveyed from its origin to our house.

Table 2. Children's drawings-criterion b

a/a	Category responses How many external cables are connected to the building of the house?	Experimental class		Traditional class	
		Pre-test (N=191) %	Post-test (N=191) %	Pre-test (N=146) %	Post-test (N=142) %
1	One cable	34	22	30,1	17,6
2	Two cables	11,5	43,5	17,1	10,6
3	Three cables	16,8	9,9	19,9	16,2
4	Many cables	6,8	6,3	7,9	14,1
5	Insufficient answer	30,9	18,3	24,7	41,5

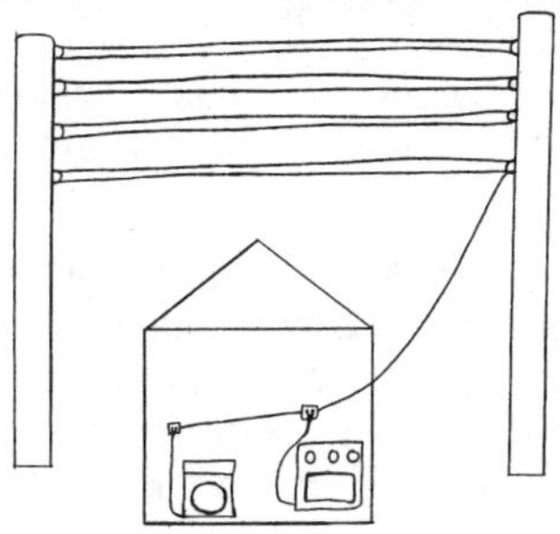

Figure 2. Students' drawing about the conveyance of electricity

Concerning the place where the external cables are connected with the wires inside the house (table 3), students believe that it is somewhere outside of the building. So, many of them (46,6% of the experimental groups and 35,5% of the traditional groups) draw the outer cables "ending" their route on the roof, on the floor or on the wall out of the house (figure 3), without showing how their way is continued inside of it.

Table 3. Children's drawings-criterion c

a/a	Category responses Which is the place where the external, outside the house, cables are connected with the wires inside of it?	Experimental class		Traditional class	
		Pre-test (N=191) %	Post-test (N=191) %	Pre-test (N=146) %	Post-test (N=142) %
1	Floor / ground / wall	46,6	13,6	35,5	19
2	Electricity board	7,9	40,3	6,2	9,2
3	Socket	10,5	22	13	4,2
4	Electric device/s	7,3	10,5	17,8	17,6
5	Other	3,7	0,5	9,6	7,7
6	Insufficient answer	24	13,1	17,8	38,7

In addition, the 24% of the experimental groups and the 17,8% of the traditional groups, give in the pre-test questionnaire an insufficient answer to this question, since they don't show in their schemes how the cables out of the house are connected to the components of the house's electric installation and they sketch the power lines "ending" their route out of the house and a little bit far from it.

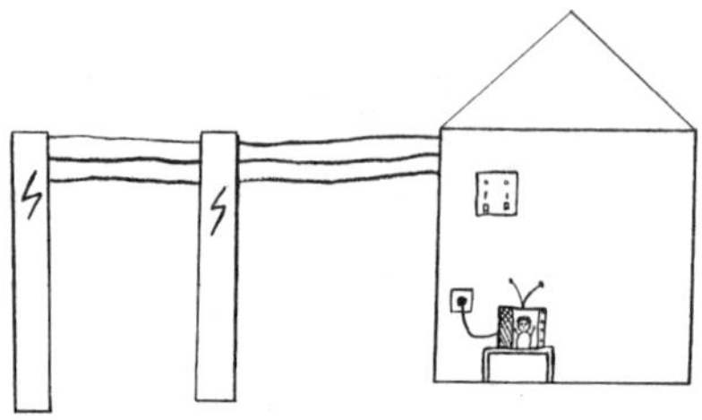

Figure 3. Students' drawing about the place where the external cables are connected with the wires inside the house

As about the connection of the electric appliances inside the house, while they are functioning (table 4), the most common idea among students before teaching (32,4% of the experimental groups and 28,3% of the traditional groups), is that each device is located away from the others, without any connection among them and operates with its own wire that is plugged in a wall socket (figure 4).

Figure 4. Students' drawing about the connection of the electric appliances

Also, many students (38,7% of the experimental groups and 42,5% of the traditional groups) give an insufficient answer to this question, since they draw the electric devices inside the house without wires and without any connection with the other components of the house's electric installation.

In the post-test questionnaire, children's drawings of the experimental classes, have been improved substantially. They have been more completed, more detailed and better structured. To be more specific, after the nine-hour innovative teaching most of the children (67,5%) present DEH or another factory as for the origin of the electric current (table 1), a lot of students (43,5%) have understood the existence of a circuit in the house and sketch two external cables to show the conveyance of the electric current from its origin to our house (table 2) and many other (40,3%) have realised that electricity board is the place where the external cables are connected with the wires inside the house (table 3).

As about the house's electric installation, most of the children depict in their final schemes three household electrical devices, whereas from their final drawings we detect four evolutionary stages to their thinking about the connection of the electrical appliances (table 4).

Table 4. Children's drawings-criterion d

a/a	Category responses How do the electric appliances are connected to each other while they are functioning?	Experimental class		Traditional class	
		Pre-test (N=111) %	Post-test (N=172) %	Pre-test (N=106) %	Post-test (N=120) %
1	One device/one wire/socket or electricity board	32,4	12,8	28,3	33,3
2	One device/one or more wires/socket or electricity board/one or more cables/pylon or factory	8,1	14	10,4	15
3	One device/two wires/socket or electricity board or DEH (circuit)	3,6	12,2	4,7	6,7
4	Parallel connection	0	33,7	1,9	2,5
5	Other	17,1	15,7	12,3	17,5
6	Insufficient answer	38,7	11	42,5	25

More specifically, after teaching intervention there has been a substantial decrease (12,8%) in the number of students who believed in the pre-test questionnaire that an electric device requires for its functioning only one wirc connected to the socket in the wall (figure 4). Many other students (14%) have understood that the electric installation is more complicated and they draw the wire/s of each electric device "continuing" its/their route behind the socket in the wall and up to the power grid's pylons or to the electricity factory (figure 5).

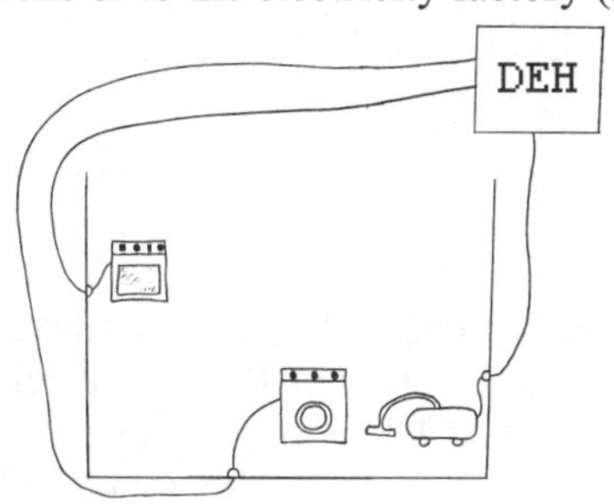

Figure 5. Students' drawing about the electric installation of the house

Moreover, the 12,2% of the children have realised that a circuit is necessary for the functioning of the electrical devices and they depict in their final schemes each household electrical appliance "having" two wires, which are connected to the power grid's pylons or to the electricity factory (figure 6).

Additionally, many students of the experimental classes (33,7%), draw, after constructivist teaching intervention, the household electrical devices connected in parallel the one with the other and with the rest components of the electric installation (figure 7).

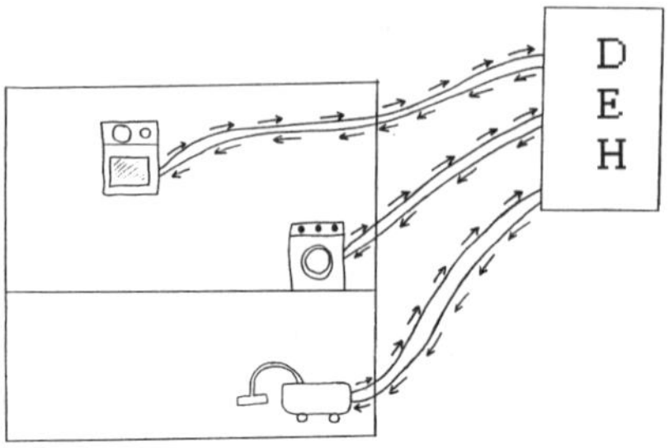

Figure 6. Students' drawing about the electric installation of the house (circuit)

Students' final drawings of the traditional classes have not really changed/improved after teaching. Children's ideas about the route that the electric current "follows" from its origin to our house have not changed and in some cases students answers in the post-test questionnaire are less satisfactory than their responses to the pre-test.

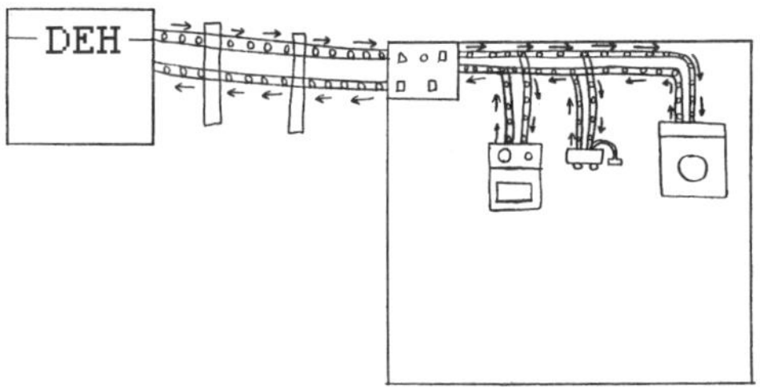

Figure 7. Students' drawing about the electric installation of the house (parallel connection)

For example in table 1, it is obvious that there has been a decrease in the number of students who believed before teaching that the electric current "comes" from DEH (24,7% in the pre-test and 20,4% in the post-test) and on the contrary there has been an increase in the number of them who draw the electric current "coming" with wires from the socket in the wall or from the electricity board.

In addition, many of them (33,3%) draw, even after teaching, every single one electric device inside the house away from the others and working independently with its own wire connected to the socket in the wall (table 4). Only a small percentage of them (6,7%) seem to have understood that a circuit is necessary for the function of the household electrical devices and the existence of a parallel circuit in the house's electric installation (2,5%).

4. CONCLUSIONS AND IMPLICATIONS

From the results of this study it is obvious that primary school students develop, before teaching and on the basis of everyday experience, a variety of alternative ideas about the mains electricity and the connection of the household electric appliances. They seem to have a narrow/ fragmentary understanding about the conveyance of the electric current from its origin to our house and apparently they can't realise that the devices inside the house are connected to each other, even if they are situated in different rooms.

Students attended the traditional groups continued to draw, after teaching, the same pictures as before. So, the traditional learning environment had little influence on their alternative ideas, which really didn't change.

On the contrary, children's drawings of the experimental classes have been improved substantially after the constructivist teaching intervention. Students developed new mental representations as for the origin and the conveyance of the electric current and about the structure and the function of the house's electric installation. Moreover, from the analysis of their final drawings it is obvious that primary-grade children have the ability to deal, successfully, with the parallel connection, since they appeared to have developed a better understanding, about these matters of electricity, only after a 3-4 hour innovative teaching.

According to these findings we can assume that it is possible for children 11-12 years old to lcarn with understanding the above issues. This conclusion is quite interesting given the fact that the parallel connection is one of the new units included in the new Greek science curriculum of the 6th grade in primary school.

REFERENCES

Bullock, J.,R., (1992). Where does electricity live? Answering the "wonder" questions in the primary grades. *Dimensions of early childhood*, 20(3), 34-36

Duit, R. & Treagust, D., (1998). Learning in science: from behaviourism towards social constructivism and beyond. In: B.J.Fraser & K.G.Tobin (Eds.), *International Handbook of Science Education* (pp.3-25). Dordrecht: Kluwer Academic Publishers,

Dupin, J.,.J. & Joshua S., (1989). Analogies and "Modelling Analogies" in Teaching: Some examples in basic electricity. *Science Education,* 73(2), 207-224

Kibble, B., (1999). How do you picture electricity? *Physics Education*, 34(4), 226-229

Newton, L. & Newton, D., (1996). Young children and understanding electricity. *Primary Science Review*, 41, 14-16

Psillos, D. & Koumaras, P. (1988). Voltage presented as a primary concept in an introductory teaching sequence on DC circuits. *International Journal of Science Education*, 10(1), 29-43

Qualter, A., (1994). Where does electricity come from? *Primary science review*, 35, 20-22

Shipstone, D.,M., (1984). A study of children's understanding of electricity in simple DC circuits. *European Journal of Science Education*, 6(2), 185-198

Solomon J. STIR Group, Black P., Oldham V. & Stuart H., (1985). The pupils view of electricity. *European journal of science education*, 7(3), 281-294

Solomonidou, C. & Kakana, D.-M., (2000). Pre-school children's conceptions about the electric current and the functioning of electrical appliances. *European early childhood educational research journa*l, 8(1), 95-111

Summers, M., Kruger, C. & Mant, J., (1998). Teaching electricity effectively in the primary school: a case study. *International Journal of Science Education*, 20(2), 153-172

DETAILED INVESTIGATION OF PROFESSIONAL PHYSICISTS SOLVING PHYSICS TASKS

Jutta Roth, Stefan Von Aufschnaiter

Institute of Physics Education, University of Bremen

ABSTRACT

It is assumed that research on expertise will help to understand what the difference of experts and novices are and how novices can be transferred into experts. However, investigations about how physicists develop and apply their knowledge to both well-known and unfamiliar tasks of their domain are rarely to be found. Such analyses would help to find out what is unique in experts' knowledge and what might differ from our expectations about how physicists act. To analyse how physicists develop and use their physics knowledge, a study with 18 German physicists has been carried out while these participants handled physics tasks on experiments. The detailed analysis of the processes occurring during the treatment of the tasks was supported by video documentation. On the background of a constructivist theory the main focus lies on the analysis of the physicists' development of meanings, in particular special skills and ways of dealing with the tasks in the dimension of time, content, and complexity. Our findings are that the physicists solve the tasks bottom-up on astonishing low levels of complexity. They both handle the given instructions and give explanations closely to the observed phenomena step by step. Competence is shown by a broad knowledge, relating physics formulae and laws to the observed phenomena (closely combining theory and practice), judging and criticising and by systematic searches for faults.

1. BACKGROUND, AIMS, AND FRAMEWORK

The investigation of expertise is an important subject for research in physics education, in particular "the processes by which novices become experts; and any advance in understanding the differences between the two is an essential prerequisite for a principled approach to transforming one into the other" (Priest & Lindsay 1992). In recent years a great number of papers on research in expertise has been published. Most of these deal with chess experts within the framework of cognitive psychology (c.f., Gruber & Ziegler 1990, Charness 1989) and other studies have been carried out within the context of artificial intelligence to model expertise (c.f., Slatter 1990). Given these various contexts for the study of expertise and the fact that there is no agreed and unequivocal definition of "expert" (Gruber & Ziegler 1996), studies about physics experts deal with participants of different experience and competence in physics. Thus, Kross & Lind (2001) studied outstanding high school students in physics, Chi et al. (1981) and Clement (1994 and 1998) studied PhD-students, whilst Snyder (2000) investigated professors of technical subjects.

D. Psillos et al. (eds.), Science Education Research in the Knowledge-Based Society, 155–161.

Apart from the work of Kross and Lind, the above studies used categorisation of physics problems to distinguish different types of expertise. The processes involved in solving physics tasks by post-graduate physicists have rarely been investigated in detail and this is the aim of our work reported in this paper. With such an investigation of the processes of construction of meaning, we aspire to describe and understand the specific characteristics of post-graduate physicists as they address physics tasks and aim to find answers to the following questions:

1. How do post-graduate physicists use their previous knowledge to solve physics tasks?
2. What approaches do these physicists typically take to dealing with physics tasks?
3. How do these physicists use physics laws and formulae?
4. How do these physicists draw upon and utilise additional available information as they solve physics tasks?
5. How can these physicists' development of meaning be described in general as they solve physics tasks?

Two essential, cognitive processes, which take place on two different time scales, can be distinguished: the construction of meaning, a short range process, and learning, a long range process (von Aufschnaiter, S. 2001). According to Pöppel (1994) the basic unit of the construction of knowledge is a mental image, synonymous with a single meaning or "Image-of-now" (von Aufschnaiter, S. 2001) which has a time range of less than three seconds. According to a theoretical framework based on a constructivist epistemology (described in von Aufschnaiter, S. 2001, von Aufschnaiter & Welzel 1999), individual constructions of meaning are inaccessible to an observer. They can only be reconstructed from the learner's utterances and actions.

Within this framework, learning can be defined as the long range development of skills (or producing meanings), which are successful in practice. Learning is a cognitive process, which can be observed only indirectly by changes in behaviour. Changes in the construction of meanings, regarding the same or similar contents, are signs that learning has occurred. Our basic approach is to describe constructions of meaning and their development in time ranges of seconds to minutes (the amount of time participants need for the treatment of a task) with respect to the dimensions of time, content, and complexity. Therefore, learning can be specified as getting "higher" in one or more of these three dimensions. Higher means faster, more complex, with more content segments and/or generating more interrelations between contents (for the Bremen Model of Complexity, v. von Aufschnaiter, C. & von Aufschnaiter, S. 2002, von Aufschnaiter & Welzel 1999).

2. METHODS AND SAMPLES

About 100 tasks, based on experiments uniformly distributed over five sessions of about 90 minutes each, were solved by a total of 18 post-graduate physicists working in pairs. The tasks were based on topics such as basics electrostatics, involving the neon bulb, capacitor and Coulomb's law, as well as electrodynamics involving the oscilloscope and passive nerve stimulation conduction. In addition, two kinds of written instructions, which contained information on different levels of complexity, were given to the participants in special cases. The first kind of intervention provided abstract information on intermediate or high levels of complexity for the participants while working on specific tasks. The second kind of intervention involved concrete information on low levels of complexity, given to the participants if they asked for it.

The leader of the investigation was present but did not interact with the participants. Only interactions between the participants were possible. The sessions were completely videotaped.

The post-graduate physicists participating in this investigation included fourteen PhD-students at the University of Bremen, a professor, two post-doctoral physicists, and a teacher. These physicists were not experts in the sense that they were specialised in solving experimental tasks about electrostatics and electrodynamics. They were more or less broadly educated graduates of various German universities who had studied physics and had additional qualifications in different areas of physics. The professor had specialist experience in presenting lectures on electrostatics and electrodynamics. One of the post-doctoral physicists was the leader of the practical courses at the Department of Physics, and four participants were specialists in neural physics, the subject of one session of our investigation. The other participants had studied electrostatics and electrodynamics as part of their basic undergraduate programme but were not studying these subjects at the time of the investigation.

We have been investigating the processes of construction of meaning in the time range from three seconds upwards. In this paper, only the amount of time the participants needed for a session, to construct a meaning or to solve a task, is taken into account. In addition, the extent to which the participants demonstrated skills such as knowing about phenomena and applying physics laws and formulae, along with technical know-how, were studied.

The participants' meanings were reconstructed from the often fragmentary and colloquially formulated verbal expressions and actions. We call such reconstructions 'ideas'. The ideas were analysed with regard to their level of complexity according to the Bremen Model of Complexity, a system of categories, which takes into account how abstract the meaning is, whether abstract properties are combined or not, and if the character of the constructed meaning is static or dynamic. Within this model, four different areas are distinguished to describe the level of abstraction of the reconstructed participants' meanings:

(I) concrete (on the levels of Objects, Aspects and Operations specific situations and objects are considered),

(II) abstract-stable (on the levels of Properties and Events stable properties of classes of objects and situations as well as connections of these properties are considered),

(III) abstract-variable (on the levels of Programmes and Principles dynamic changes in properties and their connections are considered), and

(IV) systemic (on the levels of Connections, Networks, and Systems connections of more than two variable properties are combined).

For more details see von Aufschnaiter, C. & von Aufschnaiter, S. 2002, von Aufschnaiter & Welzel 1999.

Video data have been analysed using an evaluated and well used method, developed at the University of Bremen (von Aufschnaiter & Welzel 1999). The first step was to note and to describe the succession of actions as seen on the videos. Certain sequences were then chosen to be transcribed in order to obtain longitudinal perspectives and cross sections of pairs and tasks. The meanings of the participants have been reconstructed as 'ideas' from their verbal expressions and actions. The quality of these ideas has been analysed with respect to the dimensions of content, time, and complexity as previously described.

3. RESULTS

The following results were taken from the second of the five problem solving sessions. The data from the other four sessions still need to be analysed. The longitudinal section was developed with a pair of PhD-students specialising in neural physics.

TIME:

The physicists needed on average two hours per session and about six minutes per task. This is the same amount of time required by 2nd year undergraduate university students for exactly the same tasks (Saniter 2000).

CONTENT:

In relation to the dimension of content, the physicists showed a broad knowledge when constructing general meanings about physics laws and phenomena, but they did not always know the exact formulae by heart. They were able to estimate orders of magnitude and were technically skilful. In most cases, the physicists connected formulae and phenomena successfully, which could be seen when they applied laws to describe phenomena and/or used qualitative as well as quantitative explanations. Furthermore, they drew upon their own ideas, created subtasks, and asked very specific questions. They frequently estimated results, but they often did this implicitly rather than expressing hypotheses and uttering expectations. The most obvious behaviour was that they judged and criticised the results of their measurements, their actions and arguments as well as parts of the learning environment such as the technical equipment or the formulation of the tasks. This was done by all pairs and happened several times during almost all tasks.

COMPLEXITY:

No matter what kind of previous knowledge and specialisation the physicists had, they solved the tasks on low (concrete) and intermediate (abstract-stable) levels of complexity.

About 65 % of the meanings were constructed on low levels of complexity (referring to area I). The participants constructed stable properties (first level, area II) in more than 20 % of the cases. Combining at least two properties (second level, area II) occurred in only about 5 % of the cases. The missing 20 % are statements to which a level of complexity cannot be attributed, e.g. pure emotional expressions, sighs, etc. Even for those tasks, which were intended to prompt the participants to construct dynamical meanings, they preferred to construct static formulations.

The construction of meanings on low levels was supported by formulating interrelations between different contents in more than one sentence, or with short breaks. The following transcript of the treatment of task 2.14 exemplifies the concrete levels and the breaks (table 1).

Table 1. Section of the transcript of the treatment of task 2.14

<table>
<tr><td>
Text for the task:
Connect plate 3 securely to the earth connection of the Coulombmeter.
Place plate 1 within a distance of 2 cm of plate 3, charge it with 2 kV. Measure the charge on plate 1. Q_{2cm} = ___.
Calculate the capacitance of the arrangement.
Place the plates within a distance of 0.5 cm, charge plate 1 with 2 kV, measure the charge on this plate. $Q_{0.5cm}$ = ___.
Calculate the capacitance of the arrangement now.
Explain your results!
</td></tr>
<tr><td>
HAVING CARRIED OUT THE MEASUREMENTS WITHIN 6 MINUTES, A. AND S. WENT ON TO DISCUSS THE PHENOMENON OF ELECTROSTATIC INDUCTION. THE FOLLOWING CONVERSATION THEN TOOK PLACE:
</td></tr>
<tr><td>
S: - Right, if you have them [the plates] here very, very far apart from each other, then there's nothing at all flowing onto them.
A: - Yes.
S: - Then the other plate doesn't notice that / that there is something. /7s/ And, / hm, /5s/ yes, there is an effect like that that it then leads back again. What do you think?
///
A: - Must be, because this (waves with his hand in the direction of the plates), otherwise it would be ...
//
S: - Yes.
A: - It would, it would be independent.
S: - Yes
//
A: - Otherwise, if you now consider only this one plate (points to the left plate), it is always the same now. /11s/
Otherwise, as, as much charge as we can manage to put on there (points to the plate) depends on how, how close the other one is.
S: - Exactly. ///
(Concrete levels (italics) and the breaks ("/ " refers to a break of one second) are marked.)
</td></tr>
</table>

A. and S. pointed to the concrete objects, they spoke of "here", they expressed physics terms in a colloquial, often simple language, such as "the plate doesn't notice", "there‘s nothing at all flowing onto them", or "it is always the same now". Most of the spoken sentences include breaks or gaps of one to five seconds. Breaks between sentences also occurred frequently and could take up to 11 s in the given example. Although the physicists constructed interrelations between contents, breaks prevented attributing higher levels of complexity to the physicists' meanings.

Another important result concerning the complexity is that the physicists always constructed their meanings bottom-up, from concrete to more abstract levels. This involved formulating sentences, relating to a specific question, in an increasingly exact and differentiated way, as well as interrelating more and more contents.

Although the physicists did not show construction of meanings on high levels of complexity, interventions on low levels of complexity were often regarded as containing too little information. Information, which focused on higher levels, initiated intensive discussions and was mostly judged to be helpful or not. Furthermore, the participants applied it successfully on solving the tasks. However, dialogues, which referred to this kind of information, again remained basically on low levels of complexity.

Both findings, solving the given tasks on low levels of complexity, as well as applying complex information on low levels, indicate that whilst these physicists have acquired an intuitional and implicit understanding of complex interrelations, they nevertheless minimise their effort to solve the tasks.

4. CONCLUSIONS AND IMPLICATIONS

Drawing on the findings outlined above, it is possible to give answers to the questions set out at the beginning.

1) The investigated physicists show a broad knowledge of general laws and their interrelations as well as technical skills. Instead of being able to recall the exact formulae, they deduce them from basic principles, moving bottom-up from lower to intermediate levels of complexity.

2) The way of dealing with the given physics tasks can be described as follows: the physicists engage intensely in the given learning environment. They follow all experimental and theoretical questions step by step. When confronted with the failure of an experiment, they pursue their own aims. Also in these cases, they only construct meanings on low and intermediate levels of complexity. Furthermore, the physicists spend great amounts of time in judging situations and in acting to avoid faults.

3) The physicists seem accustomed to applying laws and formulae to physics phenomena and to moving quickly and spontaneously between the “world of formulae” and “world of phenomena”. However, they construct the laws and formulae on intermediate levels of complexity.

4) The physicists prefer information on high and intermediate levels of complexity. They show that they are able to apply or to judge information without constructing meanings on high levels of complexity.

5) In general it can be summarised that the physicists develop their meanings from concrete levels of complexity to abstract levels and relate contents step by step. The main reason for staying at low and intermediate levels of complexity is not the failure of combining contents in principle but the failure of constructing interrelations in only one mental image. In addition, the use of colloquial language supported the impression of them being familiar with the physics phenomena.

REFERENCES

Charness, N. (1989). Expertise in chess & bridge. In: Klahr, D. & Kotovsky, K. (Eds.), *Complex information processing. The impact of Herbert A. Simon.* Hillsdale, NJ: Erlbaum, 183-208.

Chi, M., Feltovich, P. J. & Glaser, R. (1981). Categorization and Representation of Physics Problems by Experts and Novices. *Cognitive Science, 5,* 121-152.

Clement, J. (1994). Use of Physical Intuition and Imagistic Simulation in Expert Problem Solving. In: Tirosh, D.(Ed.). *Implicit and Explicit Knowledge: An Educational Approach.* New Jersey: Ablex, 204-244.

Clement, J. J. (1998). Expert novice similarities and instruction using analogies. *International Journal of Science Education, vol. 20, no. 10*, 1271-1286.

Gruber, H. & Ziegler, A. (1990). Expertisegrad und Wissensbasis. Eine Untersuchung bei Schachspielern. *Psychologische Beiträge, 32*, 163-185.

Gruber, H. & Ziegler, A. (1996). *Expertiseforschung. Theoretische und methodische Grundlagen.* Westdeutscher Verlag, Opladen.

Kross, A. & Lind, G. (2001). Learning biology and physics through self-explanations while studying worked-out examples. In: Psillos, D., Kariotoglou, P., Tselfes, V., Bisdikian, G., Fassoulopoulos, G., Hatzikraniotis, E., & Kallery, M. (Eds.), *Proceedings of the Third International Conference on Science Education Research in the Knowledge Based Society*. Thessaloniki: Art of Text Publications, 495-497.

Pöppel, E. (1994). Temporal Mechanisms in Perception. *International Review of Neurobiology, 37,* 185-202.

Priest., A. G. & Lindsay, R. O. (1992). New light on novice-expert differences in physics problem solving. *British Journal of Psychology, 83*, 389-405.

Saniter, A. (2000). Learning Processes of Advanced Physics Students.To be published in the Proceedings of the ESERA Summerschool, Denmark September 2000.

Slatter, P. E. (1990). Models of expertise in knowledge engineering. In: H. Adli (Ed.), *Know-ledgeengineering*. New York: McGraw-Hill, 130-156.

Snyder, J. L. (2000). An Investigation of the Knowledge Structures of Experts, Intermediates and Novices in Physics. *International Journal of Science Education, 22(9),* 979-992.

von Aufschnaiter, C. & von Aufschnaiter, S. (2002). Theoretical Framework and Empirical Evidences on Students' Cognitive Processes in Three Dimensions of Content, Complexity, and Time. *Journal of Research in Science Teaching* – accepted for publication.

von Aufschnaiter, S. (2001). Development of Complexity through Dealing with Physical Qualities: One Type of Conceptual Change? *In: Behrendt, H. et al. (Eds.). Research in Science Education - Past, Present, and Future.* Dordrecht, Boston, London: Kluwer, 199-204.

von Aufschnaiter, S. & Welzel, M. (1999). Individual Learning Processes - a Research Programme with Focus on the Complexity of Situated Cognition. In: Bandiera, M., Caravita, S., Torracca, E. & Vicentini, M. (Eds.). *Research in Science Education in Europe.* Dordrecht: Kluwer, 209-215.

LEARNING FROM WRITING IN SECONDARY SCIENCE: A CASE STUDY OF STUDENTS' COMPOSING STRATEGIES

Vaughan Prain, Brian Hand, Larry Yore

La Trobe University, Australia, Iowa State University, USA, University of Victoria, Canada

ABSTRACT

The role of writing in learning secondary science has received increasing research focus over the last decade. This study aimed to identify students' drafting strategies when they wrote for a readership other than their teacher, as well as their perceptions of the value of this task. The findings indicated that students perceived this writing process and task as valuable for their learning of science concepts and that their drafting approaches were consistent with current conceptualisations of effective strategies to enable writing to promote learning.

1. INTRODUCTION

There has been strong interest over the last decade in the role of writing as both tool and object of learning in secondary school science, with one research strand focused on the specific structural features of types of science writing (Halliday and Martin, 1993; Unsworth, 1997) and the student knowledge required to understand and reproduce these genres (Unsworth, 2000; Veel, 1996; Martin & Veel, 1998). Other researchers, such as Hanrahan (1999), Hildebrand (1998), Prain and Hand (1996) Rowell (1997), and Sutton (1992), have argued that students need to write in diverse forms for different purposes to clarify their understanding of science concepts. However, as Rivard (1994, pp. 976-978) observed, for students to learn from writing they must "possess the necessary metacognitive knowledge", and must not only know "what strategies are available for enhancing learning", but also "understand how and when to use them and why they may be useful". Sutton (1992) also argued that students need to be given a variety of opportunities to use language to explain and justify their understanding of science and scientific language.

In focusing on student perceptions and strategies when tackling diversified writing tasks this study sought to (a) identify textual strategies used by students when they attempted to explain scientific concepts in writing for younger readers, (b) identify their

D. Psillos et al. (eds.), Science Education Research in the Knowledge-Based Society, 163–171.

perceptions of the value of these strategies, and (c) determine the extent to which these strategies and perceptions were consistent with current theories of how writing supports learning. Following Keys (1999), the researchers aimed to identify the extent to which students developed conceptual knowledge when they had to elaborate understandings for novice readers. In their analysis of the nature of the language of science, Halliday and Martin (1993, p. 13) argued persuasively that many scientific terms are "grammatical metaphors", where a particular process or sequence of events entailing a verb has been nominalised into the form of a noun to stand for an "abstract theoretical entity", hence to stand metaphorically in relation to the initial process, enabling these processes to be treated as concepts for the purposes of establishing cause and effect relationships in developing scientific knowledge. Halliday and Martin (1993, p. 68) also noted this compression of meaning poses serious comprehension challenges for learners, because "experiential meanings" are "left implicit" in this new language. This research project was guided by the assumption that there were learning gains for students in being encouraged by the nature of the writing task to unpack key terminology to explain the processes or subordinate concepts represented by larger concepts in the topic. Keys (1999, p. 1046) also noted, drawing on Halliday (1985), that a major feature of science writing is expansion, defined as "the building of semantic relationships between events by the use of additional clauses that specify, define or extend the initial clause". In devising a task where students had to clarify and link key concepts in the topic, the researchers assumed that suck a task would also develop students' capacity to use expansion in science writing. The study was also informed by recent research literature on theories of how writing serves learning.

2. THEORIES OF HOW WRITING SUPPORTS LEARNING

In reviewing these theories Klein (1999) claimed there were four main explanations for how the act of writing entailed cognitive processes that promoted learning. He characterised these four explanations as follows: writing as 'spontaneous utterance'; the 'forward search hypotheses'; the 'genre-related hypotheses'; and 'backward search hypotheses'. The spontaneous utterance explanation argued that student writing promotes learning directly by shaping thought at the point of formulation, in that the expression of ideas makes understanding more explicit. The forward search explanation asserted that writers transform their ideas by ongoing analyses of their emerging texts in terms of expanding inferences, reviewing idea development, noting contradictions, and making appropriate revisions. The genre-related explanation focused on the value of the linkages between form and function of genres as a framework for learning. From this perspective knowledge of the purposes of the micro and macro structure of texts enables students to identify relationships between ideas, and hence clarify understanding of content as well as reasoning processes in science. The backward search explanation argued that through setting and addressing rhetorical and content goals students learnt through revising their writing. This research project was based on the implementation of a teaching program that required students to engage in all four of these drafting and revision processes. Previous research has highlighted that these hypotheses were not mutually exclusive.

3. METHODS RESEARCH CONTEXT AND PARTICIPANTS

The study design entailed content analysis of students' written drafts and revisions, as well as stimulated recall interviews focusing on students' composing processes and understandings of their writing. Four classes of tenth-grade biology students (N = 73) from a predominantly white and middle-class junior-senior high school in the USA participated in the project. The study took place in the context of a guided inquiry into key concepts in the topic of biotechnology, where students engaged in discussion and written tasks. This topic was chosen primarily for two reasons: the researchers considered that the use of an extended writing task was appropriate means to focus on the ethical and social implications of biotechnological developments; and this kind of task was also viewed as likely to promote students' deeper understanding of key concepts rather than the usual superficial copying of secondary sources. The research project involved collaboration between two researchers and two science teachers, entailing the development of a biotechnology unit, regular weekly group meetings, classroom observations and debriefing sessions, and the development of testing instruments to examine the learning outcomes from using writing. Six students from each of the four classes were selected for interviewing, with interviews focusing on students' perceptions of how they tackled demands of the writing task, and their perceptions of the value of this process. Past research (Hand, Prain & Hohenshell, 2000) had indicated quantitative gains in students' conceptual understandings from this approach. Students who participated in planned writing exercises demonstrated significant gains in responding to higher order conceptual questions on the end of unit tests.

4. CLASSROOM PROCEDURES

Students were assigned to write a textbook explanation on a biotechnology topic for an audience of seventh-grade students, with final drafts given to the Year 7 students to evaluate. In tackling this task students undertook the following planning and revision activities: they were given a template introduced by their English teacher to plan the writing task in terms of generic features of textbooks, such as headings, examples, and explanatory diagrams; they carried out individual research on the topic using the library, participated in small group discussion on major ideas for their topic; and they presented these ideas to the whole class, with feedback provided by peers. They also participated in small group review of peer feedback and further discussion to confirm their understanding of ideas under review.

The teachers' main role was to provide guidance and support for each stage of the process. The writing tasks in this project required students to participate in activities where they constructed and negotiated the meanings of particular concepts for themselves, for a small group of peers, the whole class, and then for younger students. The teachers provided generic templates, prompted individual student's thinking, provided a range of topics for students to select from, scaffolded students' writing attempts, offered feedback on collected research, clarified obvious misconceptions, and facilitated small and whole group discussions and presentations. First drafts of the textbook explanation were assessed for content by the science teacher and for rhetorical effectiveness by the English teacher. The students were given opportunities to discuss the feedback within their groups before completing a redrafted copy, which was subsequently scored by both teachers and used for data analysis.

5. DATA SOURCES AND METHOD OF ANALYSIS

Data sources for this study entailed 24 student drafts and final writing produced in relation to the topic including the two topic tests, and transcriptions of the interviews with these students. Key findings of the study are derived from content analysis of these students' drafts, with analysis of the interviews used to cross-reference and triangulate these findings. Final drafts of the 24 interviewed students were analysed in terms of students' perceptions of the strategies used to address the task and their sense of the value of the process to their learning.

Drawing on Keys (1999) and Halliday and Martin (1993) the students' drafts were analysed in terms of the occurrence of expansion of clauses. Clauses, as defined by Halliday and Martin (1993) are the main units of meaning in a text. These clauses consist of a subject, or implied subject, and predicate, and include processes signified by verbs, participants or objects expressed by nouns, and qualifiers of time, place, manner or condition, signified by adverbs and prepositional phrases. Halliday and Martin (1993) identify three kinds of expansion or links between ideas. These are: extension, where a second clause adds some new idea to the meaning of the first; elaboration, where a second clause describes, refines or clarifies the first; and enhancement, where the second clause qualifies the first in terms of time, place, manner, condition or cause. Enhancement is a typical feature of report writing where the author builds a chain of reasoning by specifying the circumstantial, causal or conditional links between clauses. The writing task in this project was framed so that students needed to devise various expansions to key concepts in order to make their meaning clear to younger readers. The analysis aimed to identify how students addressed this demand in revising their drafts.

Texts were analysed in terms of (a) length in words, (b) number of clauses, (c) number and percent of expanded primary clauses, and (d) changes to drafts.

6. RESULTS

Content analysis of the students' initial and final drafts indicated that all students' drafts consisted of factual propositions or primary clauses expanded through extension, elaboration, and enhancement. Student final drafts contained an average of 388 words with a range of 258 to 763 words. These drafts contained an average of 36 clauses, with an average of 29 or 80% of these clauses functioning as expansions of primary clauses. This indicates that the students generally chose a small number of key concepts/propositions to explain, consistent with the expectations required of an informative genre and addressed in the instructional program. The main kind of expansion used by students, and also consistent with this genre, was elaboration to explain key concepts and terminology, and the relationships between these concepts. The small number of expansions through enhancement evident in the final drafts can be explained in terms of the nature of the writing task, in that enhancement is more commonly used in other genres such as reports.

Several issues were identified in the analyses of the interviews of students' perspectives in tackling this writing task. These were: the perceived value of redrafting and feedback; the value of writing for a younger readership; the strategies used to meet reader needs; and the effects of engagement with this writing process. Twenty-one of the interviewed students saw value in redrafting and feedback with this task. They claimed this process enabled them to identify gaps in their writing, clarify ideas, reorder information, and check they were on the right track. They also appreciated the feedback over an extended time from two teachers, peers in a small group and the whole class, and also from their target younger reader. Over 85% of interviewed students saw value in writing for a younger readership in that it forced them to clarify their own conceptual understanding in order to explain ideas to less knowledgeable readers. They commented positively on their sense of the reality of the task because they saw its purpose as meaningful, and felt empowered to decide whether an explanation was adequate or not, given their understanding of the target readership. Over 85% of students claimed that to address the task effectively they needed to make concepts clearer to themselves, and then subsequently clearer to their target readers through changing the language and integrating various sources of knowledge. They believed that they needed to integrate different kinds of language, including the technical language of science definitions, their own everyday language and experience, and the language and experience they assumed of their target readers. Over 75% of students identified positive affective outcomes from the writing process such as an increased sense of confidence, and a willingness to participate in discussion. They also commented on increased knowledge of self-regulatory processes, of knowing if they were on track, and of achieving an expanded conceptual framework. They perceived that they needed to clarify their focus, expand inferences, review ideas, and respond to reader needs. Some perceived that being asked to transpose technical language into terms that made sense for a younger reader forced them to learn the concepts in more depth. The teachers commented that student performance in the final topic test indicated that this task and instructional program led to learning outcomes equal or better than students' usual level of performance in science.

The results of the study indicated that students used the following textual strategies to clarify major biotechnological concepts and propositions: rewording propositions in simpler non-technical language; presenting single or elaborated examples to clarify statements; making explicit the superordinate, coordinate, subordinate, or causal relationships between linked concepts; rewording propositions into extended sequences of statements to outline processes or a process; and building networks of interlocking definitions. With varying degrees of success students were able to use these strategies to explain and connect key concepts.

7. DISCUSSION

In explaining the students' performance and perceptions of the task, various possible factors can be identified as important. These include (a) students' prior knowledge of, and attitude towards, this topic (b) the specific nature of the writing task (c) the role of the planning activities including drafting, redrafting, and cycle of feedback in the instructional program and (d) students' knowledge of how to address drafting issues such as organising content and addressing reader needs and expectations. The teachers claimed that these students had very little prior knowledge of this topic, but like other past groups were generally interested in the study of genetics and biotechnology. The students perceived the planning activities, including drafting, redrafting, and repeated feedback, provided an effective space in which to devise, inspect and refine their understanding of key concepts. While the students' comments should be viewed as just one form of evidence in relation to their texts, their perceptions of the value of broad features of the instructional program appear to align with the textual strategies they employed, and support the value of the process in clarifying and organising conceptual understandings of this topic. Their comments also confirm other research findings (Chinn & Hilgers, 2000; Keys, 1999; Kelly and Chen, 1999; Prain & Hand, 1996) that claim specific aspects of task design and classroom conditions are likely to promote learning through writing. Writing tasks should entail meaningful audiences where student writers recognize task demands and have strategies to address these demands. This study confirms Sutton's (1992) claim that tasks, which require students to expand, simplify and elaborate their emerging understandings for real readerships are likely to enable students to consolidate conceptual knowledge through writing. The students' comments confirm Chinn's and Hilgers' (2000) claim that students learn from writing where there is frequent verbal and written interaction between student and teacher, participation in a collaborative learning community, clear guidelines for writing, and opportunities to write for other audiences.

However, this study entailed student engagement with one kind of writing task for a particular topic under specific classroom conditions, and should be viewed with some caution in terms of generalised application to all writing tasks or topics or all students. There is a need for far more extended analysis of the content of science, as noted by White (1994), to identify which topics and concepts might be suited to extended writing tasks of this kind, in contrast to other learning strategies and tasks. However, the students' comments on their practices and perceptions provide some evidential support for the value and effectiveness of backward and forward search hypotheses for learning through writing, as outlined by Klein (1999), especially where the nature of the task enables the higher cognitive demands of backward search to be addressed more readily. In this case the writing process supported learning by enabling writers to integrate different concepts into a single text, by providing instant feedback for reformulating and reviewing ideas, and by connecting past, present and future understandings through new textual syntheses and analysis. The specific task entailed the establishment of a global purpose, namely the production of an explanatory piece of writing for younger readers, and most students perceived this goal as valuable for guiding their subsequent focus, selection and clarification of concepts for learning about biotechnology, or participating in forward search strategies. They also claimed this process enabled them to identify gaps in their writing, clarify ideas, reorder information, and check they were on the right track, strategies clearly entailed in backwards searching of their texts. As the students' perspectives made clear, various planning experiences all contributed to their learning, not just the production of their texts; however, the students believed that the emerging text provided a clear basis for self-evaluation of their understanding and enabled them to consolidate and organize further their emerging understanding of key concepts, as well as relationships between these concepts. In this way writing can be used as an epistemological tool in promoting learning by enabling student writers to construct knowledge through developing such a framework.

8. FUTURE RECOMMENDATIONS

This research has implications for further study of task design for writing for learning as well as on instructional programs to support this process. There is the need to research the effects of different kinds of writing tasks and readerships on students' learning, as well as effective scaffolding or prompts to support this learning. There is clearly a need for programs of explicit writing instruction to enable students to understand the function/form relationships in traditional school science genres and be able to understand and construct these forms of scientific reasoning. However, there is also scope for different kinds of writing tasks that focus on clarifying understanding through writing and developing students' communication skills. While arguing strongly that the language of science enables the construction of science concepts in a language and construction often directly opposed to everyday language and concepts, Halliday and Martin (1993, p 170) also acknowledge that "common sense knowledge organizes the world in ways that can clearly be related to science understandings", and therefore there is a place for a two-way writing process between technical and everyday language and explanations and science learning. There is a need to identify which writing tasks and topics, including computer-based tasks supported by instructional programs, facilitate effective learning of target concepts or practices.

This study also suggests that future writing tasks might serve learning effectively where there is contact between writer and reader during the writing process, so that readers' needs and expectations are addressed more systematically. As noted by Morgan (1998) such contact could entail collaborative writing in different school or classroom settings using email, peer-edited written work, and collaborative problem-solving journals. As this study suggests, student writing for learning is likely to succeed where students perceive the value of the process, have strategies to address task demands, and their teachers are clear about what will count as evidence of the development of students' knowledge and skills in this writing.

REFERENCES

Chinn, P. & Hilgers, T., (2000). From corrector to collaborator: The range of instructor roles in writing-based natural and applied science classes. *Journal of Research in Science Teaching*, *37*,3, 3-25.

Halliday, M., (1985) *An introduction to functional grammar*. Baltimore, MD: Edward Arnold.

Halliday, M., A., K., & Martin, J., R., (1993). *Writing science: Literacy and discursive power*. London: Falmer Press.

Hand, B., Prain, V., & Hohenshell, L., (2000). Exploring Year 10 biotechnology students' response to conceptual questions when engaged with planned writing experiences. Paper presented at the Annual Meeting of the National Association for Research in Science Teaching, New Orleans, LA, April 28-May 1.

Hanrahan, M., (1999). Rethinking science literacy: enhancing communication and participation in school science through affirmational dialogue journal writing. *Journal of Research in Science Teaching,* 36(3) 699-718.

Hildebrand, G., (1998). Disrupting hegemonic writing practices in school science: Contesting the right way to write. *Journal of Research in Science Teaching,* 35(4)345-362.

Kelly, G. & Chen, C., (1999). The sound of music: Constructing science as sociocultural practices through oral and written discourse. *Journal of Research in Science Teaching, 36*, 883-915.

Keys, C.,(1999). Language as an indicator of meaning generation: An analysis of middle school students' written discourse about scientific investigations. *Journal of Research in Science Teaching 36* ,9, 1044-1061.

Klein, P., (1999).Reopening inquiry into cognitive processes in writing-to-learn. *Educational Psychology Review, 11*, 3, 203-270.

Martin, J. & Veel, R., (1998). *Reading Science: critical and functional perspectives on discourses of science*. London: Routledge.

Morgan, C. (1998) *Writing mathematically: The discourse of Investigation*. London: Falmer Press.
Prain, V. & Hand, B., (1996). Writing and learning in secondary science: Rethinking practices. *Teaching and Teacher Education, 12*, 609-626.
Prain, V. & Hand, B., (1999). Students perceptions of writing for learning in secondary school science. *Science Education, 83*, 151-162.
Rivard, L., (1994) A review of writing to learn in science: Implcations for practice and research. *Journal of Reseach in Science Teaching, 31*(9), 969-983.
Rowell, P.,A., (1997). Learning in school science: The promises and practices of writing. *Studies in Science Education, 30,* 19-56.
Sutton, C. (1996). Beliefs about science and beliefs about language. *International Journal of Science Education, 18*, 1-18.
Unsworth, L., (2000). Investigating subject-specific literacies in school learning. In: L. Unsworth, Ed. *Researching language in schools and communities*. London: Continuum (Cassell).
Unsworth, L. ,(1997). Explaining explanations: enhancing science learning and literacy development. *Australian Science Teachers Journal, 43* (1) 34-49.
Veel, R., (1996) Learning how to mean — scientifically speaking. In: F. Christie (ed.) *Genre and Institutions: The language of work and schooling*. London: Cassell Academic
White, R., (1994). Dimensions of content. In: P. Fensham, R. Gunstone, & R. White (eds.) *The content of science: A constructivist approach to its teaching and learning* (pp. 255-262). London: Falmer Press.

SEVENTH-GRADE PUPILS' EPISTEMIC VIEWS IN THE CONTEXT OF MODEL-BASED INSTRUCTION

Kari Sormunen, Pekka E. Hirvonen, Jouni Viiri

University of Joensuu, Finland

ABSTRACT

This paper relates to the research of personal epistemology. Different frameworks of the research area are discussed and the theoretical framework for studying young pupils' epistemic views is presented. The epistemic views deal with pupils' conceptions related both to justification and to certainty of knowledge in a science learning context. The empirical aim of this study was to discover pupils' views of their own knowledge, their physics teacher's knowledge and the scientists' knowledge. The context of the study were science 12 hours of science lessons during which the pupils were taught the modelling idea connected to the topic 'structure of matter'. The subjects in this case study were eighteen 13-year-old secondary school pupils. The results show that pupils have different views concerning the justification of knowledge and the certainty of knowledge depending on the person: a pupil, a teacher or a scientist. The results have implications for the development of science curriculula and new teaching approaches.

1. BACKGROUND

1.1 Research frameworks on personal epistemology

There has been a growing need both for psychologists and educators to discover features describing personal epistemological development and epistemological beliefs. Piaget used the term 'genetic epistemology' to describe his theory of intellectual development, initiating the interest of *developmental psychologists* in this intersection of philosophy and psychology. Having a basis in the Piagetian framework, Perry (1970) attempted to understand how students interpreted pluralistic educational experiences. Following Perry's schema, most researchers in the field have posited models that are to some degree *structural, developmental sequences* (see Hofer & Pintrich 1997 for a review). In almost all of developmental approaches (see Figure 1), it has been argued that students start out with, at best, a limited, fact-based epistemology for thinking about knowledge in their everyday life. After two or three phases an individual's beliefs might develop towards more relativistic thinking (see, e.g., Perry 1970, Kuhn 1991).

Alternatively, it has been argued that individuals may develop different epistemological stances in different domains and that some aspects of some of these stances are domain-specific (Smith et al 2000; Hofer & Pintrich 1997). This might have been the reason why *researchers of education* have been interested in the topic as well; e.g. in the fields of science and mathematics education. The bias towards

D. Psillos et al. (eds.), Science Education Research in the Knowledge-Based Society, 173–181.

personal epistemology in educational settings has mostly been either in general beliefs or in more contextual views.

In the field of science education, many of the epistemological studies are related to the nature of science research (see e.g., Lederman 1992; Kelly, Chen & Crawford 1998 for reviews) and they are not usually so committed themselves to developmental epistemology. The research on young persons' epistemological commitments or beliefs has, however, mainly concentrated on the general and formal ideas of philosophy (see Figure 1). Young people's views and beliefs have been compared to various philosophical trends like "objectivist" or "constructivist-relativist position" (Roth & Roychoudhury 1994), or "positivist" or "lakatosian" perspectives (Blanco & Niaz 1997): the target has been in categorising young people's views and beliefs according to the existing ideas of scientists or other "adults".

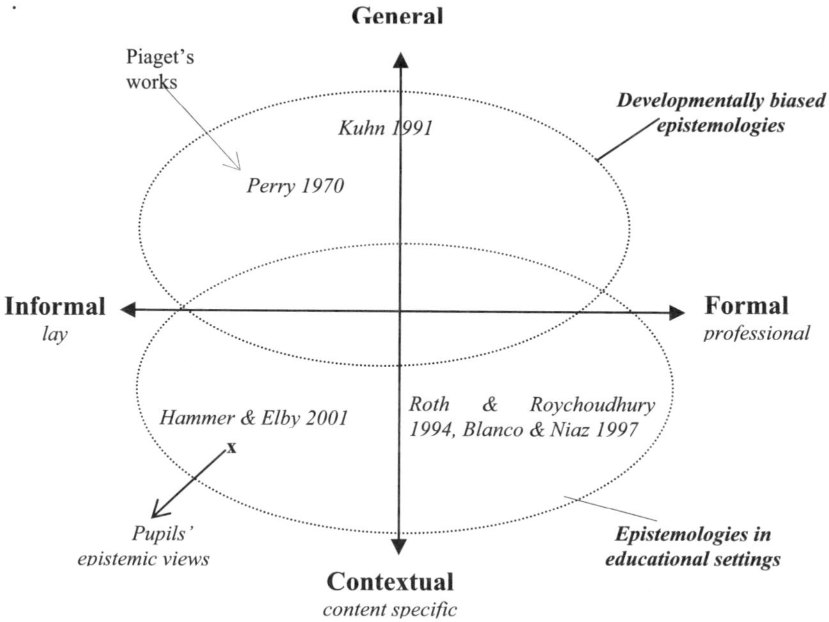

Figure 1. Different research frameworks and some exemplary studies on personal epistemological positions.

Hammer and Elby (2002) have suggested a novel, alternative framework for detecting and interpreting learners' epistemologies. They *"seek a framework of epistemological resources, smaller and more general than theories or traits, that can accommodate* <u>*contextual dependence*</u> *and provide an account of productive resources"* (p. 176). The resources here refer to cognitive resources from which learners may construct more adequate understanding. The first category of resources is for understanding the general nature of knowledge (propagation, creation, fabrication) and how it originates. The second and third categories are resources for understanding epistemological activities (accumulation, formation, checking, application) and forms (stories, rules, rule system, facts, lists, etc), which follow the notions of "epistemic games" and "epistemic forms", as developed by Collins and Ferguson (1993). In the fourth category there are resources for understanding stances one may take towards knowledge (belief / disbelief, doubting, understanding / puzzlement, acceptance).

1.2 Our framework

We assume that this approach described above has some advantages in opening new insights for evaluating, in particular, young pupils' epistemological thoughts and ideas behind them. Our model is not developmentally constrained, and the emphasis is on the context where the ideas arise (cf. Figure 1). Of course, pupils' epistemologies may not be consistent, or expressed in the terms a philosopher would use; consequently, we are using the concept *'pupils' epistemic views'*. The word 'epistemic' has fewer connotations of a worked-out intellectual position than 'epistemological' views (cf. Donnelly 1999). The pupils' epistemic views are though context-dependent and they are informal.

We are not so interested in developmental aspects in studying young pupils' epistemological thinking because it is unclear when the process of epistemological understanding begins. For instance, Boyes and Chandler (1992), using a four-stage model based on the former developmentally based studies, found all the stages represented among the high school students in their study. Our pilot study has shown the same features among 7th graders. One plausible explanation is that such development is recursive, but this is beyond of the aims of our study.

2. AIM OF THE STUDY

2.1 Research questions

The aim of the present study was to discover pupils' epistemic views in the context of the modelling of the states of matter. Our study has been based on the following research questions:

1. How do pupils describe the judgement of their own knowledge, that of their teacher, and scientists' knowledge?
2. How do pupils describe the certainty of their own knowledge, that of their teacher, and scientists' knowledge?

The research questions are developed from the two dimensions suggested to be among the main features of epistemological views: the judgement of knowledge and the certainty of knowledge (Hofer & Pintrich 1997). The questions also combine so-called *proximal* and *distal* knowledge of the nature of science (cf. Hogan 2000), because they deal both with pupils' views about their own knowledge (proximal) and with their views about professional science (distal) enterprise (scientists' knowledge).

2.2 Context

The teaching intervention, which was based on a pilot study (Saari 1997), was designed for 7th grade physics. The duration of the teaching unit was 12 hours and the teaching context of modelling was developed for this particular, case, which is described in detail by Saari and Viiri (2001). The teaching sequence had features of social constructivist pedagogy (cf. Smith et al 2000) in which pupils actively developed, tested and revised their ideas collaboratively with their peers and their teacher.

3. METHODS AND SAMPLE

As a research strategy we applied the case study and our study was explorative in its nature, because we applied the theories of epistemological views to relatively young subjects in a real school context. We chose to use interviews to obtain data, and applied qualitative analysis with an ideographic approach.

The research design consisted of a pilot study and a main study. During the pilot study, the teaching intervention was established and taught by H. Saari, one of the researchers, who had developed the intervention based on his former study (Saari 1997). The teacher of the main phase was introduced to the teaching approaches; she even followed all the lessons. The pilot phase included both pre- and post-interviews and all the lessons were videotaped. The pilot data was analysed and we continued to develop research questions as well as interview agendas. We concluded that it was better to have just post-interviews concerning pupils' epistemological conceptions because it seemed better to have a clear context for questioning about epistemic matters. The subjects were 13-year-old pupils (7th grade); fifteen of them participated in the pilot phase. In the second phase, the teaching unit was presented at another school, with eighteen pupils (7th grade as well). At the end of the teaching intervention the researcher (K. Sormunen) interviewed each pupil of the main research group (n=18).

The main questions presented in a semi-structured format were as follows:

> "How do you know that you know?"
>
> "How does your teacher know, according to your opinion, that some things are as they are?" "How sure is s/he?"
>
> "How does a scientist know that something is as it is?" "How sure is s/he?"

We did not directly ask the pupils about their opinions of the certainty of their knowledge, since our pilot study had shown that reflecting on personal certainty is very difficult. Instead, we interpreted the pupils' responses overall in the interview data to obtain an impression of each pupil's certainty of knowledge.

The interviews were transcribed and collected together with the ATLAS/ti programme. The qualitative analysis was based on the different conceptions expressed by the pupils. Firstly, each response was classified, and then the classes were combined to produce a more generalised description of the pupils' views.

4. RESULTS

The pupils' views concerning knowledge can be divided into three sections: source of knowledge, justification of knowledge and certainty of knowledge. "The source of knowledge" was added into the classification hence it emerged so clearly, although the preliminary research questions referred only to the justification and certainty of knowledge.

A considerable number of pupils described different sources of **their own knowledge**. The sources of that kind refer to knowledge as direct perception, and both as propagated and as fabricated stuff. Many pupils referred to intuition or they could not specify the justification of their knowledge, which refers to knowledge as inherent in Hammer & Elby's (2002) classification. Some of the pupils said that they knew something if they were able to answer if asked about it or if they could remember a thing; these samples refer to formation as an epistemological activity. Only two answers dealt with the evidence for an item of knowledge. Few pupils referred to rational thinking. Proofing and concluding could also be understood as epistemological activities

The certainty of the pupils' owns knowledge had seemed to be a difficult question for them in the pilot study, thus it was not directly asked in the main study. In any case, we received some implicit information about individual pupils' views of the certainty of their knowledge from their answers to other questions (which dealt with the content knowledge): they answered with considerable uncertainty. This might or might not be a sign of a relatively sound attitude towards knowledge. Some of the informants said that they "guessed" which is classified as an epistemological activity in Hammer & Elby's (2002) categorisation.

After pondering on their own knowledge the pupils were asked about **their teacher's knowledge** and how certain he/she was. Some certainty of a teacher's knowledge was the first impression we gained when analysing the responses. The pupils described the source of their teacher's knowledge as learning, studying, experience (all referring mainly to a "fabricated stuff"), communication, and education (examples for "propagation"). The justification of a teacher's knowledge is based on remembering by heart, age or proof: Only a few interviewees referred to the evidence. "Remembering" could be classified as an epistemological activity as well as proofing.

The certainty of the teacher's knowledge is described either as sure or as unsure. Many pupils referred to the teacher's position: she must be sure as she teaches them. Some pupils considered the teacher's knowledge to be very sure. A teacher's position with the conceptions of certainty refers to teachers' "epistemic authority" that might have some connections to knowledge as propagated stuff. We suggest that authoritative sources (e.g. teachers, textbooks etc; cf. Reif & Larkin 1991) could be added to resources for understanding the nature and sources of knowledge in Hammer & Elby's (2002) model.

Thirdly, the interviewees were asked **how a scientist knows** a particular thing and how sure s/he generally is. The Finnish language uses '*tiedemies*' – the concept representing '*scientist*' which literally means a '*scienceman*'. In order to avoid the male bias, the word for 'researcher' was used instead of 'scienceman'. It is, however, quite obvious that the notion of "researcher" underlines the pupils' responses to 'researching' when asked something about 'a researcher's' knowledge.

In dealing with the scientist's knowledge, the source and justification were rather mixed in the pupils' answers. According to the pupils' views, the justifications for a scientist's knowledge include intuition and rational inference – concluding as an epistemological activity, or the use of models. "Modelling" could here be interpreted as one of epistemological activities; it is like playing a game. Other bases include proof, research and different sources – e.g. propagation. One girl referred to equations, which can be classified as one of the epistemological forms.

In relation to certainty of knowledge none of the pupils considered a scientist's knowledge to be certain, though some of them considered it fairly sure in a reasonable way. Most pupils considered a scientist's knowledge to be fallible and unsure. Some responses describe the meaning of modelling: "*If s/he makes a model of an atom, it is just an idea of an atom*", "*... s/he just tries to describe it to others by using a model*".

5. CONCLUSION AND IMPLICATIONS

5.1 Findings

The idea of modelling and the constructivist approach towards teaching are not reflected in the pupils' thinking. This reflects the impression that the intentions of teachers or curricula are not always meeting the pupils' conceptions. For instance, Roth and Roychoudhury (1994) examined what they called epistemological commitment in high school physics students and identified differences between constructivistic and objectivistic beliefs. The latter commitment was predominant in spite of a curriculum that emphasised the former. Carey and Smith (1993) also discuss the difficulties of teaching a constructivist approach in science, which appears to be at odds with the common sense epistemology of seventh grade students.

The pupils consider their teacher's knowledge to be quite certain and at the same time they seem to have fairly sound views concerning scientists' knowledge. This result is quite remarkable: the teacher seems to be regarded as the origin of justified knowledge, but there exist, nevertheless, critical attitudes towards scientific knowledge. This entails, in our opinion, a challenge for school teaching: a teacher seems to be "a treasurer of knowledge". It refers to the transmission model of teaching (Newton, Driver & Osborne 1999), where the nature of science for students is seen as "a fixed body of facts primarily accessed through authoritative sources (e.g. teacher)". It may, in consequence, be hard to teach pupils to be critical, since pupils trust the teacher or the textbook.

5.2 Epistemic model

We applied in this study two aspects of theoretical frameworks. Firstly, we chose two components of the epistemological theories suggested by Hofer and Pintrich (1997) to be the main items in pupils' thinking: the justification of knowledge and the certainty of knowledge. Secondly, we were interested to see if Hammer & Elby's (2002) model of epistemic resources would have any connections with our findings.

The proposed two components (justification and certainty) attached to our research questions revealed a lot about the pupils' epistemological thinking; as a matter of fact, the third component – sources of knowledge – emerged from the analysed data quite clearly. Hofer (2000) states that the source of knowledge and the justification of knowledge belong to the same cluster. There should also be a fourth dimension – simplicity of knowledge – together with certainty of knowledge in the same cluster (Hofer 2000), but we did not recognise it. However, all the three components fit well with our ideas of epistemic views.

A quite nice amount of the resources for understanding the various aspects of epistemological items (Hammer & Elby's model) were found among our findings, and we would like to add some new ones: *authoritative sources* for the resources for understanding the nature and sources of knowledge; *proofing, concluding, remembering, guessing, and modelling* for the resources for understanding epistemological activities; and *equations* for the resources for understanding epistemological forms.

The obtained data does not directly indicate the context-dependence, but we assume that the data would have been different in another kind of context. The context always underlies teaching and interview situations.

5.3 Implications

Pupils' epistemic views make sense for themselves, they should be considered as starting points for epistemological discussions in science classrooms instead of treating the views as naïve or without any significance. Teachers should be aware of these underlying resources of their pupils, and, naturally, of their own "hidden resources". The curricula, textbooks, and standards for science education should also reflect epistemological understanding. Previous research (Sormunen 1999, Viiri 1996) conducted in Finland has revealed that teachers and student teachers have quite a poor conception of the epistemological background of physics. For teachers, it would be valuable to know some ideas of the philosophical basis of science and also the basis of pupils' epistemological ideas. It would also be crucial for teachers to recognise their own epistemic views. We suppose that then the teachers' performance would be much more efficient when teaching epistemological ideas related to science and even when teaching science per se.

In Finland, as well as in, e.g., UK, science education and science curricula have been under pressure from socio-political climates. Newton et al (1999) describe the movement as a "technocratic orientation" and there are similarities with the situation in Finland. The new criteria given by the National Board of Education (1999) were to sharpen the aims based on the former curriculum of the comprehensive schools (National Board of Education 1994). However, the descriptions of knowledge and skills related to physics and chemistry education still give a strong impression of overcrowded content where there does not seem to be any room for philosophical and historical perspectives of science (c.f. Monk & Osborne 1996). Thus, there is much to do in the field of "epistemological education".

REFERENCES

Blanco, R. & Niaz, M., (1997). Epistemological Beliefs of Students and Teachers about the Nature of Science: from 'Baconian Inductive Ascent' to the 'Irrelevance' of Scientific Laws. *Instructional Science*, 25, 203-231.

Boyes, M., C. & Chandler, M., (1992). Cognitive Development, Epistemic Doubt, and Identity Formation in Adolescence. *J. of Youth and Adolescence*, 21, 277-303.

Carey, S. & Smith, C., (1993). On Understanding the Nature of Scientific Knowledge. *Ed. Psychologist*, 28, 235-251.

Collins, A. & Ferguson, W., (1993). Epistemic Forms and Epistemic Games: Structures and Strategies to Guide Inquiry. *Ed. Psychologist*, 28, 25-42.

Donnelly, J., (1999). Interpreting Differences: The Educational Aims of Teachers of Science and History, and Their Implications. *J. of Curriculum Studies*, 31, 17-41.

Hammer, D. & Elby, A., (2002). On the Form of a Personal Epistemology. In: Hofer, B. K. & Pintrich, P., R., (Eds.), *Personal Epistemology: The Psychology of Beliefs about Knowledge and Knowing*(pp. 169-190). Mahwah, NJ: Erlbaum.

Hofer, B., K., (2000). Dimensionality and Disciplinary Differences in Personal Epistemology. *Contemporary Educational Psychology*, 25, 378-405.

Hofer, B., K. & Pintrich, P., R., (1997). The Development of Epistemological Theories: Beliefs about Knowledge and Knowing and Their Relation to Learning. *Review of Ed. Res.*, 67, 88-140.

Hogan, K., (2000). Exploring a Process View of Students' Knowledge about the Nature of Science. *Sci. Ed.*, 84, 51-70.

Kelly, G., J., Chen C. & Crawford T., (1998). Methodological Considerations for Studying Science-in-the-Making in Educational Settings. *Res. Sci. Ed.*, 28 (1), 23-49.

Kuhn, D., (1991). *The Skills of Argument*. New York: Cambridge University Press.

Lederman, N., G., (1992). Students' and Teachers' Conceptions of the Nature of Science: A Review of the Research. *J. Res. Sci. Teaching*, 29, 331-359.

Monk, M. & Osborne, J., (1996). Placing the History and Philosophy of Science on the Curriculum: A Model for the Development of Pedagogy. *Sci. Ed.*, 81, 405-424.

National Board of Education, (1994). *Framework Curriculum for the Comprehensive School*. Helsinki.

National Board of Education, (1999). *Criteria for the Evaluation of the Comprehensive School Achievements*. Helsinki. (In Finnish)

Newton, P., Driver, R. & Osborne, J., (1999). The Place of Argumentation in the Pedagogy of School Science. *Int. J. Sci. Educ.*, 21, 553-576.

Perry, W., G., (1970). *Forms of Intellectual and Ethical Development in the College Years: A Scheme*. New York: Holt, Rinehart and Winston.

Reif, F. & Larkin, J., H., (1991). Cognition in Scientific and Everyday Domains: Comparison and Learning Implications. *J. Res. Sci. Teaching*, 28, 733-760.

Roth, W. & Roychoudhury, A., (1994). Physics Students' Epistemologies and Views about Knowing and Learning., *J. Res. Sci. Teaching,* 31, 5-30.

Saari, H., (1997). *Learning the Structure of Matter at Secondary School*. Unpublished licentiate thesis. University of Joensuu. (In Finnish)

Saari, H. & Viiri, J. (2001). Teaching and Learning the Concept of the Model in Secondary School. In: Behrendt, H., Dahncke, H., Duit, R., Gräber, W., Komorek, M., Kross, A., Reiska, P., (Eds.),: *Research in Science Education – Past, Present, and Future* (pp. 259-264). Dordrecht: Kluwer Academic Publishers.

Smith C., L., Maclin D., Houghton, C. & Hennessey, M., G., (2000). Sixth-Grade Students' Epistemologies of Science: The impact of Schools Science Experiences on Epistemological Development. *Cognition and Instruction*, 18, 349-422.

Sormunen, K., (1999). Nature of Physics – Teachers' Views. In: Bandiera, M., Caravita, S., Torracca, E. & Vicentini, M., (Eds.) *Research in Science Education in Europe* (pp. 235-243). Dordrecht: Kluwer Academic Publishers.

Viiri, J., (1996). Student Teachers' Ideas of the Model Concept. In the abstract booklet of *Naturvetenskapen i skolan in för 2000*-talet (pp. 24-25). 18-22 March 1996, Kristianstad, Sweden..

NONLINEAR ANALYSIS OF THE EFFECT OF WORKING MEMORY CAPACITY ON STUDENT PERFORMANCE IN PROBLEM SOLVING

Dimitrios Stamovlasis, Georgios Tsaparlis

Department of Chemistry, University of Ioannina, Greece

ABSTRACT

We describe a nonlinear approach for studying the effect of working-memory capacity on science problem solving. The tools used are those of complexity theory and are used to test and model the phenomenon of the working-memory overload. The nonlinear method correlates the subjects' rank-order achievement scores in problem solving with the psychometric variable. From the achievement scores, rank-order sequences of the subjects, according to their scores, were generated, and in the place of each subject, his/her score was then replaced by the value of the cognitive variable. Then the sequence was mapped onto a one-dimensional random-walk model (working-memory random walk) and treated as dynamical flow. For these sequences, the nonlinear correlation exponents (Hurst exponents) were calculated. The null hypothesis was tested with surrogate data. The method is demonstrated with data from achievement scores in chemical equilibrium problems. Problems of various logical structures and mental demands were used. In problems with simpler logical structure and with low mental demand, the Hurst exponent was close to the surrogate value (demonstrating randomness), while sudden increases appeared for more complicated problems (indicating scale invariant long-range correlations). The method provides meaningful results and adds to the understanding of information processing within the frame of science education.

1. INTRODUCTION

The working memory hypothesis has provided a useful metaphor to psychological and educational research. With respect to problem solving in science education, research has shown that, in certain cases, working-memory (WM) capacity, mental (or *M*-space) capacity, and/or other psychometric variables [such as, disembedding ability (degree of field dependence-independence), and developmental level] can be predictive of the student performance (Johnstone, Hogg, & Ziane 1993; Johnstone & Kellet, 1980; Niaz & Loggie, 1993; Tsaparlis & Angelopoulos, 2000; Tsaparlis, Kousathana, & Niaz, 1998).

The working memory model states that a student is likely to be successful in solving a problem if the problem has a mental demand (M-demand or Z-demand) which is less than or equal to the subject's WM capacity (X) ($Z \leq X$), but fail for lack of information or recall, and unsuccessful if $Z > X$, unless the student has strategies that enable him/her to reduce the value of Z to become less than X. That is, as the problem increases in complexity (in terms of what has to be held and what

D. Psillos et al. (eds.), Science Education Research in the Knowledge-Based Society, 183–190

process has to be performed), there must be a decrease in achievement; moreover, if the holding/thinking space has a finite limit, the decrease of achievement may be rapid after the limit has been reached (Pascual-Leone, 1969; Scandarmalia, 1977). Note that in order for the model to be valid, a number of necessary conditions must be fulfilled (Tsaparlis, 1997; Tsaparlis & Angelopoulos, 2000).

In all aforementioned science education research, linear statistics has been used to model the phenomenon of working memory overload. Nevertheless, fostering nonlinear methods for studying such phenomena would be more appropriate. Recently, nonlinear methods have been developed to analyse the effect of WM in organic synthesis problem solving (Stamovlasis & Tsaparlis, 1999, 2000, 2001a, 2001b, 2002; Tsaparlis & Stamovlasis, 1999). The work correlates the rank order of the subject achievement scores in problem solving with WM capacity and shows how the effect of WM on problem solving can be observed with means and tools of complexity theory. The current paper consists a further exploration of the effect of WM using nonlinear methodology. It demonstrates the application of the random walk model to chemical equilibrium problem solving.

2. METHOD

The data set in the present research consists of achievement scores in chemistry problem solving and the results from the psychometric test (Tsaparlis & Kousathana & Niaz, 1998). The WM capacity of the subjects was determined through the digit backward span test (DBS). The DBS test is part of the Wechsler Adult Intelligence Scale (Wechsler, 1955), and involves both storage and processing. The chemical problems are of varying complexity: *Z*-demand and logical structure. The logical structure of the problem is specified by the number of *operative schemata* entering the problem. Nine problems of varying number of *operative schemata* as well as of varying number of *Z*-demand were used. The number of schemata varied from a minimum of two, to a maximum of four, while within each logical schema we assign a specific *Z*-demand, varying from four to six. Two separate marking schemes were followed: one for the *schemata* only and another for the *steps* only. An example of problem will clarify the method used for problem complexity estimation:

> In a vessel of fixed volume V = 24 L, a mixture of H_2, N_2, and NH_3 gases, in ratio of 1:2:2, is introduced. The mixture is heated to 1000°C and let to reach chemical equilibrium. Calculate the equilibrium constant (K_c), taking into account that at equilibrium the total pressure of the gas mixture is 41 atm, at 1000 °C.

Three schemata enter here: (1) the process of "establishment of the chemical equilibrium"; (2) the "ideal-gas equation", and (3) the "condition of chemical equilibrium". Schema (1) involves six steps for its solution, schema (2) involves five steps, and schema (3) involves three steps. For further details see Tsaparlis, *et al.* (1998). Since schema (1) involves the largest number of steps (six), we postulate a *Z*-demand of 6 for this problem.

Rank-order sequences of subjects' achievements, according to their scores, were generated, and each score was then replaced by the value of his/her WM capacity. In this way a symbolic sequence is produced for each problem with varying complexity (Z-demand and logical structure). These sequences can be treated as dynamical flows. Taking the exam can be seen as the process which generates these *dynamic* sequences. A statistical question is whether the sequences are random or there exists 'memory' in them, demonstrated by long-range correlations. In order to study the scale-invariant, long-range correlations in a dynamic sequence, we introduced a graphical representation of the sequence, which is termed a *working memory random walk* (WM-RW). We define the function $\Psi(i)$ as follows:

$$\Psi(\mathrm{i}) = -1, \text{ for } X=4; \Psi(\mathrm{i}) = +1, \text{ for } X=5 \text{ RW 4/5}$$

$$\Psi(\mathrm{i}) = -1, \text{ for } X=5; \Psi(\mathrm{i}) = +1, \text{ for } X=6 \text{ RW 5/6,}$$

With i = 1 to N representing the number of subjects. In this way, the generated sequence becomes a spatial array of -1s, or +1s, so that it is possible to map the sequence onto a one-dimensional random walk. The nonlinear Hurst exponent is calculated for each sequence (Sprott & Rowlands, 1995). If $H = 0.5$, the sequence is a normal random walk or Brownian motion. On the other hand, if $0.5 < H < 1$, there exist scale-invariant, long-range correlations with persistent behavior, characteristics of black noise. Such correlations reject the null hypothesis that the sequence is random ($H = 0.5$). There is "memory" in the sequence, which then possesses higher *order*/higher *entropy* or *disorder*/lower *entropy*.

Calculations showed that the value of Hurst exponent for a sequence depends on the complexity (schemata and Z-demand), of the problem. To test the null hypothesis of randomness, that is if the sequences or series are random, surrogates were developed for each sequence of symbols, by taking the original data set and randomizing it completely by shuffling. More details, about the method and the mathematics employed, are given in Stamovlasis and Tsaparlis (2001).

3. RESULTS AND INTERPRETATION

The method correlates the rank order achievement scores with the WM capacity of the subject. This correlation is a function of the Z-demand of the problem and the logical structure. Figure 1 shows graphs of the Hurst exponent (H) as a function of Z-demand and the logical structure of the problem for random walk RW 4/5. We observe that for the problem with Z-demand 4, the value of H is close to surrogate data. This means that all subjects solve (more precisely are *capable* of solving) the problem and there is randomness in the rank-order achievement sequence. At $Z = 5$, we observe long-rage correlations ($H = 0,8$) because subjects with WM = 5 outscore subjects with WM = 4. The correlations still persist in the problem with $Z = 6$. The same pattern was observed for organic synthesis, where the problems have a simple logical structure (Stamovlasis & Tsaparlis, 2001).

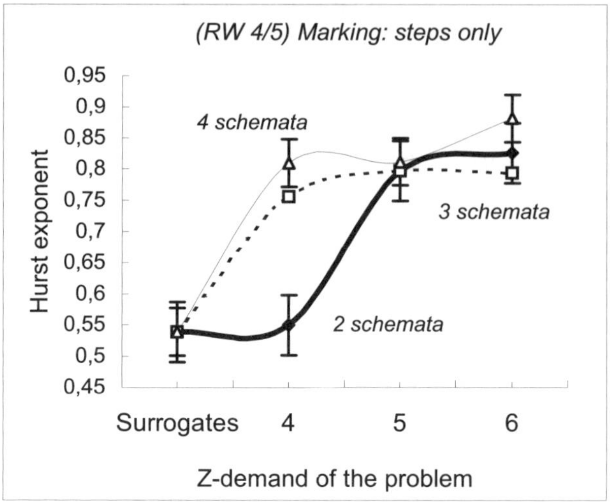

Figure 1. Hurst exponent as a function of Z-demand and the logical structure of the problem. Random walk RW 4/5

A logical structure of two schemata is not very different from one with just one-schema; as a result, WM correlates with rank-order achievement sequence at the same *Z*-demand value. In the case of three and four schemata, long-rage correlations appear already from $Z = 4$, that is, a small increase in logical structure of the problem causes mental load capable to differentiate subjects with WM = 4 from those with WM = 5. This demonstrates a degree of interaction between *Z*-demand, logical structure and WM capacity.

Figure 2 has the data for random walk RW 5/6. We observe that for the problem of two schemata, no long-rage correlations appear for all *Z*-demands. Subjects with WM = 5 and WM = 6 come in a random way in the rank-order achievement sequence. The two phases separate in the problem with three schemata and a *Z*-demand of 6. In other words, we observe a shif*t* towards larger number of schemata, that is, long-rage correlations do not appear at the two-schemata problem, but with a delay of one additional schema. In the case of four schemata, long-rage correlations follow the same pattern observed for the three-schemata case. The above deviations from the previously observed pattern of the simple logical case (Stamovlasis & Tsaparlis, 2001) demonstrate dynamic interactions between the cognitive variable and the logical structure (Pascual-Leone, 1987; Niaz, 1991).

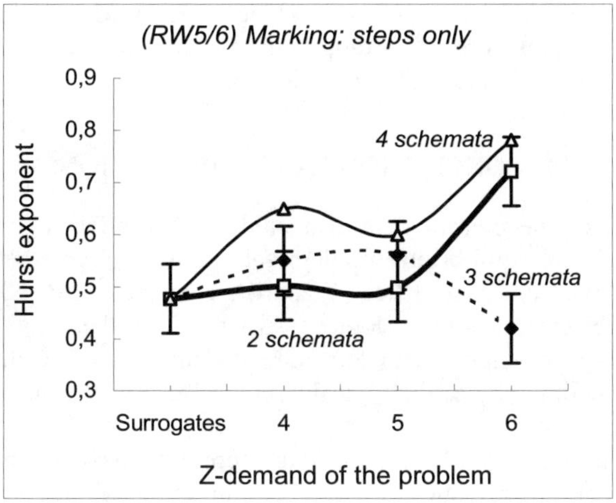

Figure 2. Hurst exponent as a function of Z-demand and the logical structure of the problem. Random walk RW 5/6.

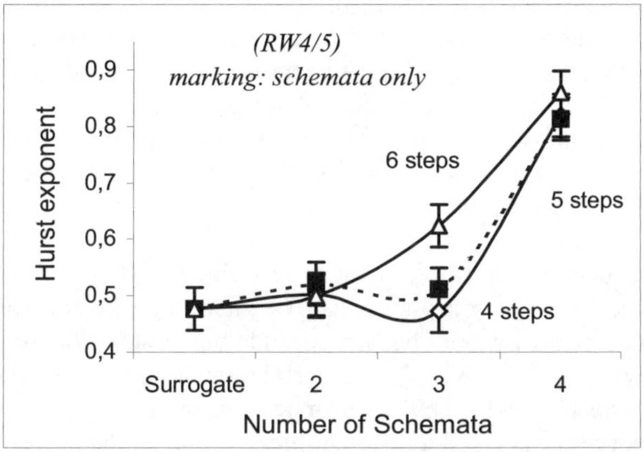

Figure 3. Hurst exponent as a function of Z-demand and the logical structure of the problem. Marking in schemata only and for random walk RW 4/5

Figure 3 shows the Hurst exponent as a function of the complexity of the problem for the marking of schemata only. This marking is an estimation of subject's ability to retrieve correctly operational schemata from long-term memory before applying the steps of problem solution. We observe that only in problems with logical structure of four schemata, subjects with WM = 5 can outscore those with WM = 4. There, we observe strong long-rage correlations. In the problem with three schemata and six steps, such correlations also appear.

4. CONCLUSIONS AND IMPLICATIONS

The findings of the present nonlinear study of the effect of working memory capacity on chemical equilibrium problem solving are in agreement with previous work that employed linear statistics (Tsaparlis, Kousathana, & Niaz, 1998). They demonstrate the effect of the logical structure of the problem on student performance. Even problems with low mental demand can be difficult for some students, as the complexity of the logical structure (that is, the number of schemata) increases.

An important finding is that long-range correlations exist between working memory and rank-order achievement scores in the marking of schemata. That marking is a measure to what extent the subject has learnt an algorithmic procedure. It seems that high processors recall the algorithm better than low processors, even though the mental task does not require high processing capacity. The material seems more familiar to the high processors. According to a general information-processing model, familiar or interesting catching components are admitted through the sensory filter (Johnstone & Selepeg, 2000). However, the judgment of what is familiar must be controlled by what has already been established in long-term memory. Thus, there exists evidence that what has been established in long-term memory (learnt material) is correlated to information processing capacity. During the learning period the material (input) to be learnt is recoded/changed in working memory and stored in long-term memory. This procedure is more efficient for high processors and the constructed knowledge network is likely to be more permanent and less rote and easily forgotten (Johnstone & Selepeg, 2000). Our findings supply empirical evidence to support a theoretical connection between constructivism and information processing.

Correlations between working memory capacity and students' achievements can also be observed by employing the (linear) Pearson coefficient. Linear correlation has an analogous behavior, but does not provide the context for rigorous patterns to be revealed: evidence is provided of a correlation between two variables, without them being connected in fact by a linear relationship. The current application of complexity theory has revealed a different picture, that is, the power law correlation; in addition, it has demonstrated that the nonlinear methodology has proved efficient and insightful.

An important issue, which may constitute a shortcoming of the presented methodology, is the problem of *nonstationarity*, in both its notions: *reproducibility* of the measurements (in our case, of working memory capacity), and the *size* of the available data set (Kantz & Schreiber, 1999). We have discussed elsewhere this problem (Stamovlasis & Tsaparlis, 2001).

In conclusion, we maintain that this work makes an important contribution towards understanding the role of working memory capacity in problem solving. A research tool has been developed, which is suitable for modeling nonlinear phenomena. It is such phenomena which science education research predominately deals with. The implication for instructional practice is that we can design better strategies for teaching problem solving. Guided by the findings of research, we are able to construct a series of problems with varying mental demand (*M*- or *Z*-demand) and logical structure. We can then facilitate student success first to problems with simple logical structure and varying *Z*-demand, up to the point where the students are able to modify *Z* by *chunking* within their working memory capacity (*X*) (that is, by treating together bits of information). Problems with increased complexity could be left for later use, when the students have acquired more experience and motivation.

REFERENCES

Johnstone, A., H., (1984). New stars for the teacher to steer by? *Journal of Chemical Education,* 61, 847-849.

Johnstone, A., H. & El-Banna, H., (1986). Capacities, demands and processes - A predictive model for science education. *Education in Chemistry*, 23, 80-84.

Johnstone, A., H., Hogg, W., R., & Ziane, M., (1993). A working memory model applied to physics problem solving. *International Journal of Science Education,* 15, 663-672.

Johnstone, A., H. & Kellet, N., C., (1980). Learning difficulties in school science - Towards a working hypothesis. *European Journal of Science Education, 2,* 175-181.

Johnstone, A., H. & Selepeg, D., (2000). A Language Problem Revisited. *Chemistry Education: Research And Practice In Europe* (vol. 2, pp. 19-20). (CERAPIE). [http://www.uoi.gr/conf_sem/cerapie]

Kantz, H. & Schreiber, T., N., (1999). *Nonlinear time series analysis.* Cambridge UK: Cambridge University Press.

Niaz, M. & Loggie, R., H., (1993). Working memory, mental capacity, and science education: Towards an understanding of the 'working memory overload hypothesis'. *Oxford Review of Education,* 19, 511-525.

Pascual-Leone, J., (1969). The encoding and decoding of symbols by children: A new experimental paradigm and neo-Piagetian model. *Journal of Experimental Child Psychology,* 8, 328-355.

Scandarmalia, M. (1977). Information processing capacity and the problem of horizontal decalage: A demonstration using combinatorial reasoning task. *Child Psychology,* 48, 301-345.

Sprott, J., C. & Rowlands, G., (1995). *Chaos data analyzer*. Physical Academy Software.

Stamovlasis, D. & Tsaparlis G., (1999). An application of complexity theory to an information model in science education. Paper presented at the *9th International Conference for Chaos Theory in Psychology and Life Sciences*. Berkeley University, CA, USA: July 23-26.

Stamovlasis, D. & Tsaparlis G., (2000). Nonlinear analysis of the effect of working-memory capacity on organic-synthesis problem solving. *Chemistry Education: Research and Practice in Europe* (CERAPIE), 1, 375-380 [http://www.uoi.gr/conf_sem/cerapie]

Stamovlasis, D. & Tsaparlis G., (2001a). Application of complexity theory to an information-processing model in science education. *Nonlinear Dynamics in Psychology and Life Sciences,* 5, 267-286.

Stamovlasis D. & Tsaparlis G. (2001b). Nonlinear analysis of the effect of working memory capacity on student performance in problem solving: the case of chemical equilibrium problem solving. *Proceedings of the third International conference of the European Science Education Research Association* (ESERA), August 21-27, 2001, Thessaloniki, Macedonia, Greece.

Stamovlasis, D. & Tsaparlis G., (2002). A complexity theory model in science education problem solving: Random walks for working memory and mental capacity. *Nonlinear Dynamics in Psychology and Life Sciences.* Accepted for publication.

Tsaparlis, G., (1998). Dimensional analysis and predictive models in problem solving. *International Journal of Science Education,* 20, 335-350.

Tsaparlis, G., & Angelopoulos, V., (2000). A model of problem solving: Its operation, validity, and usefulness in the case of organic-synthesis problems. *Science Education*, 84, 131-153.

Tsaparlis G. & Stamovlasis D., (1999). A working-memory random walk for analyzing problem-solving data: The case of organic-synthesis problems. P*roceedings of the 2nd International Conference of the European Science Education Research Association* (pp. 226-228), (ESERA). Kiel, Germany.

Tsaparlis, G., Kousathana, M., & Niaz, M., (1998). Molecular-equilibrium problems: Manipulation of logical structure and of M-demand, and their effect on student performance. *Science Education,* 82, 437-454.

Wechsler, D., (1955). *Wechsler adult intelligence scale manual.* New York: Psychological Corporation.

THE NATURE OF GROWTH IN CHILDREN'S SCIENCE UNDERSTANDINGS: INSIGHTS FROM A LONGITUDINAL STUDY

Russell Tytler, Suzanne Peterson

Deakin University

ABSTRACT

This paper uses selected data from a longitudinal study of fourteen young children, concerning their changing views of evaporation over a three year period, to argue that children's science learning pathways are highly complex, fundamentally contextual and framed within a network of personal stories. It is argued that the longitudinal methodology opens up a more complex and nuanced view of children's conceptual learning in school settings than is afforded by cohort studies, and that the focus on individuals over time forces a very different construction of the learner than is represented in mainstream conceptual change literature.

1. BACKGROUND, AIMS AND FRAMEWORK

The last few decades has seen an enormous amount of research into students' learning in science, that has involved a theoretical shift away from a Piagetian presumption that children's learning is fundamentally determined by the developmental growth of logical structures, to a greater focus on the conceptual content of children's ideas; their ways of making sense of phenomena. Despite this shift, there are many commonalities between Piagetian and individual constructivist views of learning. In much of the conceptual change work, cross age cohort studies are used to describe developmental sequences of conceptualisations through which children pass (Driver, Leach, Scott & Wood-Robinson 1994). The presumption concerning stability in students' thinking has been retained, this time in terms of quasi-theory-like 'conceptual frameworks', replacing the more general notion of stages in the development of logical structures.

Questions have been raised about the coherence and stability of individual students' conceptions, and the inconsistency with which individuals use conceptions has been demonstrated in a number of studies (eg. Tytler, 1998a). Research exploring the stability of individuals' ideas across time has been mostly limited to time scales within a year (Engel Clough, Driver & Wood Robinson, 1987; Tytler, 1998a,b). The detailed nature of individuals' growth in understanding over longer periods has rarely been explored. In the absence of such studies, the presumption that learning occurs as a developmentally determined sequence of conceptualisations continues to dominate the framing of science curricula and advice concerning science teaching and learning.

D. Psillos et al. (eds.), Science Education Research in the Knowledge-Based Society, 191–199.

White (2001) has called for some time for longitudinal studies tracing the emergence and development of conceptions. Black and Lucas (1993) argue that a longitudinal research program based on classroom learning "will be necessary before we have a thorough understanding of the way(s) in which pupils progressively build their understandings" (p. 235). Despite these arguments, such studies have been rare in science education because they are time-consuming and difficult.

2. RESEARCH METHODS

The study this paper is drawing on is following 14 children through their primary school years. We worked with the children in one class in the preparatory year (Age 5), and they are at the time of writing in their fourth year of schooling (Grade 3). Each year we have worked with the children and their teachers in at least two classroom sequences on different topics, gathering data from written responses, tapes of class discussions, teacher interviews, and interviews of the children. In the interviews we explore, among other things, their learning from the sequence and their views of previous or related activities. Each interview has been transcribed and the NUD*IST software analysis program used to assist in developing interpretive categories. Underlying the research approach is a presumption that learning and development are inextricably linked. We take the view that over the time scales we are studying it is not useful to make the distinction between growth in understanding based on experiences inside or outside of the classroom, and more general epistemological or cognitive development. The aim of the study is to investigate the interactions between children's learning in science, their personal conceptual histories, and their lives in classrooms. The methodology is providing more fine grained insights into individual children's construction of science knowledge than was possible in cross age studies we have previously been involved in. In the paper we focus on two children's changing views of evaporation to illustrate the ways in which longitudinal studies can provide insights into children's learning not accessible to cross age studies.

3. CHILDREN'S VIEWS OF EVAPORATION

There has been a substantial amount of research carried out on evaporation (Osborne & Cosgrove, 1983; Russell, Harlen & Watt, 1989; Bar & Galili, 1994; Tytler, 2000).

The topic is interesting partly because of the iconic imagery of the water cycle that is culturally pervasive and universally promoted in primary schools, and partly because an understanding of the scientific view requires a view of air, of matter, and of interrelationships of entities such as atmosphere and clouds, the sun, heat, and surfaces. Bar and Galili's (1994) analysis represents a good example of the conceptual trajectory approach of children's understandings. They claim to identify four distinct stages in children's progression of understanding of evaporation and condensation; water disappears, water is absorbed into surfaces, water is transferred upwards, and water disperses into air. They associate the transitions between stages with particular ages, with the highest stage occurring at about age 13.

Within the present study we have substantial information about the 14 children's responses to a range of evaporation phenomena, from an evaporation unit they did in their first (preparatory) year, when they were age 5, to various activities in class and during interviews in grades 1, 2 and 3. During a grade 2 interview, for instance, we asked them to put a wet hand on a slate and draw round it with chalk, then explain what happens. We then showed them their work from 2 years previous and asked them to comment on any changes in their thinking. The transcripts presented in the next section represent selections from the sequence of explanations of different evaporation phenomena, for two children, Amy and Chad. Taken as a whole, the transcripts illustrate the complexity and multi-dimensionality of individual development of understandings.

4. FINDINGS: PATHWAYS OF GROWTH IN UNDERSTANDING EVAPORATION

Amy and Chad's cases (and those of the other children) illustrate the contextual dependence and instability of their conceptions of evaporation. Their explanations are quite different for different phenomena. Amy, in the preparatory grade and grade 1, oscillates between explanations involving water soaking (or melting) into the ground (*Prep., March:* 'The water went down to the soil and the flowers grow .... It went down to the soil and dried up'; *Prep, September:* '... in the soil. It makes worms feel nice ...'), going into drains and to the sea, going up to the sky, and to the clouds as part of a water cycle. She is unable to link the idea to water boiling (*Grade 1, November:* (The water) 'just bubbled up onto the side'). Amy's ideas, from the start, were peppered with water cycle imagery but she seemed to have little control over the application of the idea (*Prep, March:* 'The water goes up to the clouds. It goes round in cycles until it finds a shady spot and then goes up to the clouds').

Children in the preparatory year consistently failed to see the classroom activities – puddles, clothes drying, or boiling water – as linked (Tytler & Peterson, 2000). The children were using a number of ideas quite differently for different phenomena. More than this, they would oscillate between different conceptions for the same phenomenon.

Chad did in fact build a consistency of sorts into his preparatory year explanations, based around the preposition 'into'. In March he applied this to clothes drying ('The water sinks *into* the grass'), water boiling ('The water dried up *into* the metal'), and puddles ('I think the water went *into* the concrete'). This linguistic turn was also applied to upward movement of the water, transcending Bar and Galili's stages. He explained a cloth-drying experiment: 'In the sun the water will drink up into the sun. In the wind the water will go onto the fan's blades and dry up when the fan stops'. A year later (Grade 1, September) he has not shifted in his view of boiling:

Chad:The salt sucked it (the water) up and then the salt dried up.

But later in the same interview, concerning the fog inside a terrarium, he was able to construct a view of evaporation:

> Chad:… it's got so hot in there, that from the soil, um, water has evaporated … 'cause it's really warm on the inside and it's cold on the outside, and it makes a fog, and … the fog turns into water.

He linked this understanding to a family discussion concerning fog on glass.

> Chad:Um, my Dad told me that um, if it's warm on the inside and there's glass or something like that, it…and it's cold on the outside, it makes fog, and fog turns into water…. 'cause one morning um, we were in the car, Dad was taking me to school, and there was fog on the window… And I asked him what makes fog.

And in November, being asked to remember the puddle, he extended this view to include a role for air:

> Chad:umm... from the sun it got dehydrated. It evaporated … umm... what happens, it just sort of... just sort of like... um, the sun is really hot and it dries it all up and it goes into air.

However, in the same interview he was unable to translate this into a coherent view of clothes drying. He retains an 'into' view of evaporation.

> Chad:… (the water) just evaporates. It soaks into the ... umm the shirt or whatever it is and it just, um...sort of evaporates … Um … it goes in and the sun dries it up.

In Grade 1, neither Chad nor any of the grade 1 children understood water boiling as evaporation, and did not clearly associate the generation of steam with dropping water levels. They could not be said to have had an understanding of evaporation independent of context, and they commonly constructed entirely different explanations for the different phenomena. Nevertheless, over the years the children grow in their understandings of evaporation. In grade 1 Chad begins to use the word 'evaporation' and in grade 2 he is able to use it with increasing power and consistency in describing phenomena. In May of Grade 2 he explained a wet handprint on a blackboard evaporating into the air, by linking it with a beach.

> Chad:If it was on a beach and say it was high tide .... let's just say the sand wasn't there maybe concrete.... .and it's high tide and when the tide goes out if there be wet patches like that it would evaporate and go back into the air and there's a cycle of rain. It rains and then ...

By grade 3 (June), Chad has a coherent story to tell about what evaporation is, and seems to actively look for links. Unlike previously, he talks naturally of the water evaporating when a salt solution is boiled. He explains what he means by 'evaporate'.

> Chad:It's like when a puddle evaporates and it goes into the cloud but there's not clouds cos we're in a room here, so it just becomes a part of the air…. water that's evaporated. And it rises up.

In the same interview, he describes what happens with a damp tea towel:

> Chad:umm… the sun would have evaporated it and just like the puddle… um, the same thing would have happened and since it would've been outside on the line, it would've gone up to the clouds and then go down as the rain.

The growth relates both to his extension of this idea to more phenomena, but also in the sophistication of the way he expresses the idea.

In grade 2, following a classroom unit in which evaporation and condensation were covered, Amy uses the word 'evaporate' and becomes increasingly aware of the sun's role, which she interprets first as 'taking', then 'attracting'. In Amy's class the unit included discussion of the water cycle in terms of 'tiny droplets' of water rising into the air. In explaining the evaporating handprint:

> Amy: It's drying up....Because the sun's shining in through that window... Because the sun takes water.... I forgot the reason why it takes water but the sun takes water up and umm .... and it dries up... It's evaporating.

She relates this to an experience she had, spilling a drink on the pavement.

> Amy: Because it had been in the sun for a few minutes it came up, it went.... it came through the air in tiny droplets that you can't see them and ..um.... they evaporated up into the air and the sun took them.

This idea of droplets persists into grade 3 (June), when she explains what happens with a damp tea towel.

> Amy: It evaporates into the sky. ... the sun makes it evaporate..... because it's so hot, um, tiny microscopic droplets go into ... um, because they're attracted by it, I think.

5. DISCUSSION

In grades 2 and 3 Amy and Chad can both apply the word evaporation with some confidence to an expanded range of phenomena, and they both use water cycle imagery. Perhaps on a standard outcomes test they would show up as equivalent. However, despite similar teaching sequences on evaporation in Grade 2, they express their understandings very differently.

5.1 Characterising the children's learning over three years

Amy, from the beginnings of the prep year, seemed to have access to a 'water cycle' version of evaporation, as did many of the children. In this first, and indeed in subsequent years, we can identify a tension between the more perceptually accessible idea of water 'soaking in', and the frequent intrusion of this water cycle notion which is quite prevalent, even iconic, in the primary school curriculum tradition. Amy's thinking oscillates between a number of conceptions across the first two years. Her growth in understanding could be characterised as increasing control over the idea of a water cycle. She resolves her competing versions of the evaporating puddle with the idea that water goes into the ground and finds its way to the sea through drains whence it goes up to the sky. Later, armed with the idea of 'tiny droplets', she is able to characterise evaporation more directly, and she can talk about different evaporation phenomena in these terms. Her problem remains, however, in not being able to put together the ideas of air, sky, clouds and sun. She seems not to have developed a coherent mental model of what happens to the water, such that she can vary her account in a convincing way and apply it to details of phenomena. This is evident in the relatively 'schooled' nature of her explanations.

Chad, from the beginning of his school life has worked at exercising control over his explanations. In grade prep he enforces a partial coherence on his ideas through metaphorical language; prior to and during the evaporation unit he consistently referred to water going 'into' surfaces. In grade 1 he uses the word 'evaporation' for the first time, and embellished this with a condensation story featuring his father. That story, like Chad's later beach story, indicates a narrative control lacking in Amy's accounts, in that Chad uses them to illustrate the general evaporation principle. The narrative structure is subservient to the ideas that Chad is promoting (fog on a cold surface, evaporation from a non porous surface). They have a 'for instance … ' quality about them. This narrative control allows Chad to extend the concept of evaporation to new situations in a coherent way, and perhaps this is the essence of understanding. Chad seems concerned to construct a mental model which explains in some detail the pathways of the evaporated water. In grade 2 he responded confidently to the question of whether the evaporated water might still be in the room rather than the clouds, and a year later anticipated this question and incorporated it into his account. He seems to have a more confident grasp than Amy of the role of air as a carrier of the evaporated water, and indeed a much more confident grasp of the material nature of phenomena.

These insights into children's developing understandings differ in a number of respects from standard accounts of 'conceptual progression' (eg. Bar & Galili, 1994) in understanding evaporation. Firstly, the assumption that moving from a view of displacement of water 'into surfaces' to one of 'upwards' displacement represents a developmental shift is contradicted by the layeredness and contextual dependence of the younger Amy and Chad's notions and their ability to play with multiple ideas (see also Tytler, 2000). Secondly, while the movement to a conception of evaporating water dispersing into the air does indeed represent a significant advance in understanding, it's attainment is very context dependent, and once attained the understandings surrounding it are very different. We would argue that the notion of conceptual trajectories, while useful to some degree, fails to capture major features of Chad's and Amy's growth in conceptual understanding.

By grade 2 both Chad and Amy are both beginning to acknowledge the transcendent nature of 'evaporation' as an idea. The quality of their understandings are, however, quite distinct, related to the associations they draw upon, and the way they reason with them. Woven throughout the children's interpretations we can discern some fundamental aspects of how they see themselves as learners, and their commitments and orientations to schooling. We see these as narratives through which children explore their developing subjectivities.

5.2 Children's narratives of self

5.2.1 Amy

Amy's facility with language, and concern with finding the 'right' technical word, has shown up in many interview contexts. Her display of the word 'evaporation' (she was very concerned in one earlier interview when she could not remember the word), and maintenance and embellishment of her reference to 'tiny microscopic droplets' over a full year seems to us to be part of this commitment to technical language as representative of understanding. It may be that this is part of a 'pleasing adults' narrative she constructs in class and in interviews.

Over the prep year Amy increasingly became preoccupied with friends, and gendered occupations and games. Her concern for flowers and worms seemed to us to be part of a general 'other'-oriented focus that also showed in her continual weaving of her friends into her responses. When asked if these activities remind her of anything, she invariably talks of herself or her schoolfriends, as in the evaporating handprint discussion in Grade 2:

> Amy: I spilt my whole drink on the pavement and in few minutes I came outside and it had nearly all gone … I can think of something Evelyn did at school!. It was the same thing I did, except she filled her lunch box up with water and she tipped it onto the pavement and she made a line.

Amy is very aware of her own ideas and their status. In grade 1, she was scathing about her previous, somewhat confused description of cycles in relation to puddles, and imposes a strong water cycle story.

> Amy: I think it's a very silly idea.... It dries up... the water goes underground... and it goes through the ground and it goes into the sea... and then the clouds take it up in the sky and then it rains down again.

In grade 2 she was spontaneously and surprisingly judgmental about her drawing, 2 years earlier, when asked if she remembered the puddle event.

> Amy: Mmm.... my picture was silly.... All the people were stick people.

She is surprisingly aware of her previous thoughts. In Grade 3 (June), when asked whether the damp tea towel reminded her of anything, she referred (accurately) to what she had written in prep and reviewed in Grade 1.

> Amy: Yep, in Prep when I said that the raindrops went into the ground. They fell off and they went into the ground… some of them do.

5.2.2 Chad

Chad is a child who seems to impose strong discipline over his thinking. His thoughtful responses in class and in interview, often accompanied by long pauses, and often including an 'I don't know', illustrate the seriousness of his commitment to ideas. He is very quiet in class, and does not shine in language activities. He has always seemed to enjoy talking to us. His stories relate more to his life of the mind than to the personal and social matters that occupied Amy's attention. His anecdotes refer to events outside school, and often to textual accounts from books, or expert individuals. His stories of a conceptual conversation with his father, and of sea water evaporating from a beach, supported a significant advance in his view of evaporation in different contexts. Chad seems less concerned than Amy with technical language, but rather throws his explanatory energies into weaving a coherent account of events. He is much less interested than Amy in his previous explanations, and more concerned with explicating his current understandings.

6. CONCLUSION: CHILDREN LEARNING SCIENCE

Amy and Chad are only two of the children we are following. The fourteen children weave themselves into very individual narratives as seekers of knowledge, as strongly private thinkers, as learners of technical terms, or as seekers after technical insights. When these children were asked if particular phenomena reminded them of anything, they invariably recounted a story that contained narrative elements of personal involvement, sometimes social commitments and concerns, but most often recognisable aspects of their identities as we have come to understand them. We view these stories as much more than noise against the backdrop of a definable conceptual position. They are part of their understandings of the phenomena, and a fundamental aspect of knowing. They are related, we believe, to the personal contexts identified by Hellden (in press), and the personal framing of idioms described by Holgersson (2001). Science understanding cannot be described simply in abstracted conceptual terms, but is rooted in context, and related to episodic memory (White, 1988), which is itself related to children's construction of identity. The task for longitudinal researchers remains; to tease out the nature and interconnectedness of these elements.

While it would be foolish to extract a definitive statement about teaching from this account of two children's journeys of understanding, the study overall is moving to a position concerning how these findings relate to effective teaching and learning in practice. Even from the evidence presented in this paper the following implications can be drawn: that assessment of children's understandings needs to be more layered than simply making judgments about the achievement of particular outcome levels; that probing a child's prior understandings should involve not simply a notional conceptual 'position' but also their approach to using their understandings to construct explanations; that if understanding can only be adequately described with reference to multiple contexts, then teaching sequences should involve children responding to a range of contexts; and that if children's understandings are bound up with their 'narratives of self' then an important part of teaching must involve responding to these, as we suspect happens intuitively with effective teachers.

This study has shown the complexity of the relationship between children's science ideas and their ways of reasoning, their wider lives, and their sense of themselves. We would argue the longitudinal study provides a metaphor for learning that is more powerful for teachers than the developmentalist perspectives promoted by cohort studies. The notion of children's narratives of self could provide insights for teachers into the variations in children's learning orientations, that cut more deeply than ideas of intelligence, or learning styles.

REFERENCES

Bar, V. & Galili, I., (1994). Stages of children's views about evaporation. *International Journal of Science Education*, 16(2), 157-174.

Black, P. & Lucas, A., (Eds.) (1993). *Children's informal ideas in science*. (pp. 209-229). London, Routledge.

Driver, R., Leach, J., Scott, P. & Wood-Robinson, C., (1994). Young people's understanding of science concepts: implications of cross-age studies for curriculum planning. *Studies in Science Education*, 24, 75-100.

Engel Clough, E., Driver, R. & Wood-Robinson, C., (1987). How do children's scientific ideas change over time? *School Science Review*, 69(247), 255-267.

Hellden, G., (in press). Longitudinal studies - providing insight into individual themes in science learning and students' views of their own learning. In D. Psillos, P. Kariotoglou, V. Tselfes, G. Fassoulopoulos, E. Hatzikraniotis, M. Kallery (Eds.), *Science Education Research in the Knowledge Based Society*. Dordrecht, The Netherlands. Kluwer Academic Publishers.

Holgersson, I., (2001). Young children and molecules - examples from a longitudinal study on children's views of matter. *Proceedings of the Third International Conference on Science Education Research in the Knowledge Based Society* (pp. 89-91). Thessaloniki: European Science Education Research Association.

Osborne, R. & Cosgrove, M. (1983). Children's conceptions of the changes of state of water. *Journal of Research in Science Teaching,* 20(9), 825-838.

Russell, T., Harlen, W. and Watt, D. (1989). Children's ideas about evaporation. *International Journal of Science Education,* 11, 566-576.

Tytler, R. (1998a). The nature of students' informal science conceptions. *International Journal of Science Education*. 20(8), 901-927.

Tytler, R., (1998b). Children's conceptions of air pressure: Exploring the nature of conceptual change. *International Journal of Science Education*, 20(8), 929-958.

Tytler, R., (2000). A comparison of Year 1 and Year 6 students' conceptions of evaporation and condensation: dimensions of conceptual progression. *International Journal of Science Education*, 22(5), 447-467.

Tytler, R. & Peterson, S., (2000). Deconstructing learning in science: Young children's responses to a classroom sequence on evaporation. *Research in Science Education*, 30(4), 339-355.

White, R.,T., (2001). The revolution in research on science teaching. In: V. Richardson (Ed.) *Handbook of Research on Teaching*. New York: Macmillan.

White, R.,T., (1988). *Learning Science*. Oxford: Blackwell.

HOCS PROBLEM SOLVING VS. LOCS EXERCISE SOLVING: WHAT DO COLLEGE SCIENCE STUDENTS PREFER?

Uri Zoller

Faculty of Science and Science Education – Chemistry, Haifa University – Oranim

ABSTRACT

The acquisition of higher-order cognitive skills (HOCS) such as problem (not exercise) solving, decision making and critical thinking, by our students, constitutes a major instructional goal in contemporary science teaching. The driving force for this research has been to provide supporting research-based evidence that will encourage the use of HOCS-promoting science teaching strategies and examinations by science teachers at all levels. The main findings of two of our relevant studies were: (1) college science students consistently perform best on algorithmic (ALG), compared to their performance on lower order cognitive skills (LOCS) which, in turn, is higher than on HOCS exam questions; (2) given a free choice, LOCS questions are pragmatically preferred on HOCS questions by the students; (3) students' "ideological" preference is HOCS, compared to ALG/LOCS which are referred to as 'computational questions'. The implications concerning the current HOCS-oriented reform in science education will be critically discussed.

1. BACKGROUND/RATIONALE

The acquisition of higher-order cognitive skills (HOCS) by our students, beyond the so-far dominating lower-order cognitive skills (LOCS), is a major instructional goal in contemporary science teaching. The HOCS capabilities such as question asking (Zoller, 1987), problem solving (Gabel & Bunce, 1994), decision making (Scholz & Zimmer, 1997), and critical thinking (Ennis, 1989) – all of which require evaluative thinking (Zoller, 1999) are considered by many to be the most important learning outcomes to which good teaching should aim. These skills constitute the core of the assembly of performance processes needed for coping with previously unencountered complex problem situations, conceptual understanding, transfer (Solomon & Perkins, 1989), and system (inclusive) thinking with respect to both the science disciplines and real-life problems (Zoller, 1990; Scholz & Tietje, 2002).

Two major convictions are driving current efforts to reform science education: (1) it is vital for our students to develop their HOCS capability; and (2) students should construct a deep conceptual understanding of any scientific topic they study rather than simply learn to apply algorithms to problem sets (Rutherford & Ahlgren, 1990; Zoller, 1993; Zoller, Lubezky, Nakhleh, Tessier & Dori, 1995). Both objectives should be targeted by teachers and students alike as partners. Any progress towards their attainment would require the application of new teaching

D. Psillos et al. (eds.), Science Education Research in the Knowledge-Based Society, 201–207

methods that would mesh with these desired learning outcomes. Examinations as well as other assessment means must also be consonant with these goals (Zoller, 1993; 2001). Therefore, the development of (a) HOCS-promoting examinations; (b) innovative assessment procedures of performance on HOCS questions; and (c) means for evaluating the effectiveness of HOCS-oriented courses and teaching strategies will facilitate the acquisition of HOCS by all students (Zoller, 1999; 2001; Zoller, Dori & Lubezky, 2002).

In the above context it should be emphasized that HOCS extend far beyond the skills of analysis, synthesis and evaluation *per se*. In addition to these and the HOCS of critical system (lateral) thinking, decision making, problem solving, evaluative thinking and more, they also include the synergistic interactions/ integration of these skills (in varying proportions) in different situations and contexts. While Bloom taxonomy is, in essence, an hierarchical 'level-reference' model, focusing traditionally on disciplinary "*knowing*, understanding, application by students, of known algorithms in solving exercises [algorithmic (ALG)-type questions], etc." (LOCS), the "HOCS teaching and learning" is focusing on interdisciplinary *transfer* capability. In short: passive "knowing" vs. active evaluative thinking.

The focus of the research here reported was three-fold: (1) college science students' performance on ALG/LOCS vs. HOCS chemistry exam questions; (2) students' attitudes to ALG-LOCS vs. HOCS-type exam questions; and (3) these students' actual preferences – of LOCS vs. HOCS questions in real (chemistry) exam situations. All of the above within traditional chemistry teaching.

2. AIMS

1. To assess college science students' performance on ALG/LOCS vs. HOCS chemistry exam questions.
2. To assess students' attitudes towards each of these categories of exam questions in chemistry and in the science disciplines in general
3. To identify the science students actual LOCS vs. HOCS preference under real exam situations.

3. METHODS & RESEARCH POPULATION

The following six chemistry examination questions were constructed and categorized as ALG(A), LOCS(L), HOCS(H) and mixed type, by 6 expert science/chemical educators.

Q1. Calculate the maximum weight of SO_3 that could be obtained from 1.9 moles of oxygen and an excess of sulphur in the reaction $2S+3O_2 \rightarrow 2SO_3$.

Q2. Potassium, vanadium, and iron crystallize in a body-centered cubic unit cell. Given the lengths of the unit cell edges and the atomic weights listed below, which of these elements has the highest density?

Potassium: a = 5.250A   Vanadium: a = 3.024 A   Iron; a = 2.861 A

AW = 39.098 AW = 50.942 AW = 55.847

Q3. The atomic number of the element magnesium is 12 and its molar mass is 24.3 g/mole. The mass numbers of its three natural isotopes are 24, 25 and 26. Which of the following statements is incorrect? (Circle the appropriate letters).

a. The three isotopes have the same chemical properties.

b. The three isotopes have the same nuclear charge.

c.* The mass number of the most abundant isotope is 26.

d. A portion of the nuclei of magnesium atoms contain 14 neutrons.

e.* All elements have two or more natural isotopes.

(* denotes correct item response)

Q4. Is PH_3 or BH_3 a base or an acid? Explain and defend your answer.

Q5. Ionization potential refers to the energy required to remove an electron form an atom. The first ionization potential refers to the energy required to remove the first electron, the second potential refers to the removal of the second electron, etc. Which of the following two would you expect to have a higher ionization potential: a sulphur atom (S) or a phosphorus atom (P)? Explain.

Q6. Adding an electron to an oxygen atom is a reaction which is associated with emission of energy. Adding a second electron to the resulting O^- ion is associated with energy absorption. What is your explanation to these phenomena?

The solving of both ALG and/or LOCS questions required just the application of memorized algorithms and/or low level of understanding. The response to HOCS questions required the dealing with unfamiliar situations and ill-defined problems and/or analysis, synthesis and problem-solving capabilities as well as to identify/make connections and think critically (Zoller, 1993; Zoller & Tsaparlis, 1997; Zoller et al., 2002). This six-question exam was administered as a mid-term general chemistry exam to 64 freshmen - biology (70%) and physics mathematics (30%) majors at a liberal art-oriented university. The corresponding students' responses were scored on a 0-100-point scale for each question in accordance with pre-established criteria.

A stratified sample (N=14) of the student research population was later interviewed. A mid-term take-home examination, within a general chemistry course for a different group of freshman biology majors (N=22) was administered. These students were asked to choose just two questions out of the ten A, L, H, or mixed (M: A/L/H, etc.; i.e., questions containing algorithmic and/or LOCS-and/or HOCS-type components simultaneously) included, to work them out at home, and to submit their 'final product' to their professor – for grading.

4. RESULTS

In response to the first aim of this research, the target population's LOCS and HOCS performances were assessed and averaged, for each of the three categories, to facilitate delineation and comparison.

The summary of the students' grades on the ALG, LOCS and HOCS questions of the chemistry exam is given in Table 1. The highest scores were on the ALG questions, significantly different from their average score on the other LOCS-type questions, suggesting the high proficiency of science students on applying algorithms to solve exercises. Furthermore, the particularly high performance of the students on the ALG questions points to the centrality of algorithmic teaching and assessment practices in contemporary chemistry (and science) education. The lowest scores were on the HOCS questions, while those on the LOCS questions fell between the two extremes. Thus, students performed consistently on each of the three categories in the order ALG>LCOS>HOCS.

Table 1. Means of students' performance on examination questions by question type (Zoller et al., 2002)

Question Type		ALG		LOCS		HOCS	
University	N	Q1	Q2	Q3	Q4	Q5	Q6
Haifa	64	87.2 (SD=20)		68.2 (SD=32)		55.9 (SD=29)	

The above results are in agreement with the conclusions of our previous studies (Zoller et al., 1995) and those of others (Pickering, 1990; Nakhleh, 1993; Niaz, 1995) with respect to algorithmic and *conceptual* questions, which are not necessarily 'HOCS questions'. Significantly, however, they not only corroborate the previous findings, but also extend their scope form the conceptual category – to the HOCS category level. Thus, our findings suggest that: (a) success in algorithmic exam questions does not imply success on HOCS questions; (b) success in algorithmic questions does not imply success even on LOCS questions; and (c) HOCS questions are the most difficult for students to cope with (Zoller et al., 2002). Yet, although 'HOCS learning' is our goal in science teaching, the algorithmic/ LOCS-type examinations serve as the major (inclusive, in most cases) assessment tool of *learning* in science.

Guided by the research questions the interviewees were asked – among others – to respond to the following questions:

- Which type of the exam questions (e.g., Q_1, Q_4, Q_6) do you prefer?
- What, in your opinion, are the main difficulties that you have confronted in your dealing/coping with them?

The interviewees were not asked, however, to explain or rationalize, any differences between their declared preference and actual choice.

Selected students' responses are given below. They represent typical responses, not necessarily in accordance with the "quantitative" weighted preferences of the interviewer.

S1: 'I like questions of the type of Q3 which requires much deeper understanding than Q1… It still does not deal with numbers but it requires to 'introduce' into the answer a lot of data which are often associated with different aspects of the same problem…It is not just to look at a certain aspect like in Q1 and then, according to intelligence, to decide. At any rate, when I study for a test I do not simply learn to use formulas and numbers, 'cause I like in a test to sit and look at what I have learnt...since part of the test is to review what I have achieved in the study and what I know. This, in fact, what is important; what will remain later…a question of type Q3 requires *thinking* and it is much more difficult to explain…'

S2: '…This is a tough question…both…actually… [I prefer] *computational* questions because I am used to this style of questions…the other type (referring here to the HOCS Q5) is difficult for me…'cause I need to analyze and actually search for reasons…I have difficulties in how to approach the question, how to *think* about it and what should be looked for…. With the first type (referring to Q1), I don't have a problem since it is computations and in general it is easier for me…'

S5: '…Computational questions (referring here to the ALG-type questions) definitely not, because such questions do not testify for anything…. I prefer Q6-type questions (referring here to the ALG-type questions). They are much easier for me because I can express myself and explain to myself what I think…'

S11: 'This is a tough question…I don't know…sometimes I prefer *theoretical* questions (referring here to the HOCS' type questions) when I know that I understand the material well…It depends on the material…there are computational things that I know and theories also; namely, if I'll know the theory I will know to answer computational questions. My difficulty with thc computational questions is that I think that may be sometimes I do not *remember* the way of computation…'

S12: '…The easiest for me are questions that I can find out if my solution is correct…then I also feel safer. There is no difference if these are computational or thinking questions…the main thing is that I would know if my answer is correct…The truth is that in order to achieve a good grade it would be good for me to have computational questions so that everything then is simple and…that's it…but I think that we should have of both types, so that the evaluation would be fair and also that it will be balanced…'

Thus, the essence of the interviewed students' responses concerning their exam questions – type preference was, that they appreciate and have favorable attitude towards HOCS questions. In practice, however, the majority (about two thirds of the interviewees) would prefer ALG and LOCS questions. The students' main criteria for this pragmatic preference is the level of difficulty, the HOCS-type questions considered to be the most difficult to "solve".

It is *noteworthy*, that the highest scored students indeed selected, in the take-home exam, only algorithmic and LOCS questions to respond to. Apparently, the drive to succeed on exams, at all cost, is the decisive factor, which determines the students' pragmatic preference (Table 2). The scores and the LOCS/HOCS distribution of the low achievers are not provided since they have no direct bearing on the issues here studied and discussed.

Table 2. Students' performance (80-100 score range) versus the LOCS/HOCS distribution of their selected take-home exam questions (N=22; F=16, M=6)

Score	Number of students		(%)	Type of Questions	Level
96-100	4		(18)	A&L	LOCS
91-95	6	2	(9)	A&L	LOCS
		2	(9)	A&A/H	LOCS & MOCS[2]
		2	(9)		LOCS & MOCS
80-90	4	2	(9)	A&A/L/H or A/H	LOCS & MOCS
		2	(9)	A/L/H & A/H or L/H	MOCS

[2] MOCS: Mixed LOCS and HOCS questions

5. CONCLUSIONS AND IMPLICATIONS

Our research findings – (1) college science students perform, consistently, on each of the exam questions, in the order of ALG>LOCS>HOCS; and (2) although students' "ideological" preference is HOCS>ALG/LOCS, their pragmatic preference under real exam situations is the reverse; i.e., preferring exercise (not problem) solving. Apparently, this "pragmatic preference" reflects the students' wish to 'guarantee' a high grade with minimum effort (... and aspirations). Our findings suggest that traditional instructional strategies of science (chemistry) teaching and even more so – assessment, are not compatible with the development and fostering of students' HOCS; i.e., promoting of HOCS learning. Our study supports the effort being made worldwide to ensure the LOCS-to-HOCS paradigm shift in science education at all levels. It is time to move science instruction from an algorithmic and factual recall orientation dominated by ALG/LOCS to a decision-making, problem-solving and critical thinking approach, dominated by the HOCS orientation.

SELECTED REFERENCES

Ennis, R.H. (1989). Critical thinking and subject specificity: Clarification and needed research. *Educational Researcher*, 18(3), 4-10.

Gabel, D.L. & Bunce, D.M. (1994). Research on problem solving: Chemistry. In D.L. Gabel et al. (eds.). *Handbook of Research on Science Teaching and Learning*. New York: Macmillan Publishing, pp. 301-326.

Nakhleh, M.B. (1993). Are our students conceptual thinkers or algorithmic problem solvers? *Journal of Chemical Education, 70*(1), 52-55.

Niaz, M. (1995). Relationships between student performance on conceptual and computational problems of chemical equilibrium. *International Journal of Science Education, 17*, 343-355.

Pickering, M. (1990). Further studies on concept learning versus problem solving. *Journal of Chemical Education, 67*, 254-255.

Rutherford, F.J. & Ahlgren, A. (1990). *Science for all Americans*. New York: Oxford University Press

Scholz, R.W. & Tietje, O. (2002). *Embedded case study methods: Integrating quantitative and qualitative knowledge.* Thousand Oaks, Sage.

Scholz, R.W. & Zimmer, A.C. (eds.) (1997). *Qualitative aspects of decision making*. Lengerich. Pubst. Science Publisher.

Solomon, G. & Perkins, D. (1989). Rocky roads to transfer: Rethinking mechanisms of a neglected phenomenon. *Educational Psychologist, 24*, 113-142.

Zoller, U. (1987). The fostering of question-asking capability – A meaningful aspect of problem solving in chemistry. *Journal of Chemical Education, 64,* 510-512.

Zoller, U. (1990). Environmental education at the university: The "Problem solving-decision making act" within a critical system-thinking framework. *Higher Education in Europe, 15*, 5-15.

Zoller, U. (1993). Are lecturing and learning compatible? Maybe for LCOS: Unlikely for HOCS. *Journal of Chemical Education, 7*(3), 195-197.

Zoller, U. (1999). Teaching tomorrow's college science courses – Are we getting it right? *Journal of College Science Teaching, 29*(6), 409-414.

Zoller, U. (2001). Alternative assessment as (critical) means of facilitating HOCS- promoting teaching and learning in chemistry education. *Chemical Education Research and Practice in Europe, 2,* 9-17 [http://www.uoi.gr/conf-sem/cerapie].

Zoller, U., Lebezky, A., Nakhleh, M.B., Tessier, B. & Dori, J. (1995). Success on algorithmic and LOCS vs. Conceptual chemistry exam questions. *Journal of Chemical Education, 72*, 987-989.

Zoller, U., & Tsaparlis, G. (1997). Higher and lower cognitive skills: the case of chemistry. *Research in Science Education, 27*(1), 117-130.

Zoller, U., Dori, Y.J., & Lubezky, A. (2002). Algorithmic, LCOS and HOCS (chemistry) exam question: Performance and attitudes of college students. *International Journal of Science Education*, 24(2), 185-203.

PART III

Teaching and Communicating Science

SCIENCE AND TECHNOLOGY EDUCATION: A HIGH PRIORITY POLITICAL CONCERN IN EUROPE

Invited Key Note Address Esera, Thessaloniki, Aug 2001

Svein Sjøberg

University of Oslo, Norway

ABSTRACT[i]

This paper attempts to describe and analyse some challenges facing S&T education by relating these issues to a wider social setting. The paper suggests a series of underlying reasons for why we have problems with recruitment and public image. This is followed by a similar analysis of who needs S&T – and for what purposes. The point here is that the challenge may be perceived differently from different perspectives and interests. Hence, there may also be different views on suitable strategies to solve 'the problem.' A critical description of S&T in *schools* is offered, and some international *recent trends* are briefly described. These trends may provide ideas for possible ways ahead and may underlie discussions of research priorities for associations like ESERA.

1. THE CHALLENGE

Our societies are dominated and even 'driven' by ideas and products from science and technology (S&T). The workplace and the public sphere are increasingly dependent on new as well as the more established technologies. Knowledge and skills in S&T are crucial for most of our actions and decisions, as workers, as voters, as consumers etc. Meaningful and independent participation in modern democracies assumes the ability to judge evidence and arguments in the many socio-scientific issues that are on the political agenda. One might expect that the increasing significance of S&T should be accompanied with a parallel growth in the interest in these subjects as well as increasing understanding of basic scientific ideas and ways of thinking. This does, however, not seem to be the case.

In many countries, the *recruitment* to S&T studies is falling – or at least not developing as fast as expected or planned for. This lack of interest in science often manifests itself at school level at the age where curricular choices are made. In many countries there is noticeable decrease in the numbers of students choosing (some of) the sciences. This trend is further enlarged in the enrolment to tertiary education. A similar trend occurs in some areas of engineering and technology studies. It should, however, be noted that there are large (and interesting!) differences between the various countries and between the different areas of S&T.

D. Psillos et al. (eds.), Science Education Research in the Knowledge-Based Society, 211–220.

Statistical data and most surveys do, however, not shed much light on the underlying causes. *Why* has S&T apparently lost its attraction among young people? Unless one has some ideas about this, intervention programs to increase the interests in S&T are not likely to have success. The following points are attempts to suggest explanations, although some of claims can be backed up with research evidence.

2. DISENCHANTMENT WITH S&T? 13 POSSIBLE REASONS....

Reasons for the doubt in and dissatisfaction with S&T have to be found in the youth culture and in society at large. The decline in recruitment must be understood as a social and political phenomenon that occurs in many (but not all!) highly industrialized countries. This means that the current situation can hardly be explained fully by events or reforms in each individual country. One should seek for more general and common trends found in different countries. The following is an attempt to suggest underlying reasons for the present situation. The first point refers to schools, the other are related to wider social trends.

Outdated curriculum. Many studies show that pupils perceive school science as lacking relevance. It is often described as dull, authoritarian, abstract and theoretical. The curriculum is often overcrowded with unfamiliar concepts and laws. It leaves little room for enjoyment, curiosity and a search for meaning. It often lacks a cultural, social and historical dimension, and it seldom treats the contemporary issues (see later paragraph)

Science: Difficult and 'untrendy'? Scientific knowledge is by nature abstract and theoretical. It also often contradicts 'common sense'. It is also often developed through controlled experiments in artificial and 'unnatural' and idealized laboratory settings. Learning science often requires hard work and intellectual efforts (although school science could and should be tailored to better meet the needs and abilities of the pupils!) Concentration and hard work is not part of present youth culture. In a world where so many 'channels' compete about the attention of young people, such subjects become untrendy.

Lack of qualified teachers S&T are often poorly treated in teacher preparation for the early years. Moreover, the students who choose to become primary school teachers are often those who did not take or did not like science themselves in school. The present decline in recruitment of science teachers is now being felt also in secondary schools.

Anti- and quasi-scientific trends and 'alternatives'. In many Western countries there is an upsurge of 'alternative' beliefs in the metaphysical, spiritual and supernatural. These movements are often labelled 'New Age', and comprises a rich variety of world-views, therapies etc. They include beliefs in UFOs, astrology, several forms of healing. A common denominator is often the rejection of scientific rationality, which they often characterize as being mechanistic, reductionist etc. Although most 'alternatives' reject science, some do, however, also base their ideas on misinterpretations of ideas taken from modern science, like quantum mechanics.

Postmodernist attacks on S&T This may be seen as the more 'serious' and academic version of the critique imbedded in the above mentioned 'alternative' movements. Many postmodernist thinkers reject basic elements of science, and reject notions like objectivity and rationality. The more extreme versions assert that scientific knowledge claims say more about the researcher that about reality, and that all other 'stories' about the world have the same epistemological status. In this tradition, notions like 'reality', 'truth' etc are seldom used without inverted commas! These postmodernists attacks on scientific thinking have even been called "Science War" in the US, and book titles like "The flight from science and reason" and "Higher superstition" indicate the tone of the 'debate' and how these trends have been met by the scientific community.

Stereotypical image of scientists and engineers. Many research projects indicate that the perceived image of the typical scientist and engineer is stereotypical and problematic. The image of the 'crazy scientist' is widespread, possibly supported by cartoons, plots in many popular movies and in media coverage. Scientists (especially in the hard, physical sciences) are by pupils often perceived to be authoritarian, closed, boring – and somewhat crazy. They are *not* perceived to be kind or helping and working to solve the problems of humankind.

Disagreement among researchers perceived as problematic. Scientists debate and disagree on many contemporary socio-scientific issues (like causes of global warming, effects of radiation, possible dangers of GM food etc.) Such discussions are the normal processes for the healthy development of new scientific knowledge. Recently, such debates are also taken to the mass media and are not (as before) confined to professional conferences and journals. The disagreement in public may, however, confuse and disappoint people who are acquainted with 'school science', where scientific knowledge is presented as certain and uncontroversial.

Problematic values and ethos of science The traditional values of science are meant to safeguard objectivity, neutrality, disinterestedness and rationality. Taken to the extreme, however, these values may seem to justify absence of ethics, empathy and concern for the social implications of science. The search for universal laws and theories may lead to an implied image of science as abstract and not related to human needs. For many people, science is cold and lacks a human face.

Dislike of an over ambitious science? The achievements of science may call for admiration, but also unease. Many people dislike the image and ambitions of modern biotechnology. They have emotional and rational fear about scientists who are 'tampering with Nature', and 'Playing God'. Similarly, many people react emotionally when physicists talk about their quest for 'The Final Theory', also called 'The Theory of Everything' or the search for 'The God Particle' (the title of a book by Nobel laureate Leon Lederman). Such perspectives may attract some young people, but it is not unlikely that the ambitions of modern science will scare others. Many people feel that science intruding areas that they consider sacred – and they do not want a world where science can explain everything. Many people like to think of Nature as sacred and mystical – not as explainable, controllable and rational. Their avoidance of science is a deliberate choice of values – not something that may be 'corrected' by more information from the scientists!

The new image: Big Science and techno-science Science used to be seen a search for knowledge driven by individual curiosity. Scientists have historically rightfully been described as radicals and revolutionaries who often challenged religious and

political authorities.
Present science is different in fundamental ways. We have in the last decades seen a fusion between science and technology into what is called techno-science and Big science (NASA, CERN, The Human Genome Project etc.) The scientists and engineers of today often work close to national, industrial or military interests. The earlier image of scientists being dissidents and rebels has been replaced with a less exotic image of scientists being loyal workers in the service of power and authority. Scientists and other 'experts' are often on the pay-roll of industry, military or the State. Hence, their role as neutral defenders of objectivity and truth is questioned by the media, by many scholars – and also pupils in schools.

Scientists and engineers: No longer heroes? Not very long ago, scientists and engineers were considered heroes. The scientists produced progressive knowledge and fought superstition and ignorance, the engineers developed new technologies and products that improved the quality of life.

This image is, however, 'history' by now, at least in the more developed countries. For many young people in rich, modern societies, the fight for better health and a better material standard is an unknown history of the past. They take the present way of living for granted, not something that has been developed by S&T. They do not see the fruits of S&T, but are more able to see the present evils of environmental degradation, pollution, global warming etc. Forgetting the victories of the past, many put the blame of the current problems on S&T.

The new role models: Not in S&T We live in a world that is in part created by the media. Football players, movie stars and pop artists are exposed and earn fortunes. The lives of journalists and other media people seem interesting and challenging. Although rather few young people enter these careers, the new role models on either side of the camera create new ideals. The young people also know that lawyers and people at the stock exchange earn ten or hundred times more money than the physicist in the laboratory. They also know that lack of physics or mathematics knowledge is no hindrance to such careers.

A white-coated hardworking (and not very well paid) scientist in a lab is not the role model of young people of today! This social climate does not create an atmosphere where it is easy to convince young people that they should concentrate on their science learning!

Communication gap between scientists and the 'public'? The S&T establishment is often confused and annoyed when met with critique. In the past, they have enjoyed enormous popularity, increasing budgets and excellent recruitment. They are not used to face distrust, and they have not been in need to justify their research in public debates. The immediate reaction to the new situation is the search for scapegoats, often found in the schools and in the media.

The problematic situation is often seen by the S&T establishment as a problem of information. Critique and scepticism are often interpreted as based on 'misunderstandings' and lack of knowledge from the public. In some cases this may of course be justified, but the new situation does also call for a form for self-critique within the S&T community. Communication works 'both ways' and a lack of mutual understanding cannot be blamed on one part only.

At least some of the points above may be valid explanations for the disenchantment with S&T. The importance of each point will, of course, also vary between countries. Some of these points may be actively addressed, others are more deep-rooted and outside the direct influence of political decisions.

3. WHO NEEDS SCIENCE AND TECHNOLOGY – AND WHY?

The problematic situation for S&T can be seen from different perspectives and different interests. These range from industry's concern about national, economical competitiveness to a concern about empowerment at the grassroots level for the protection and conservation of nature. Different conceptions of 'the crisis' may possibly lead to different solutions. Here is an indication of possible stakeholders and their arguments for learning S&T.

- **Industry** needs people with high qualifications in S&T. Modern industry is high-tech and often called 'knowledge industry'. This industry is in need for highly qualified scientists and engineers to survive in a competitive global economy. This aspect is of importance for the economy of the nation.
  But one should keep in mind that young people do *not* base their educational choices on what is good for the nation!
- ***Universities and research institution*** have similarly a need for researchers (and teachers) to maintain research at high international level and to provide good learning possibilities for coming generations of experts, researchers and teachers.

The above-mentioned two groups constitute a highly skilled elite. But the actual number of such people may not necessarily be very high. It would also be a mistake to have mainly these groups in mind when reforming S&T in *schools*. A policy based on this perspective could even further decrease the proportion of young people who find S&T interesting, and who would choose to continue with S&T. The next perspective is one of high importance for a much larger group, the teaching profession:

Schools need qualified teachers in S&T. The decline in recruitment has already hit the teaching profession. Well-qualified and enthusiastic teachers constitute the key to any improvement of S&T in schools -- and for the further development of knowledge, interests and attitudes of ordinary citizens when they have left school. The S&T teachers also play a key role in the recruitment of people to the S&T sector.

The long-term effects of a lack of good S&T teachers could be very damaging, although the effects are not so immediately observable as the lack of qualified people in industry and research. The S&T teachers need a broad basis for their activities. A solid foundation in the academic discipline is important, but not enough. They need broader perspectives and skills in order to cope with challenges of the sort outlined earlier in this paper. In short: S&T teachers do not only need a foundation *in* S&T, they also need to have perspectives *on* S&T in a historical and social context. This may require reforms in teacher training.

The next points, although different, are of importance for more or less *all* citizens.

- **A broader labour market needs S&T competencies** People in general need qualifications in S&T to compete on the modern labour market. The need is great and growing fast, as knowledge and skills based on science and technology become prerequisites in new areas and new parts of the labour market. Not only doctors, pharmacists, engineers and technicians need S&T. Health workers handle complicated and dangerous equipment, secretaries and office staff need good computer literacy etc. New as well as more traditional technologies often dominate the workplaces, and those with skills in these areas may have a competitive advantage for their further career. Many countries have also identified a need for people with S&T skills to replace those retiring in the near future.
  There is also a general need to become flexible and able to learn. A foundation in S&T as well as mathematics is of great importance to develop such learning skills. Besides, most of the changes are likely to be related to technological innovations, and people with basic S&T skills may be better equipped to cope with changes and innovations.
- ***S&T for citizenship and democratic participation:*** As stated in the introduction, our modern society is dominated by S&T. Many aspects of life have a dimension related to S&T. All citizens are confronted with such issues as consumers and as voters. As *consumers* we have to take decisions about food and health, quality and characteristics of products, claims made in advertisements etc. As *voters* we have to take a stand and be able to judge arguments on all sorts of issues. Many of these political issues also have an S&T dimension. In such cases, knowledge of the S&T involved has to be combined with values and political ideals. Issues relating to the environment are obviously of this nature, but also issues relating to energy, traffic, health policy etc. have S&T dimensions. It is indeed hard to think of any contemporary issue that does not have some aspects relating to S&T.
  Social and political issues should not be seen as 'technical' – and left in the hands of the 'expert'. A broad Public understanding of science and technology may in fact be a democratic safeguard against 'scientism' and a domination of experts.
  The above 'democratic argument' does not only assume that people have some grasp of the *contents* of S&T. It also requires some public understanding of the *nature* of S&T and the role they play in society. Among other things, people need to know that scientific knowledge is based on argumentation and evidence, and that statistical considerations about risks play an important role.
  Everybody cannot become 'experts', but everybody should have tools to be able to judge which 'expert' and what kind of arguments one should trust.

The points raised here call for some caution, since some of the above aims may possibly compete. The concern for recruitment of possible Nobel Prize winners and researchers at CERN may not coincide with the concern for a broad public understanding of science – or for the protection of wildlife and natural resources.

4. PRESENT CURRICULA – THE CRITIQUE

Science curricula are key factors to sustain and develop the interest in science. Science education in most countries is criticized for being traditional and rather old-fashioned. The implicit image of science is that it is mainly a massive body of authoritative and unquestionable knowledge. Most curricula and textbooks are overloaded with facts and information at the expense of concentration on a few 'big ideas' and key principles. Many new words and 'exotic' concepts are introduced on every page of most textbooks. Although very few pupils will pursue further studies in science, preparation for such studies seems to be a guiding curriculum principle. There is often repetition, where the same concepts and laws are presented year after year. Such curricula and textbooks often lead to rote learning without deeper understanding.

Moreover, this textbook science is often criticized for its lack *relevance* and deeper *meaning* for the learners and their daily life. The content is often presented without being anchored to social and human needs, neither present nor past. The historical context of discoveries is often reduced to biographical anecdotes. Moreover, the implicit philosophy of textbook science is by most scholars considered a simplistic and outdated form of empiricism.

It should also be noted (as in point 2 in the previous listing) that science often is seen as *demanding* and *difficult*. The ideas are not always easy to grasp, and their understanding often requires concentration and hard work over a long period of time. The youth of today are not used to cope with such demands. If one shall hope for such efforts, the pupils will need a strong motivation, and they need to find something that is seen as very valuable. This is not always the case. Although science *per se* can be seen as difficult, the demands of *school science* may of course be adopted to suit the age of the learners!

This science curriculum has to 'compete' for popularity and attention with other school subjects. Many of these subjects have qualities that meet the students' needs for *meaning* and *relevance*. The content of such subjects is less authoritarian, and there is a place for opinions and feelings of the learners. This is seldom the case in school science.

5. SCIENCE AND TECHNOLOGY IN SCHOOLS – RECENT TRENDS AND RESPONSES

The challenges for S&T education outlined above have been met in different ways. Many countries have introduced more or less radical reforms, and there has been support to curriculum development and experiments.

A general trend is that there seems to be less influence from the (traditional) academic organization of curricula and contents. An underlying concern is that S&T should contribute to more general aims of schooling in a situation where 'everybody' attends school for 12-13 years. The general tendency is a widening of the perspective and a gradual redefinition of what counts as valid school science. Social and ethical aspects of S&T are often becoming part of the curriculum. The following is a listing of some trends. Many are related, but still mentioned separately. Not all these trends are found in all countries, but together they represent a series of ***identifiable tendencies***:

- **Towards "Science for all"** More weight on aspects of science that can be seen to contribute to the overall goals of schooling. Key concern: liberal education ('allmenn dannelse', 'allmänn Bildning' Bildung, Formation.....) Hence; there is less weight on traditional academic contents and science as mainly as preparation for tertiary studies in science. Specialisation is postponed to the last few years of school.
- **Towards more subject integration.** In the early years of schooling, S&T is usually more or less integrated with other school subjects. Only later are the sciences presented as separate disciplines. The level where this specialization starts varies between countries. It is a general trend that separate science subjects are taught only at a late stage. (e.g. in Norway, only the two last years of upper secondary school have single science subject.)
- **Widening perspectives** More weight on cultural, historical and philosophical aspects of science and technology. S&T are presented as human activities. These aspects may also appeal to the pupils that are in search for 'meaning', not only factual information and the accepted correct explanations.
- **NOS: The Nature of Science** The 'Nature of science' has become an important concern in the curriculum. This often means a rejection of the often stereotypical (and false) image of science as a simple search for objective and final truths based on unproblematic observations. The weight on recent understanding of the nature of science also implies a stress on the social, cultural and human aspects of science. Science is presented as knowledge that builds on *evidence* as well *arguments* in a creative search for meaning and explanation. This aspect also strengthens that human and social relevance of science, and may attract pupils who value such aspects.
- **Contexts become important** More weight on putting science and technology in meaningful contexts for the learner. This often implies examples from everyday life and current socio-scientific issues. These themes or topics are by their nature interdisciplinary, and require teacher cooperation. Such issues often requires methods like *project work*. (For which teachers have to be adequately educated.)
- **Concern for the environment** Towards more weight on environmental questions as part of the S&T curriculum. Environmental issues are often of the socio-scientific nature mentioned above, and their treatment often requires *project work* in *interdisciplinary* settings.
- **Weight on Technology** Technology has recently been introduced in many countries as a subject in its own right, also in the general part of the education system. In other countries, it has received a broader place within the science curriculum, not only as interesting concrete examples to illustrate scientific theories and principles. The curricular definition of 'technology' is, however, often confusing and incoherent. In some countries technology is placed in a context of 'design and technology' (as in the UK). In other countries the term technology implies modern information technology and ICT.

In some places, the stress is on the technical (and underlying scientific) aspect of technology. In other countries the weight is put on human relations to technology, society and technology etc.
The stress on technology, utility and practical examples is often used to build confidence in the children. Through technology they may experience that S&T is also about *doing* and *making things work.*

- **STS: Science, Technology and Society** STS has become an acronym for a whole 'movement' within S&T education. The key concern is not only the Science and the Technology content, but also the relationship between S&T and society. The trends described in the preceding points (relevant contexts, stress on the environmental and the role of technology) can also be seen as belonging to an increase of the STS perspective.
- **Inclusion of ethics** When S&T issues are treated in a wider context, it becomes evident that many of the topics have ethical dimensions. This is of course the case when dealing with socio-scientific issues. But ethics is also involved in discussions relating to 'pure' science, like what sorts of research one ought to prioritise (or even allow), and the moral dilemmas in e.g. using animals in research. Again, this ethical dimension may contribute to giving S&T a more human face. It is also likely to empower future voters on important political issues on which they are invited to take a stand.
- **"Less is more"** This has become a slogan for curriculum development. More weight is put on 'great stories' of S&T and on presentation of key ideas and their development, often in an historical and social context. These key ideas replace (the impossible) attempt to give an encyclopaedic coverage of the whole of science. One hopes to avoid the curse of the overcrowded curriculum that leaves so little time for reflection and search for meaning.
  By choosing 'typical' and important stories, one hopes to convey an understanding of the nature of S&T. One also hopes to nourish curiosity and respect for S&T – and to inspire some students to pursue S&T.
  'Narratives' have become a key word for this development.
- **Information technologies as subject matter and as tools**
  Information and communication technologies (ICT) are products that by their definition 'belong' to the S&T sector. (The 'hardware' is science-based technologies; the 'software' builds on basic mathematics etc.) Hence, the underlying physical and technical ideas are to an increasing extent treated as important subject matter on their own right in S&T curricula.
  Besides, ICT provide new tools that are very suitable for teaching and learning in S&T. The whole range of 'ordinary' software is used, including databases, spreadsheets, statistical and graphical programs. In addition, modelling, visualization and simulations of processes are important. ICT is also used for taking time series of measurements for a wide variety of parameters ('data logging').
  S&T subjects are likely to be key elements in strategies to develop ICT to become a better educational tool. It is also likely that S&T teachers are better

educationally equipped for this task than most other teachers – although they are also in need for ways to be updated and retrained.

6. WAYS FORWARD?

As indicated in this paper, 'the problem' has many dimensions, and different interest groups may understand and conceive the challenges in widely different terms. The perspectives of industrial leaders are often different from those of the environmental activists. It has also been argued that the problems related to the interests in and attitudes to S&T cannot only be perceived as merely *educational* challenges. They have to be understood and addressed in a wider social, cultural and political context. Hence, 'solutions' may be as different as the way in which the challenge is understood.

If one accepts that the problems of recruitment to and attitudes to S&T are embedded in a wider social context, one will also need a broader approach than only to address school reforms, curriculum reform, reforms in teacher training and in higher education. If the challenges are of a deeper social and cultural nature, as argued here, then there is no easy one-shot solution. One will need to look beyond the education system, and involve different stakeholders. There is a need for reforms that are context specific, that require multiple approaches and are implemented of long periods of time. Initiatives will also have to be monitored, and the development and results will need continuing discussions, informed by evidence and careful analysis.

7. CONCLUDING REMARK

The above perspectives are now a high priority in the European Union (and other OECD countries). A special Action Plan for Science and Society is part of the newly (Nov 2002) adopted 6th Framework Programme in the EU for the period 2002-2006. As science education researchers, we should seize this opportunity. It is up to us to show that our research and our reflections are of value for a proper understanding of the challenges – and for the formulation of viable action to face what is seen as a growing crisis.

---

[i] This paper is based on an invited contribution to Meeting of Ministers of Education and Research in the European Union held in Uppsala, Sweden 1-3 March 2001. An enlarged and fully referenced version will be published in Jenkins, Edgar (ed) (2002) *Innovations in Science and Technology Education Vol VIII* Paris, UNESCO. It is also part of the publication Sjøberg, S (2002) *Three contributions to science education*, Acta Didactica 2/2002 Dept. of Teacher Education and School Development, University of Oslo, Norway (can be obtained from svein.sjoberg@ils.uio.no)

A Mesoscopic Model Of Liquids For Teaching Fluid Statics

Ugo Besson[1], Laurence Viennot[1], Jacques Lega[2]

[1]*LDSP University Paris 7, France,* [2] *UCL, Louvain La Neuve, Belgium*

ABSTRACT

After a study on students' conceptions concerning pressure in fluids, we elaborated a short teaching sequence on the statics of liquids, implemented and evaluated for two years among science students in the first university year. The sequence makes use of a model of liquids at the mesoscopic level through an analogy with a set of sponge balls, treated first as mechanical entities, then as a model of elements of fluids. The aims and the results of this experimentation are presented and discussed.

1. BACKGROUND AND AIMS

Many studies have been conducted on pupils' ideas and difficulties concerning gases and liquids (Piaget & Inhelder 1955, Engel & Driver 1985, Séré 1986, Giese 1987, Kariotoglou & Psillos 1993, Besson 1997 and 2001). Investigations on students' conceptions regarding liquids (Engel & Driver 1985, Giese 1987, Kariotoglou & Psillos 1993) show that most pupils correctly think that pressure increases with depth, but for many of them it also depends on the total volume of liquid and "acts" only, or most strongly, downwards – two ideas which often go with a non-differentiation of pressure and force (Kariotoglou et al. 1995). Some models of fluids have been proposed in teaching sequences (Séré 1990, Chomat et al. 1988, Méheut 1996, Borghi et al. 1996), but none takes into account both compressibility and weight.

In this research, we study students' reasoning and elaborate a teaching sequence, focusing on their representations of fluids. The aim is to study the pressure of a liquid in a static situation in the presence of gravity, without taking into account the thermic and kinetic aspects.

We feel that, to promote understanding, it is not enough to say what a given quantity *has to be*; it is also necessary to explore how the state has "come about". To connect a *global* and *formal* reasoning, based on learnt rules or formulae, for example $\Delta p=-\rho g\Delta h$, and a *local, causal* reasoning, based on what occurs and changes in a given spot of fluid, it is necessary to activate a *systemic* form of reasoning, based on the interactions between the parts of the fluid and on a transmission of changes that can explain the *mechanism* which makes it possible to establish a new equilibrium. One should also avoid asserting that liquids are incompressible, as that seems to exclude any local modification in a fluid, and therefore any difference in its action on a wall or on surrounding fluids. Variations

D. Psillos et al. (eds.), Science Education Research in the Knowledge-Based Society, 221–229.

in the volume of fluids are often negligible in calculations, but we think they are essential to an understanding of what is happening.

2. STUDY OF STUDENTS' CONCEPTIONS ON FLUIDS

We questioned a total of 229 Italian and French school pupils in grades 9 to 12, and 458 science students in the first university year in Belgium; the latter had just been given a class on fluid statics. After preparatory interviews, paper and pencil questionnaires were used. In one of these, the "Vessels" questionnaire, we asked them to compare the forces exerted by the water on the bottom of three containers of different shapes (/__\ , |__|, __/), all the containers being filled to the same level, and all the bottoms being of equal size. In another, the "Fish" questionnaire, we asked them to compare the pressure for two fish at the same depth, one fish being in the open sea and the other in an underwater cave (This question is taken from an Italian high school textbook, Pugliese Jona 1984. A similar question was used in Kariotoglou & Psillos 1993).

As regards the "Vessels" questionnaire, a majority of school pupils and university students (60% and 59% respectively) answered that the force is greater in the largest receiving container __/; only 14% of pupils and 33% of the students answered correctly, that the forces are equal. The justifications indicate that there is a strong tendency to equate "more water" with "more force on the bottom".

For the "Fish" situation (fig.1), the great majority of the school pupils considered that the pressure was different, i.e., greater in the open sea (36%) or in the cave (45%). The majority of the university students (68%) answered correctly, that the pressure was equal for the two fish; but, even after instruction, 32% considered that the pressure was different (for 24%, it was greater in the open sea). In their justifications, they often refer to the fact that pressure depends on the depth, quantity, or weight of the water above the considered point. This is the "overhead fluid" conception, according to which what causes pressure is the fluid above – its *weight, acting vertically downwards*. Upward or horizontal actions are very seldom considered. Moreover, some students, who think that the pressure in the open sea is higher than in the cave, consider that the rock does not exert a force on the water, at the roof of the cave, because its ceiling "holds up on its own", and does not need to be *supported by the water: "The rock does not rest, does not push on the water: if the water dries up, the rock does not fall* ". This idea may cause students to give the wrong answer: that the pressure is smaller inside the cave.

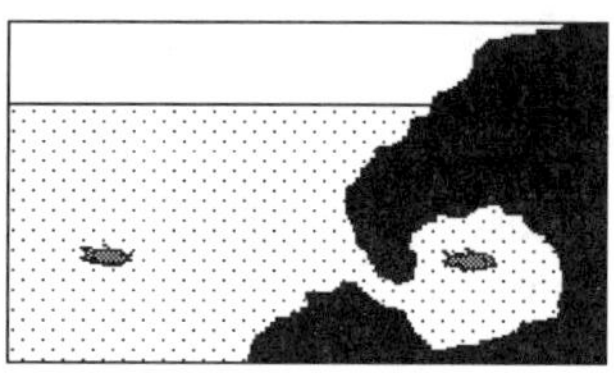

Figure 1 The two fish

Thinking that there are forces acting between the water and the rock at the roof of the cave can initiate a positive change in students' reasoning, so that they see the proposed situation as more complex. To encourage this form of reasoning is one of the objectives of our sequence.

3. DESIGN OF A TEACHING SEQUENCE

The difficulties encountered led us to elaborate a short teaching sequence (two and a half hours), which was implemented and evaluated for two years among science students in the first university year at Louvain-la-Neuve (Belgium), after a traditional lecture on the subject. The theoretical class on fluid statics being unchanged, an experimental group (N=71 the first year, N=169 the second year) participated in the sequence while a control group (N=77, N=94) did classical exercises on the same topic. The sequence is a *micro sequence*: we also want to test whether it is possible to produce a change in students' conceptions concerning specifically targeted points, through apparently small modifications in the teaching of traditional content. In particular, no extra time, or additional teachers, was devoted to the experimental sequence. In this respect, there was no additional cost.

The common analysis being centred on vertical downward actions, and almost never on upward or horizontal actions, it was necessary to activate a horizontal analysis as a preparation to an analysis of isotropy, using the idea of a modification of the state of the fluid associated to variations of pressure. This implies both the idea of global reaction of the system to a disturbance and that of a local change in the fluid. The sequence proposes a model of liquids, based on a mechanical analogy at the mesoscopic level. The objects chosen as analogues of mesoscopic units are sponge balls, treated first as mechanical entities, then as model of elements of fluids.

The mesoscopic character of the model is essential. Indeed, to describe the effects of weight statically, the units of the model must be in contact with one another. Scattered balls, such as molecules, do not work in this respect. Any microscopic model requires a consideration of kinetic aspects. A static model has to be mesoscopic, so as to hide within the unit of the model (here, the sponge ball) the kinetic aspects whose effects are manifested by elastic resistance and pressure forces. The sponge balls are not, then, the analogues of molecules, but rather of *elements, droplets* of fluid. Indeed, an analysis in terms of elements of fluids, at a mesoscopic level, is the standard method used in physics to deal with both the statics and dynamics of fluids, molecular kinetics being hidden and represented globally, in its effects, by quantities such as pressure and temperature. But this decomposition into small parts is made in an abstract and mathematical way, as an intermediate procedure, with the aim of returning at once to a continuous representation.

We wanted to make this mesoscopic decomposition *concrete*, to appeal to students' intuition, by evoking real macroscopic objects with which to establish an analogy, as regards the limited field of the statics of fluids in the presence of gravity.

The mechanism of the sponge balls, with its rules and its limits of application, constitutes the syntax which transforms intuitive analogy into a simple but operating model through which to clarify certain aspects of fluid statics, to activate specific forms of reasoning, with passages between the model and fluids.

Obviously, this model cannot go very far, yet we suggest that it can be a useful didactic tool for specific and limited objectives.

4. DESCRIPTION OF THE SEQUENCE

The sequence begins with a purely mechanical consideration of the sponge balls. The sponge balls can be deformed and compressed, they react to deformations with elastic forces acting in the same way in all directions. They are somewhat rigid, the deformations are small in relation to their size, but that is taken into account in the reasoning. Three situations are studied, involving first a series of balls aligned horizontally and pushed against a vertical barrier, then in a vertical position with a push of the hand downwards, then vertically with an upward push.

In the horizontal situation (fig. 2), a series of balls is placed on a table. At one end there is a barrier, at the other end someone pushes on the first ball. In a situation of equilibrium, the forces of the interaction between adjoining balls *must have the same magnitude all along the horizontal line.* The teacher leads the students to represent the forces as vectors on a graph, making it clear that the forces on the right-hand side have the same magnitude as those on the left, and then to draw a graph of the function *F(n)* representing the magnitude *F* of the forces according to the order number *n* of the contact points between two adjoining balls.

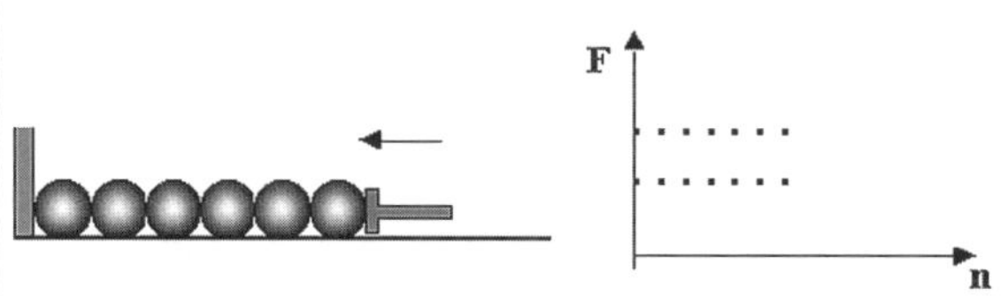

Figure 2 Model of sponge balls. Horizontal situation.

Pushing harder increases the intensity of the forces, but *equally*, so that the graph is another straight horizontal line, only set higher up.

Two vertical situations are studied (fig. 3), in which the balls are placed in a column. In the first situation, the ball at the bottom rests on the ground and someone pushes down on the top ball. In the second one, the top ball presses against the ceiling and the bottom ball is pushed upwards. In these situations, the forces between adjoining balls do not all have the same magnitude: they increase towards the bottom, each in a value equal to the weight of one ball. As in the horizontal situation, the teacher helps the students to draw a vectorial graph of the forces (stressing that, for each pair of balls, the upward force has the same magnitude as the downward one), then to make a graph of the function *F(n)* representing the magnitude *F* of the forces according to the order number *n* of the contact points between two adjoining balls. The points of this graph are now on an oblique straight line. Pushing harder on the balls causes small deformations, and the forces increase, but *equally*, so that the

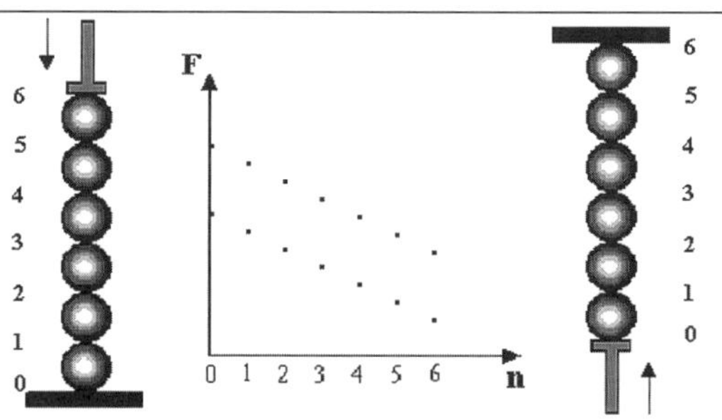

Figure 3 The sponge ball model. Vertical situations.

difference between one force and the next remains unchanged, and equal to the weight of each ball. The points of the graph are still on an oblique straight line, parallel to the first and higher up.

Then, the teacher proposes the analogy between the sponge ball model and liquids. Horizontal and vertical situations are discussed once more, an analogy being made between the force diagrams for the sponge balls and that concerning pressure in fluids, for each situation. Students analyse pressure in a closed container, which is filled with water, with a piston to push on the water horizontally (fig. 4); questions like the following (provided here with some elements of the answers) are discussed:

– What are all the points where the pressure is equal to that of a point near the piston, or of points A, or B, or C?
– If one pushes harder on the piston, what happens? As with the sponge balls, the diagram showing the pressure *p(x)* for points *x* on the same horizontal line gives a higher horizontal line. This means that all the pressures change, while remaining equal among themselves.
– Do the changes in pressure near the piston have effects in a vertical direction? Yes, because, in this model, when p_m increases, balls on the same horizontal line push harder against adjoining balls, even in the vertical direction, upward and downward.

Indeed, this is a weak point in the model; it can help students understand that the push is transmitted in all directions, but it cannot justify the isotropy of the forces of pressure. *One should assume* that this isotropy exists, but the model alone does not lead to this conclusion.

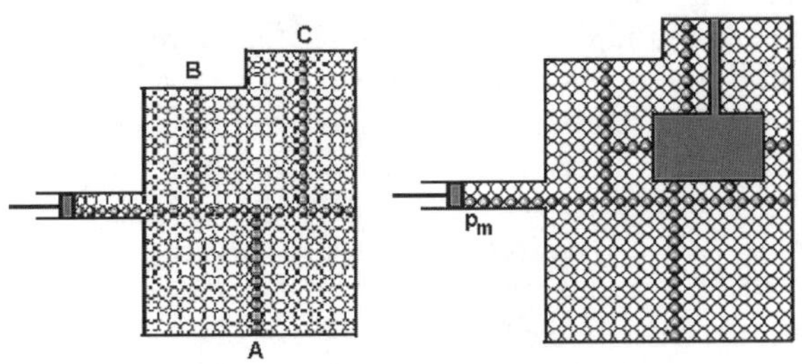

Figure 4 Analogy between liquids and the sponge ball

The teacher can propose other shapes of containers, to study the interactions of the water against the sides of an immersed object or an enclave.

In the final session of the sequence, the students, working in pairs, were asked to return to the first question of the “Vessels” pre-test, and compare the forces exerted by the water on the bottoms (of equal areas) of three containers that are shaped differently (/__\, |__|, __/), and are filled to the same level. The students discuss this, the professor moving from group to group.

5. EVALUATION OF THE SEQUENCE

We conducted two types of evaluation. The first was an external evaluation, consisting of a pre-test and post-tests. Before the beginning of the sequence, the same questionnaire, "Vessels", was given to both the experimental and control groups, as a pre-test. About four weeks after the sequence, three questionnaires were given to both groups as a post-test. In the second, internal, type of evaluation (only in 2000) we analysed the students' discussions in the final session, to understand how and to what extent the proposed didactic elements affect students' reasoning. The sessions observed took place a week after the session in which the model and the analogy with fluids were introduced. We recorded the discussions of 9 groups of volunteers over periods of 15 to 20 minutes. We also have 14 answer sheets.

In this paper, we shall give only a few results concerning the experimentation conducted in 2000, which was more complete than in 1999, from two questionnaires, "Fish" (fig. 1) and "Ball in water" (fig. 5), and from the discussions.

5.1 The questionnaires

For the "Fish" questionnaire, the students in the experimental group give a higher percentage of correct answers (that the pressure is equal): 81 %, versus 70 % in the control group. The khi-two test indicates that this difference is significant (at the significance level p=1.4%). This first result indicates a moderate positive effect of the sequence.

Some justifications given by the students indicate that the sequence has had a definite influence. Several students in the experimental group refer to the sponge ball model and many reason in terms of horizontal actions, often referring to the model.

> " The pressure is equal... if you consider the model of the sponge balls, the one at the far left is subjected to forces, which reproduce from ball to ball till they reach those in the cave ".

The interaction between the water and the rock at the roof of the cave is more often recognized in the experimental group. Among those who answered correctly, 29 students in the experimental group mention the force exerted by the roof of the cave on the water to justify that the pressure is equal for the two fish, even though there are different heights of water above them.

> "The pressure is equal, because the difference in height is compensated for by the force which the wall exerts". "The force exerted by the roof of the cave on the water is equal to the force which would be exerted if there were water above instead of rock ".

These justifications suggest that some students have made considerable progress by using the model. This aspect is one of the stated objectives of the sequence: to promote an image of fluids based on the presence of reciprocal actions between the mesoscopic elements and the walls, which are transmitted from one element to the next, propagating local modifications, in a systemic vision of the physical situation.

For question 1 of the "Ball in water" questionnaire (fig. 5), there was a very clear difference in favour of the experimental group. The percentage of correct (or nearly correct) answers is much better in this group: 64% as opposed to 45% in the control group. The khi-two test indicates that this difference is very significant (at the significance level p<0.1%).

Unlike question 1, question 2, which bears on buoyancy, did not yield significantly different results between the two groups. Indeed, due to external

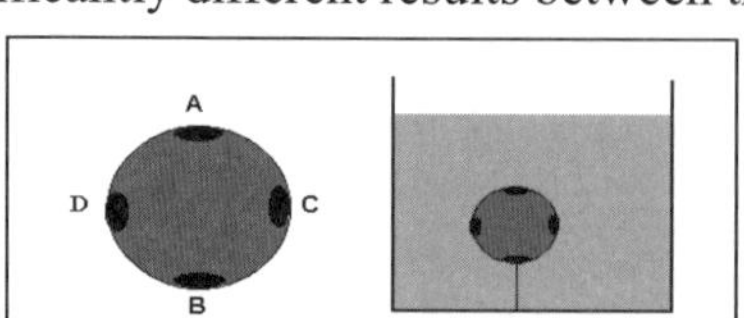

Question 1
Does the water exert a force on A, B, C, D?
If yes, say whether these forces have the same magnitude or list them in order of magnitude.
Question 2
Do these forces have anything to do with buoyancy?

constraints, the available time was very short, and we could not include the subject of buoyancy in the sequence. The great majority of students consider the buoyant force as different from pressure forces, both in its origins and characteristics. This distinction is often made explicit. Many students stress the

A ball is immersed in water. Four identical discs are drawn on it, at the top A, at the bottom B, to the right C, to the left D.

Correct answer: $F_A<F_C=F_D<F_B$; Nearly correct: $F_A< F_B$ and $F_C=F_D$

Figure 5 "Ball in water" questionnaire

Correct answer: $F_A<F_C=F_D<F_B$; Nearly correct: $F_A< F_B$ and $F_C=F_D$

Figure 5 "Ball in water" questionnaire

"global" nature of the buoyant force; it is considered as a volume force, which acts upon the whole ball, is applied at the centre of the ball and is directed upwards, in contrast with pressure forces, which are the contact, *local*, forces, acting upon each little part of the ball. This contrast is so radical that the connection between the two phenomena becomes difficult to grasp, without specific didactic intervention.

5.2 The final discussions

The students' participation in the discussions was lively; they generally showed good commitment, as is apparent from their frequent objections, the manner in which they showed resistance to other points of view, or changed theirs. We find four main types of reasoning.

1. In a smaller, narrower space, the pressure is higher, because the fluid is "compressed", "crushed".
2. In the recipient containing more fluid there is more weight, so the force on the bottom is stronger. In short: "more water = more force on the bottom".
3. Pressure depends on the weight of the column of fluid directly over the considered point. In short: "column of water above", "overhead fluid".
4. Pressure depends on height; at a same depth or height, the pressure is equal. In short: "same height = same pressure", "formula $\Delta p= -\rho g\Delta h$ ".

Reasoning 3 is the most common and the one most used in the discussions. It is often used with the model of the sponge balls, in terms of "columns of balls", or "balls", above. It can also be considered, in most of the formulations, as a mixed local - global reasoning and is often the first step towards a complete, systemic form of reasoning. It leads, more or less quickly, to the equality of forces for containers *a* __/ and *b* |__|, but clears up few difficulties for case *c* /__\ , because the columns of water are of different heights at the sides of the container.

One notices above all that the simple idea of equal pressure at the same height is not enough for the students to solve the problem, and does not convince them. Sometimes it is accepted at once, but then questioned or rejected, because it does not seem compatible with a closer analysis in terms of forces. Further work is needed to bring the students to accept that the forces are equal in the three containers.

In this respect, the proposed model appears to be an effective tool, on which the students rely to a great extent. Often, reactivating some of the elements of the sequence leads students to overcome their blockages in the discussions.

Even when it does not solve all the difficulties of a problem, the model encourages students to approach these on a higher level, giving them a different and more probing view of the proposed physical situation. It may take students beyond the naive idea of "just weight", or "weight of the fluid above". The use of the model activates better articulated, more powerful and productive reasoning in the discussion among the students and provides them with more convincing arguments.

The model becomes effective only when interactions between balls are considered. Indeed, as long as the balls are drawn or thought of as scattered, as having no contact with one another, the model is not really considered in the reasoning of the students, it is a visual memory rather than an *operating model*. But as soon as touching balls are considered, another reasoning is activated and sets off new types of problems and solutions.

6. CONCLUSIONS AND IMPLICATIONS

After the sequence, an appreciable number of students have reached a more satisfactory understanding of the physical situation, taking into account the interactions between the elements of fluids. This seems to confirm that the choices underlying the sequence were indeed relevant for the stated objectives. In particular, the implementation of a mesoscopic model proved effective as regards the development of systemic reasoning and the reconciliation of local and global approaches. The intuitive and qualitative character of the model turned out to provide students with a shared language to discuss efficiently the physical situations they were confronted with.

In view of the progress observed concerning pressure forces exerted on an immersed object, it is probable that completing this sequence with an explicit study of buoyancy would significantly improve students' understanding of this point.

The next step in modelling should make it possible to "enter" the mesoscopic unit represented earlier by a sponge ball, which can lead to a study at the microscopic level of the particles that constitute a fluid, paving the way for a subtler approach of isotropy and a study of thermo-elastic properties.

REFERENCES

Besson, U. (1997). *La pression dans les fluides*, Mémoire DEA, LDSP, University of Paris "Denis Diderot" (Paris 7).

Besson, U. (2001). *Une approche mésoscopique pour l'enseignement de la statique des fluides. Etude des raisonnements des apprenants, élaboration et expérimentation d'une séquence d'enseignement.* Ph.D. thesis, University of Paris "Denis Diderot" (Paris 7).

Borghi, L., De Ambrosis, A., Invernizzi, C., Mascheretti, P., (1996). Un modèle pour la compréhension des propriétés des liquides. *Didaskalia*, 8, 139-153.

Chomat, A., Larcher, A. Méheut, M. (1988). Modèle particulaire et activités de modélisation en classe de quatrième. *Aster*, 7, 143-184.

Engel, E., Driver, R., (1985). What do Children Understand about Pressure in Fluids ? *Research in Science and Technological Education*, 3, 133-144.

Giese, P, A. (1987). Misconceptions about Water Pressure. *Proceedings of 2nd International Seminar* (vol. 2, pp.143-148). Ithaca, Cornell University

Kariotoglou, P., Psillos, D. (1993). Pupils' Pressure Models and their Implications for Instruction. *Research in Science and Technological Education*, 11, 95-108.

Kariotoglou, P., Koumaras, P., Psillos, D. (1995). Différentiation conceptuelle: un enseignement d'hydrostatique fondé sur le développement et la contradiction des conceptions des élèves. *Didaskalia*, 7, 63-90.

Méheut, M. (1996). Enseignement d'un modèle particulaire cinétique de gaz au collège. *Didaskalia*, 8, 7-32.

Piaget, J., Inhelder, B. (1955). *De la logique de l'enfant à la logique de l'adolescent.* Paris, PUF.

Pugliese Jona, P. (1984) *Fisica e laboratorio.* Vol.1, Loescher, Turin, 1984.

Séré, M.G. (1986). Children's Conceptions of the Gaseous State. *European Journal of Science Education*, 8, 413-425 (see also *European Journal of Science Education*, 4, 1982, pp.299-309).

Séré, M. G. (1990). Passing from one model to another: which strategy ? In: Lijnse, P.L. et al. (eds), *Relating macroscopic phenomena to microscopic particles* (pp.50-66). Utrecht, CD-β press, 1990.

The Importance Of Weightlessness And Tides In Teaching Gravitation

Igal Galili, Yaron Lehavi

Science Teaching Department, The Hebrew University of Jerusalem, Jerusalem 91904, Israel

ABSTRACT

Presentation of weightlessness and tides in university level physics textbooks was examined through the perspective of the relationship between weight definitions and phenomena explanations. In addition, the pertinent conceptual knowledge of physics teachers and students (physics majors) was probed. Introductory instruction of gravitation, as reflected in the textbooks, often tends to omit tidal effects, and the explanation of weightlessness never incorporated tidal phenomena (as a reliable indicator of gravitational force). Explanations of tides in textbooks often miss the mutual free gravitational motion of the objects involved and the symmetry of tidal deformation. The elicited shortcomings of students' and teachers' explanations of weightlessness and tides were related to their presentation in textbooks. It was suggested that an approach emphasizing weight definitions, as well as synergetic treatment of weighing and tidal effects in the state of weightlessness, may contribute to the presentation of gravitation in a contemporary correct perspective. Implications for teachers' content and pedagogical-content knowledge are also considered.

1. BACKGROUND, AIMS AND FRAMEWORK

Despite both being related to gravitation, weightlessness and tides are traditionally studied independently within a regular general physics course. Several studies reported serious difficulties with respect to the understanding of both phenomena by students and teachers (Galili and Kaplan, 1996; Viri, 2000; Galili, 2001). These reports also mentioned unsatisfactory presentation of weightlessness and tides in introductory physics textbooks. In the light of previous studies, it seems plausible to check the relationship between these deficiencies and the concepts of weight and gravitational forces, as exploited in the explanations of weightlessness and tides. Commonly in physics instruction, the gravitational definition of weight is used and the teacher invests a great deal of effort to prevent students from the literal interpretation of weightlessness as zero gravitational force (a misconception). This situation can be demonstrated schematically, by establishing correspondence between weight definition and the inference regarding weightlessness in a free fall:

Weight = Gravitational force [gravitational definition] ⇒ “Weightlessness” [Apparent, fictitious]

Weight = Weighing result [operational definition] ⇒ Weightlessness [Real, gravitation-independent]

D. Psillos et al. (eds.), Science Education Research in the Knowledge-Based Society, 231–239.

Despite their historical importance (Viri, 2000), tides and tidal phenomena appear only rarely in physics courses. This, when prominent physics educators advocate teaching this subject and provide explanations appropriate for introductory physics courses (Arons, 1979; Whithers 1993; Hartel, 2000).

Usually, tides present an optional topic and often serve solely as an illustration of the universal law of gravitation. This pedagogy, which regards weight, tides and weightlessness as conceptually unrelated, motivated the present study that explored physics textbooks as well as the knowledge of teachers and students with regard to these subjects.

2. METHODS AND SAMPLES

The definitions of weight and explanations of weightlessness and tides were examined in a representative sample of 25 standard physics textbooks[a].

In parallel, an open questionnaire was administered to two groups: practicing high school physics teachers (Group-T, N=75)[b] and university students majoring in physics at the Hebrew University of Jerusalem (Group-S, N=10)[c]. The subjects were asked to answer the following questions:

> Q_1. Define the concept of weight.
>
> Q_2. Explain the weightlessness experienced by astronauts in a spaceship orbiting the Earth.
>
> Q_3. A passenger in a sealed free-falling elevator simultaneously releases a steel ball and a helium filled balloon (*the situation was demonstrated by a drawing*). Explain the passenger's observation regarding the motion of the two objects.
>
> Q_4. When do ocean high tides attain their maximum: during (a) full moon, (b) new moon, (c) half-moon (*the situations were demonstrated by a drawing*). Explain your answer.

The responses of the teachers and students along with the presentation of the subjects in the textbooks were analysed and categories of weight definitions, explanations of weightlessness and tides were established. The examination explored interrelation and coherence of concept definitions and phenomena explanations. It compared between the textbook presentation and the subjects' knowledge.

2.1 Results and interpretation

Definitions of weight. All the textbooks in our sample treat gravitation (including weight) in two separate contexts: 1) when elaborating the laws of motion and 2) in the chapter presenting Newton's Universal Law of Gravitation.

With regard to the employed definitions of weight, we found a prevailing use (76% of the textbooks) of the nominal gravitational definition (Definitions 1 and 2, Table 1) and its mathematical form W = mg. Only five books adopted operational definition (directly related to weighing, Definition 3), and only one (Halliday et al., 2001) mentioned both types.

Table 1. Frequencies of weight definitions (multiple responses are allowed)

#	Types of Weight definition	Textbooks	Group-T	Group-S
1	Weight is the gravitational force exerted on the body [without introduction of apparent weight]	48% [12]	46% [35]	[10]
2	Weight is the gravitational force exerted on the body [with introduction of apparent weight]	28% [7]	1% [1]	–
3	Weight is the result of standard weighing by a calibrated scale	20% [5]	16% [12]	–
4	Weight is equal to the supporting force	–	53% [40]	–
5	Weight is a resultant force of gravitational and inertial forces	–	3% [4]	[1]
6	No definition	4% [1]	–	–

Notably, about one half of the textbooks (Category 1) did not distinguish between weighing result and gravitational force. Only some (Category 2) developed the required distinction, by introducing "apparent weight", as being equal to weighing results (reserving "true weight" for the gravitational force itself). The approach of Category 1 overlooks the essential difference between "apparent" and "true" weights, most salient in the state of a free gravitational movement. Thus, measuring g for use in the formula W=mg, one obtains solely its apparent quantity (French, 1995). Despite the fact that weighing results are not always determined by gravitation alone (in accelerated motion) one finds, without reservation, statements like:

> "The weight of the body can be determined experimentally by balancing the force of gravity against a second calibrated force [spring] acting on the body" (Fishbane et al., 1993, p. 127)

> "In practice the corrections needed to obtain the true weight from the [spring] scale reading are of the order 0.1% and can usually be ignored" (Benson, 1991, p. 86)

This approach ignores the factual uncertainty in the interpretation of weighing results, which makes Definitions 1 and 2 incompatible with Definition 3. A salient shortcoming is that the way to measure the "true weight" (gravitational force) is normally missing. This lack of relation to measurement can be striking, if not misleading:

> "The weight of a body is the total gravitational force exerted on the body by all other bodies in the universe." (Young and Freedman, 1996, pp. 383)

The distribution of weight definitions in Group-T was rather complex (Table 1) with the prevalence of the operational definitions 3 and 4. The university students (Group-S) kept with the textbooks and provided nominal (gravitational) definition. Notably, about 17% of the teachers adopted a "complementary" approach, providing the gravitational *and* operational definitions, thus ignoring their conflicting nature. The term *apparent* in addressing weight, introduced to avoid just this particular difficulty, appeared only in one response.

Explanations of weightlessness

Examining explanations of weightlessness in the sample of textbooks, we found that only 12 textbooks explicitly explained weightlessness (Table 2), meaning that 60% (!) of the textbooks did not provide any explanation for this state.

Table 2. Explanations of weightlessness (some textbooks provided more than one)

#	Type of explanation/treatment	Textbooks	Group-T	Group-S
1	All the objects share the same acceleration	28% [7]	19% [14]	–
2	Fictitious (Inertial) force cancels gravitational force	–	27% [20]	–
3	Small (or no) gravitation	–	22% [17]	[7]
4	No force of support is required to prevent falling	12% [3]	11% [8]	–
5	Scale reading (apparent weight) is zero	8% [2]	–	–
6	"Objects are influenced solely by gravitation"	–	5% [4]	–
7	Not explained, but suggested as an end-of-chapter problem	16% [4]	–	–
8	Not mentioned/Not explained	44% [11]	–	[3]

Nine of the textbooks warned their readers not to interpret weightlessness literally since it is *fictitious* and does not truly indicate the absence of weight (within the framework of the gravitational definition of weight):

> "The condition we call weightless does not mean "no weight". Instead it means no "apparent weight." ... The man in the elevator still has weight mg - that has not changed - but he has no sensation of weight, as he free-falls. The term weightlessness is a very poor one, almost guaranteed to cause confusion because it seems to imply that objects have no weight." (Knight, 1997. pp. 168)

Note that the reference was made to "sensation" and not measurement, as it could (and should) be. This essentially reduces the importance of the reservation made by the author.

Some textbooks distinguish between the "apparent" (really observed) and "true" (solely hypothetical) weightlessness. Although using the same mathematical formula they provide the student with different interpretations of g (calculated versus measured). Textbooks that defined weight operationally were free of this subtlety, which escaped the understanding of most students. Such texts regarded weightlessness as zero weighing results (e.g. Lerner, 1996).

Teachers and students addressed weightlessness by considering astronauts and buoyancy in a free gravitational motion (Q_2 & Q_3). Tables 2 and 3 present the major categories of the accumulated responses. Importantly, many teachers explained weightlessness by using inertial forces (Type-2). Such an explanation, which did not appear in the textbooks, presents a subject out of the standard curriculum for colleges and high schools. This account for weightlessness was found frequent in the intuitive knowledge of school students (Galili and Bar, 1992), who explained weightlessness by neutralizing the gravitational force by centrifugal one. The relatively popular reasoning by diminished gravitation (Type-3) is a common misconception of high school students and was interpreted as induced by the gravitational definition of weight (Galili and Kaplan, 1996). Explanation of Type-6 merely rephrases the situation, presenting a verbatim contradiction to the gravitational definition of weight. The fact that no physics freshmen could provide a correct answer indicates a real problem, even among the good students. Finally, responses 2, 3 and 6 overlooked gravitational effects due to the tidal forces.

With regard to the steel ball and helium-filled balloon in a free-falling elevator (Q_3) both groups exhibited much confusion (Table 3). Although the subjects recognized the state of weightlessness with regard to the steel ball (91% in Group-T), they were much less unanimous (48% in Group-T) with regard to the helium balloon. The following response is illustrative:

> "The ball falls together with the man and there is no relative velocity between them. The balloon will **rise up because it is lighter than the air**." (Our emphasis)

Table 3. Subjects' predictions and major types of reasoning regarding objects dropped in a free falling elevator (Q3)

#	Responses	Group-T	Group-S
1	The steel ball floats	91% [68]	[10]
2	The steel ball falls	5% [4]	-
3	The helium balloon floats	48% [36]	[3]
4	The helium balloon ascends	45% [34]	[6]
	Types of reasoning		
1	"The balloon is lighter than the air (buoyancy)"	30% [23]	[1]
2	"They all share the same acceleration/motion"	21% [16]	[3]
3	"There is no buoyancy during free falling"	10% [8]	[1]
4	"The resultant force on any object in a free fall is zero"	6% [5]	[1]
5	"In a free fall all objects are weightless"	6% [5]	[1]
6	"Gravitation is absent in a free fall"	4% [3]	-

Only 10% of both groups satisfactorily explained the floating of the balloon by the weightlessness of both the air and the balloon (absence of buoyancy) in a free gravitational fall. The others were confused regarding the relationship between the two phenomena (Table 3). Even most of those who predicted correctly the floating of the helium balloon did not explain it properly. One may also mention that the conception of "relative lightness" of helium ("helium is lighter than air" - 30% in Group-T), which causes the unconditional buoyancy, apparently revives the Aristotelian concept of intrinsic "lightness" (levity), in all aspects conflicting weightlessness in its modern sense.

Explanations of tides

Tidal phenomena rarely appear and even more seldom are explained in the textbooks. In five, out of thirteen, textbooks that did mention these phenomena, tides appeared merely as an end-of-chapter exercise. Furthermore, tides are never mentioned in conceptual relation with weight and weightlessness and commonly the separation is large (e.g., Ch. 5 and Ch. 13, Benson, 1991). This despite the fact that tidal forces present an important gravitational effect, which becomes particularly observable in the state of free-fall (Ohanian and Ruffini, 1994). So, the rather common claim that in a free fall *all* effects of gravitation cease is not accurate.

Only four authors (Benson, 1991; Fishbane et al., 1993; Hewitt, 1998; Ohanian, 1998) provide a conceptually complete treatment of the phenomenon, addressing the dynamic nature of tides (the essential role of the Earth-Moon mutual free-fall). Only such an approach can account for the symmetry of the tidal bulges on the opposite sides of the Earth (a historically important feature of tides that puzzled generations of scientists). Other important characteristics of tides (such as dependence on the geographical latitude) are mentioned extremely rarely. Only in Ohanian (1998) one can learn that tidal effects are a measurable phenomenon also at a scale of a laboratory (ocean tides are the only tidal phenomenon addressed in the rest of the textbooks).

Generally, it appears that most textbooks present tides (if at all) as "another phenomenon explained by Newton's law of gravitation". They ignore the important role of tidal phenomena in probing gravitation, in contrast to weighing, which cannot reliably do the same. With regard to the knowledge of teachers and students, the responses to Q_4 clearly indicate a serious flaw. All the subjects related the maximum of the high tides (spring tides) solely to the "new moon" arrangement, Sun-Moon-Earth, and not to the equally appropriate setting of the "full moon" (Sun-Earth-Moon):

> "The maximum is on a new moon **for the sum of forces** is then the greatest" *(our emphasis)*

Only one (teacher) mentioned the symmetry (equality) of the tidal bulges. The responses yield a straightforward interpretation: spring tides were mistakenly explained by summing the Sun's and the Moon's gravitational forces exerted on a particular point of the Earth and not by the spatial difference in these forces. Evidently, our subjects missed the symmetric contributions of both, the Sun and The Moon, to the phenomena. This symmetry, as shown qualitatively by Mach and in details by others (e.g. Arons, 1979), arises from the free fall of the Earth. The tides might be treated in terms of either forces or relative accelerations of the Earth's fragments. Notably, the symmetry of tides and its origin did not appear in the great most of the textbooks.

Only one teacher suggested a dynamic explanation of tides, concluding however, that tides (but not spring tides, as it should be) appear on both new and full moon. This teacher mentioned also the influence of friction between the water and the solid crust on the timing of the phenomena:

> "Theoretically, if friction were absent, tides had to appear on a new and full moon. But, to my best recall, there is a considerable difference [in timing] with the reality. Tides on the earth result from two factors together: the moon's attraction in different points on the Earth's surface plus the circular motion around the common centre of mass."

This response, although addressing the important feature of equality of tidal bulges, possibly exhibits confusion with regard to the periodicity of the phenomena.

The students, showing even greater confusion, could not explain tides correctly. For instance, similar to the above response, one predicted a monthly period of tides (see the last sentence):

> "The high tide appears when the moon succeeds in pulling the sea towards it or when the sea is being "thrown away" from the earth in its orbit around the moon. If I stand on the beach during high tide then the moon is above me (pulling the sea) and I see him full. Or it is in the opposite position and then I stand at the outer side of the circle, the sea is thrown and again I see high tide. Low tide appears when I see half moon."

Generally, the results display a highly fragmentary and incomplete knowledge of both teachers and students with regard to tidal phenomena.

Considering the central role of textbooks in the process of science instruction, it is instructive to juxtapose both the frequent features of teachers-students' knowledge and the correspondent contents of the examined textbooks.

Weightlessness

Physics textbooks: Often do not address weightlessness; Introduce "apparent weightlessness" for zero weighing results and "true weightlessness" for zero gravitational force, providing no means to distinguish between them; Often argue by equal accelerations of the body and its support to account for weightlessness; Rarely address other phenomena related to weight (e.g. buoyancy) in free fall.

Teachers/students: Often argue by small or no gravitational force to account for weightlessness; Often cancel gravitational force by centrifugal (inertial) force; Are confused with regard to what is and is not measurable in a free fall; Often use equality of acceleration of the body and its support to account for weightlessness; Exhibit difficulties in explaining buoyancy in the state of free fall.

Tides

Physics textbooks: Rarely consider tidal effects and almost never in small-scale; Normally use the difference of gravitational forces on the extremes of objects; Rarely employ relative gravitational accelerations, by which parts of the considered body would fall separately; Rarely mention equality of tidal bulges; Rarely mention spring tides.

Teachers/students: Fail to explain tidal effects and use the sum of gravitational forces to explain spring tides; Do not consider the difference of forces along the body of the Earth; Do not consider accelerations, by which parts of the considered body would fall separately; Miss the equality of the tidal bulges.

This juxtaposition shows that appearance of deficiencies in teachers'/students' knowledge of the subject, corresponds to incomplete or missed explanations of weightlessness and tides, as provided by the textbooks.

3. CONCLUSIONS AND IMPLICATIONS

This study extends the evidence for the unsatisfactory knowledge of weight and weightlessness to university students and physics teachers, and added to that even a stronger claim regarding tides. It was found that physics textbooks provide little support for the construction of a better knowledge. The findings imply that the reasons for this situation might be the general approach to teaching gravitation – the subject matter contents of the instruction (for which textbooks serve an important, although not exclusive, resource). Accordingly, three remedial actions are suggested.

First, we recommend that attention should be given to the weight definition. The present study implies a clear correspondence between the unsatisfactory knowledge of definitions of weight and the low quality of explanations of weightlessness and tides provided by both teachers and students. The difference between nominal and operational definitions of weight should be elaborated.

Secondly, if the gravitational definition of weight is adopted, the concept of "apparent weight" should be introduced in order to relate the notion of weight with people's experience with weighing. The fact that "true weight" (the gravitational force) is not a subject of local weighing should be explicitly discussed. The instruction based on the operational definition of weight seems preferable, being free of these subtleties. It naturally matches students' tactile experience and spontaneous knowledge regarding weighing. The merits of the operational definition appear to be especially salient when accounting for the state weightlessness.

Thirdly, the pedagogical potential of tidal phenomena greatly exceeds their mere illustrative use. We suggest contrasting weight and tides (at both large *and* small distance scales) in the context of weightlessness. Such a synergetic instruction of these concepts may encourage students' appreciation of the important epistemological question - what can and what cannot be measured with regard to gravitation. It may change the common perception of tides as being relevant solely at astronomical distances, also preventing the mistake of explaining tides by the gravitational pull itself, instead of its gradient along the considered body. In addition, the knowledge of tidal effects can contribute to the understanding of the meaning of Einstein's principle of equivalence, currently considered to be a prerogative of advanced university courses (e.g. Misner *et al.*, 1973) or proper scientific texts (Dicke, 1964). Weight and tides can become mutually supporting topics, facilitating the construction of a meaningful and contemporary adequate concept of gravitation already within an introductory physics course.

Understanding of weightlessness and tides illustrates the general aspect of science education, the importance of the concept definitions, rooted in the philosophy of science (Margenau, 1950). Clarification of the interrelation between alternative definitions of the same concept (in our case, weight) and their role in the explanations of related phenomena (in our case, weightlessness and tides) may essentially contribute to the construction of this concept as well as to the understanding of the phenomena. Thus, it promises to facilitate both the “content” and the “pedagogical-content” knowledge of physics learners and teachers.

REFERENCES

Arons, A., B. (1979). Basic physics of the semidiurnal lunar tide, *Am. J. Phys.* 47(11), 934-37.

Benson, H. (1991). *University physics* (Wiley, NY,).

Dicke, R., H. (1964).The many faces of Mach, in: Chu H. and Hoffmann W. F. (Eds.), *Gravitation and Relativity,* Benjamin, NY.

Fishbane, P., A., Gasiorowicz, S. & Thornton, S., T. (1993). *Physics for scientists and engineers,* Prentice Hall, Englewood Cliffs, NJ.

French, A. (1995). On weightlessness. *Am. J. Phys.* 63, 105-106.

Galili, I. and Bar, V. (1992). Motion implies force. Where to expect vestiges of the misconception? *Int. J. Sci. Ed.*, 14(1), 63-81.
Galili, I. & Kaplan, D. (1996). Students' operation with the concept of weight. *Sci. Ed.* 80(4), 457-487.
Galili, I. (2001). Weight Versus Gravitational Force: Historical and Educational Perspectives. *Int. J. Sci. Ed.*, 23 (10), 1073-1093.
Hartel, H. (2000). The tides: a neglected topic. *Phys. Ed.* 35(1), 40-45.
Halliday, D., Resnick R., & Walker J. (2001). *Fundamentals of Physics,* Wiley, NY.
Hewitt, P. (1998). *Conceptual Physics,* Addison-Wesley, Reading, Mass.
Knight, R., D. (1997). *Physics Contemporary Perspective.* Addison-Wesley, Reading Mass.
Lerner, L., S. (1996). *Physics for Scientists and Engineers,* Jones and Bartlett, Sudbury, Mass.
Margenau, H. (1950). *The Nature of Physical Reality.* McGraw-Hill, NY, pp. 220-244.
Misner, C., W. Thorne K. S. and Wheeler J. A. (1973).*Gravitation,* Freeman, San Francisco.
Ohanian, H., C. (1998). *Physics,* Norton, NY.
Ohanian, H., C. & Ruffini, R. (1994). *Gravitation and spacetime* (ПП. 35-54). Norton, NY.
Viri, J. (2000). Students' Understanding of Tides. *Phys. Ed.* 35(2), 105-109
Whithers, M. ,M. (1993). Why do tides exist?. *Phys Teach.* 31, 394-398
Young, H., D. & Freedman, R., A. (1996). *University Physics,* Addison-Wesley, Reading, Mass.

---

a The university physics textbooks of the sample were published in the USA during the last five years and are well known to physics educators teaching physics in many countries, including the country of the study, Israel.

b The teachers attended special workshops in different regions of the country.

c All were at least second year students who had completed the introductory courses. In Israel there are relatively few students who major in physics, thus our sample is more representative than the number may imply.

MAKING DECISIONS ABOUT BIOLOGICAL CONSERVATION ISSUES IN PEER GROUP DISCUSSION

Marcus Grace, Mary Ratcliffe

University of Southampton, UK

ABSTRACT

An understanding of biological conservation is an essential component of education for sustainable development, and much of this understanding is initiated and cultivated through the school science curriculum. This study explores the value of group discussion among 15-16 year old pupils, and some aspects of their interaction, while making decisions about conservation issues. The great majority of pupils welcomed the opportunity to discuss the issues. Most pupils modified their solutions to the conservation management problems as a result of peer group discussion, and the activity appeared to have a long-lasting modifying effect on their attitude, particularly with regard to killing (culling) animals. Pupils' views are context-dependent, as are the underlying values and biological concepts, which indicates that pupils should engage in peer group discussion of a range of conservation scenarios in order to make informed decision-making about biological conservation issues.

1. BACKGROUND

Global recognition of biological conservation as a precondition for sustainable development impelled 157 political leaders to sign the Convention on Biological Diversity at the Earth Summit in 1992 (Glowka et al, 1994). Professional conservation biologists have identified a large number of biological concepts as essential underpinnings for school education programmes (Grace and Ratcliffe, 2002), but some believe that cultural, aesthetic or utilitarian factors are more important in deciding conservation priorities (Spellerberg, 1996). Environmental issues are socially constructed, and even professionals are not always able to agree on approaches to conservation.

Education for sustainable development is now part of the statutory National Curriculum for Science in England and Wales, which states that '*...science provides opportunities to promote... education for sustainable development, through developing pupils' skills in decision making on the basis of sound science...*' (Qualifications and Curriculum Authority, 1999, p.9). Despite the powerful rhetoric from central government in support of the Convention on Biological Diversity, and the importance of biodiversity in underpinning sustainable development, it is curious, and in our view regrettable, that the science curriculum does not include the terms 'conservation' or 'biodiversity'. It is therefore interesting and useful to explore how English pupils, nearing the end of their compulsory schooling, handle discussions and decision-making about conservation issues.

D. Psillos et al. (eds.), Science Education Research in the Knowledge-Based Society, 241–247.

2. AIMS AND FRAMEWORK

The aims of this study were to gain an insight into whether peer group discussion affects 15-16 year olds' views towards biological conservation, and how pupils interact while involved in a conservation decision-making activity. When referring to group interaction in this paper, we use the term 'discussion' as meaning verbal interaction focusing on the resolution of a controversy, and recognise that some researchers may prefer to call this 'argument' (Newman *et al*, 1999). The tasks given to pupils here are ones in which they are encouraged to develop 'informed opinions', i.e. where commitment is not necessarily present (attitude formation), rather than 'informed decisions' which imply finality and action (behaviour). There is considerable evidence that attitude and behaviour are not necessarily correlated (Fishbein and Ajzen, 1975), and particularly when considering environmental issues (e.g. Newhouse, 1990; De Jager and van der Loo, 1990).
Even though people may favour certain conservation actions, they may not demonstrate them personally. However, it may be possible through informed discussion of socio-scientific issues to clarify and modify attitudes, even if behaviour remains unaffected.

The study builds on the use of decision-making frameworks in youngsters' consideration of socio-scientific issues (Ratcliffe, 1997), which draws on extensive research on decision-making in theory and practice using normative and descriptive decision-making models.

3. METHODS AND SAMPLES

Decision-making exercises lasting 30 to 40 minutes were given to 131 fifteen year old pupils at four different co-educational schools in the south of England, after piloting the tasks with four small groups. These pupils were all above average achievers in science. The tasks related to one of two real conservation issues, one concerning the conservation of elephants in Africa, the other the conservation of puffins (a familiar British seabird) in competition with rabbits. Elephants, puffins and rabbits were chosen for the study as species that these pupils would have a relatively strong desire to conserve. They are all familiar species to the pupils, and are the kinds of animals that receive the most positive attitudes among young people (Greaves *et al.* 1993; Grace and Sharp, 2000). Brief written information and colour photographs were used to introduce the issues, and an effort was made to avoid explicit mention of relevant scientific concepts at this stage. Although puffins and elephants are both endangered species, the scenarios differ considerably. Elephant conservation is a social issue, affecting local economies through tourism, crop destruction and the ivory trade; whereas the puffin conservation scenario is presented as a less complex issue with no significant impact on the local human population. Puffins nest in burrows in the soil on cliffs and islands around the coast of Britain. Expanding rabbit populations compete for burrow space and cause soil erosion.

The pupils (47% girls) worked in 'peer groups' familiar to them as part of their normal science lessons (4-6 per group). The data required to address the research questions was collected using the following methods:

1. *A 'pre-test' questionnaire* stating the scenarios, and asking pupils for their personal view of what should be done and why. To ensure that pupils' answers best reflected their own real views, pupils were asked not to confer, and the questionnaire was administered during normal class time, under 'examination conditions' and under the supervision of their usual class teachers.
2. *Audio-taping in conjunction with a decision-making framework.* The steps in the framework were: listing options; identifying important criteria; considering scientific information available and required; weighing up the pros and cons in relation to the options; making a decision; and reviewing the decision-making process. The authors did not intervene in discussions other than managing procedural aspects of the activity. No other teachers or classroom assistants were involved. Pupils were asked to consider one of the scenarios using the framework, and think about any factors they thought important in making these decisions, but to focus particularly on the scientific information needed. Pupil discussions were audio-taped, listened to several times, and transcribed.
3. *A 'post-test' questionnaire* immediately after the discussion, asking pupils what they thought should be done about the situation and why, and how they felt about tackling problems in groups like this, rather than on their own.
4. *A semi-structured interview* with each group one year later, to explore memories of the activity in terms of issues discussed, decisions made, the decision-making framework, views of peers and science used.
5. *A final questionnaire* at the end of their compulsory schooling having completed the whole science national curriculum, asking what they thought should be done about the elephant/ puffin situation and why, which areas of science are important to understand when making decisions about conserving animals, and asking about their interests and experiences in relation to wildlife conservation.

4. FINDINGS AND DISCUSSION

For both scenarios, pupils provided a range of solutions in the pre and post-tests and in the final questionnaire. A comparison of individual choices at these three stages may indicate the possible impact of the group discussion on their views. A fundamental question influencing their overall decision is whether we should kill (cull) individual animals in order to protect people or other animals, and this aspect accounted for a substantial part of the conversation in all discussion groups.

Most adolescents at this age view species extinction as an unacceptable consequence of human activities (Grace and Sharp 2000). However, girls are known to be more environmentally aware and active than boys, tending to express a more 'sympathetic' view, and stronger positive attitude towards conservation (Morris and Schagen, 1996). It might seem to follow that girls are less favourably disposed towards the killing (culling) of individual animals, although species extinction and killing individual animals are very different issues in a scientific context. Table 1 shows that discussion resulted in an overall increase in the number of boys and girls advocating culling as a management solution. Both before and after the discussion, boys favoured culling more than girls, and although more pupils advocated culling rabbits than elephants, the gender difference was fairly consistent for both scenarios.

There were some stereotypical male comments at the start of discussions, recommending that the rabbits and elephants be shot for sport and human pleasure, but this tended to be brief. In some groups, pupils had very strong and opposing views about killing animals at the start, but their final group decision was a compromise between the two, indicating that the activity had a tempering effect on their attitudes.

It is worth noting that the final group decision was not always reflected in pupils' post-activity private written decisions, suggesting that some individuals may go along with the general consensus to avoid further discussion and we should acknowledge that unguided discussion may suit certain personalities better than others. However, most pupils did show signs of modifying their personal decisions subsequent to the discussions.

The majority of pupils chose another solution which did not involve culling (e.g. constructing fences, translocating or sterilising animals), but after discussion many pupils (over 70%) changed their views or elected for a combination of ideas as a solution to the problem. A minority proposed the same solution before and after discussion, with slightly more girls than boys giving a consistent response (for rabbits: 27% girls, 21% boys; for elephants: 29% girls, 26% boys).

Personal choices given in the final questionnaire one year later, were surprisingly similar to the post test results, particularly with regard to culling (Table 1).

There was little connection between attitude and experience/interests. In questionnaire responses, 17% of pupils said they were 'very interested' in wildlife, and the majority (75%) said they were 'quite interested' in wildlife; but only 25% claimed to watch programmes or read articles about wildlife at least once a week. 4% of the pupils belonged to some kind of wildlife group. 30% of these 'wildlife enthusiasts' supported culling of rabbits and 26% supported culling of elephants. In this respect they did not differ much from the rest of the pupils in the study.

Observation of the pupils in the classroom, and analysis of audio-tapes so far has shown that pupils at this age can remain fully engaged when discussing these issues in a semi-structured way following the decision-making framework. Very few groups stuck rigidly to the framework, but used it as an aid to keeping on task. The majority of pupils employed some elements of a decision-making strategy, such as mentioning some consequences of their decisions, or suggesting the need to seek further information. However, few pupils took this further by arguing for consideration of the advantages and disadvantages of various courses of action.

Off-task behaviour was apparent in all groups, although much of this was stimulated by, and related to the conservation issue. For example, the elephant issue led to brief discussions about personal experiences connected with fox-hunting, and products from rare animals used in 'oriental' medicine.

All groups used some scientific concepts and values in deciding about the conservation issues, but more weight appeared to be given to values, particularly as the discussions proceeded. The only concepts raised and discussed, by all groups for both scenarios, were 'species', 'rarity' and 'culling'. The number of concepts and amount of use differed according to the scenario. Most groups discussed 'ecological concepts', but only one group discussed concepts relating to genetics and inheritance. More details about these concepts and values used are given in Grace and Ratcliffe (2002).

The values discussions were mostly biocentric and ethical, centring around the belief that animals have 'an intrinsic right to live'. This supports the work of Greaves *et al* (1993) who found that, ethical considerations were the commonest reasons given by children for preventing animal extinction. Each group also raised the anthropocentric view that the animals are worth conserving because they are aesthetically appealing to humans. Individual pupils did not necessarily make the same value judgements about culling elephants and rabbits. For example, one girl stated: "elephants deserve more respect because they are more intelligent."

5. CONCLUSIONS AND IMPLICATIONS

The findings in this study indicate that pupils can fully engage in decision-making about conservation issues. Post-test questionnaires asked pupils to rate how useful it was to discuss these issues, and the great majority of pupils (87%) gave a positive response (i.e. 'useful' or 'very useful'). Interviews also confirmed that pupils preferred to work in small groups rather than on their own.

As a result of the discussions, over 70% of pupils modified their solutions to the conservation issues. Although those advocating culling as a solution remained in the minority, this number increased after discussion and remained noticeably consistent one year later. This change in attitude supports Solomon's (1992) suggestion that group discussion can assist attitude change, and is consistent with Zoller et al's (1990) assertion that an STS (Science-Technology-Society) course can "substantially change the viewpoints/position of senior high school students..." (p33-34). It contrasts, however, with Aikenhead's (1989) hypothesis that, in resolving conceptual conflicts, group decisions emerge from members' original choice preferences, rather than from their interactions during discussion. Aikenhead indicates the difficulty in identifying factors, which contribute to viewpoint change, but in the present study the peer group friendship seemed sufficiently robust to allow disagreement without much personal conflict. There are signs in this study that discussion of the issues reduces the rigidity of views and brings pupils towards a compromise view. This is noticeable for example when considering gender differences. As a result of the discussions, girls appear to become more favourably disposed towards culling, and boys become more prepared to investigate alternative options to culling.

Interviews and questionnaires a year after the study indicated that pupils had little memory of the decision-making framework, but remembered the scenarios and the decisions they made. However, they still brought very little science into the discussion. We believe that conservation decision-making can only be properly informed if pupils are aware of the essential basic biological concepts underpinning conservation management, i.e. those regarded as essential by conservation biologists. The inclusion of values considerations in conservation education is also very important. Pupils' use of values and scientific concepts is context-dependent, indicating a need for them to discuss a range of conservation scenarios in order to maximise their understanding of the linkage between the underlying values and biological concepts. Selection of appropriate issues will depend on the pupils' interests and experiences. Witherspoon (1994) identified three factors influencing environmental values: 'social values', 'rational perceptions' based on scientific evidence, and 'romantic world views' based on scepticism about scientific and economic progress. Despite the noticeable change in attitude during this study, the majority of adolescents seemed to demonstrate a tendency to take an emotional or 'romantic' stance over biodiversity conservation issues.

This research begins to provide information about how pupils working in peer groups attempt to make decisions about conservation matters. Involving pupils in the peer group discussions appeared to have a long-lasting modifying effect on their attitude towards specific conservation issues. We have a long way to go if we are to encourage concern among adolescents for conservation of diversity between and within species, but we suggest that peer group discussion about the relative importance of conserving a range of organisms is crucial in developing an ability to make decisions about biological conservation issues. In our opinion, this ability is a prerequisite to making informed decisions about wider issues of sustainable development.

Table 1. Percentage of girls and boys advocating culling as a conservation management solution

	Pre-test attitude			Post-test attitude			Final attitude		
	% of girls	% of boys	% of all pupils	% of girls	% of boys	% of all pupils	% of girls	% of boys	% of all pupils
Cull rabbits	16	25	21	28	36	32	26	37	32
Cull elephants	12	17	15	20	26	23	21	26	24

REFERENCES

Aikenhead, G.,S. (1989). Decision-making theories as tools for interpreting student behaviour during a scientific enquiry simulation. *Journal of Research in Science Teaching*. 26, 3, 189-203

De Jager, H. & Van der Loo, F. (1990). Decision-making in environmental education: notes from research in the Dutch NME-VO project. *Journal of Environmental Education,* 22, (1) 33-42

Fishbein, M. & Ajzen, I. (1975). *Belief, attitude, intention and behaviour: An introduction to theory and research.* Reading, Mass. Addison-Wesley.

Glowka, L., Burhenne-Guilmin, F., Synge, H., McNeely, J.A. & Gundling, L. (1994). *A guide to the Convention on Biodiversity*. Cambridge: International Union for Conservation of Nature and Natural Resources

Grace, M. & Sharp. J. (2000). Young people's views on the importance of conserving biodiversity. *School Science Review*, 82 (298), 49-56.

Grace, M. & Ratcliffe, M. (2002). The science and values that young people draw upon to make decisions about biological conservation issues. *International Journal of Science Education, 24, (11) 1157-1169*

Greaves, E., Stanisstreet, M., Boyes, E. & Williams, T.,R. (1993). Children's ideas about animal conservation. *School Science Review*, 75 (271), 51-60

Morris, M. & , Schagen, I. (1996). Green Attitudes or Learned Responses? *Global Environmental Education*. Berkshire: National Foundation for Education Research.

Newhouse, N. (1990). Implications of attitude and behaviour research for environmental conservation. *Journal of Environmental Education,* 21 17-32

Newman, P., Driver, R. & Osborne, J. (1999). The place of argumentation in pedagogy of school science. *International Journal of Science Education*, 21 (5) 553-576

Qualifications and Curriculum Authority (1999) *The National Curriculum Handbook for Secondary Teachers*. London: Her Majesty's Stationery Office.

Ratcliffe, M. (1997). Pupil decision-making about socio-scientific issues within the science curriculum. *International Journal of Science Education*, 19 (2) 167-182.

Solomon, J. (1992). The classroom discussion of science-based social issues presented on television: knowledge, attitudes and values. *International Journal of Science Education*, 14 (4), 431-444

Spellerberg, I., F. (1996). *Conservation Biology*. Harlow: Longman

Witherspoon, S. (1994). The greening of Britain: romance and rationality. In Jowell, R., Curtice, J., Brook, L. & Ahrendt, D. (Eds) (1994) *British Social Attitudes: the 11th Report*. Aldershot: Dartmouth Publishing Company.

Zoller, U., Ebenezer, J., Morley, K., Paras, S., Sandberg, V., West, C., Walthers, T. & Tan, S.,H. (1990). Goals attainment in Science-Technology-Society (S/T/S) education: expectations and reality. A probe into the case of British Columbia. *Science Education* 74 (1), 19-36

DISCOURSE IN THE LABORATORY: QUALITY IN ARGUMENTATIVE AND EPISTEMIC OPERATIONS

Jiménez Aleixandre, M.P[1]., Reigosa Castro C[2]. & Díaz De Bustamante J.[1]

[1] *University of Santiago de Compostela, Spain,* [2] *High School A Pontenova, Spain*

ABSTRACT

The conversations and actions of 9th & 10th grade (14 to16 years old) students are analysed while working in small groups in the laboratory (Physics, Chemistry and Biology) on open tasks. The paper focuses on 1) argumentative operations, in particular warrants, justifications of claims and actions; and 2) epistemic operations, as data construction, progression from data to explanations. Students were audio and videotaped and the recordings analysed by means of Toulmin's layout and Habermas' frame for communicative and dramaturgical actions. The quality of argumentative operations is explored through the analysis of justifications used along a sequence of 5 tasks and 12 sessions; and the quality of epistemic operations drawing on a frame by Duschl & Erduran.

1. LABORATORY WORK AND SCIENTIFIC CULTURE: FRAMEWORK OF THE STUDY

Laboratory work is essential in science learning but its potential has yet to be fulfilled. Not always do students and teacher share the same meaning for experiences, in our opinion due, among other, to teacher expectations focused on results, without paying attention to the processes, to the heuristics or paths which students may follow, or to the difficulties (cognitive, procedural or other) which they encounter in the process. This reflection is needed to illuminate the obstacles that hinder the use of laboratory to introduce students to scientific practices.

In our proposal, practical work is designed as problem solving, intending to enculturate students into scientific culture. As scientific inquiry is in itself ill-structured, open tasks, situated and ill-defined problems seem appropriate to this enculturation. By open, here, we do not always mean a variety of solutions, but having to decide (and justify) which steps to take in order to solve the problem. This approach, where the process is given pre-eminence over the results, is contrasted to standard laboratory work, where tasks are structured as recipes and it is the result, which matters. But it is not devoid of problems, for instance, students have been socialized in a role, which does not demand them to take decisions and may resent what they perceive as too little guide from the teacher (Reigosa & Jiménez, 2000).

D. Psillos et al. (eds.), Science Education Research in the Knowledge-Based Society, 249–257.

Our focus is the laboratory discourse, the communication while students are solving open tasks. Communication should allow participants to construct shared meanings and knowledge yet, as Kelly & Crawford (1997) note, students can share tasks and not knowledge, having different access to it, depending on the cognitive resources available to them. The objective of the paper is to explore the knowledge construction by students, with two foci:

- Argumentative operations, in particular the type of justifications for the claims;
- Epistemic operations, data construction, pattern searching, commitments.

Argumentative operations are related to generation and justification of knowledge claims; argument is essential both in doing science and in communicating scientific claims. The argumentation is analysed with Toulmin's (1958) model, an adequate tool for discourse analysis as he focuses in the practice of arguments in linguistic –formal or informal– contexts.

We seek to identify operations and practices related to the culture of science practitioners in the students' discourse. These practices are better viewed not as stereotyped "steps", but as operations related to doing science (Jiménez, Díaz & Duschl 1998a). Of particular relevance are the epistemic operations, related to knowledge construction in the science domain, for instance explanation procedures, analogies, comparisons, and data construction. Some are illustrated below, and a detailed discussion is found in Jiménez *et al* (1998a). For the analysis of epistemic operations, Duschl & Erduran (1996), establish a sequence of steps from data to scientific explanation. This model is used to explore how students cooperate to progress along the path.

raw data – hard data – patterns – pattern explanations

The first step is data selection, when from the available data some are chosen as part of data construction. The second is interpretation of data in order to identify patterns, and the third explanations, which account for the patterns.

Not all the operations and justifications of the students belong to scientific culture. Some are better placed under a category, sometimes labelled school culture, and here in terms of Habermas' (1987) dramaturgical actions. For Habermas, in *dramaturgical* actions participants constitute a mutual audience for a performed role, as opposed to *communicative* actions, oriented to understanding.

The focus of the paper is identifying argumentative and epistemic operations in the students' discourse and exploring their quality. Some criteria for quality are:

- In the argumentative operations: warranted claims are considered of a better quality, and warrants supported either in background knowledge or in empirical evidence of better quality than those supported, for instance, in authority. Communicative actions belong to scientific culture and so are of better quality than dramaturgical.
- In the epistemic operations: in Duschl and Erduran (1996) model from data to explanation, the last steps are considered more difficult, and its performance a sign of better quality. Also unassisted performance is considered of higher quality.

2. METHODS AND CONTEXT OF THE STUDY

Participants: six small groups of secondary school students (9th & 10th grade, corresponds to 14-15 & 15-16 years old); four for the Physics tasks and two (F, G) for the Biology task. The tasks were part of their regular coursework with their teachers, (in Physics, the second author). The Physics task was carried in different contexts: Groups 1 and 2 were in the evening shift, being older than the average tenth grader; it was the first time that they worked with open tasks. Group 3 and 4 were part of the same class in the day shift, for them "Springs" was the last of five open Physics & Chemistry tasks, whose content can be seen in table 3.

The tasks: 'An explorer in the field' and 'the tracks of the thief'

– *An explorer in the field* (Springs): (part of the handout)

> 'An explorer wants to measure the mass of small stones but he doesn't want to carry scales as it would take too much space. He needs to know whether he could use, instead of the scales, a spring or an elastic band and a measuring tape. He asks you to study if they would suit him and why, if it is possible to use any one or how could he know that a particular one is adequate for his purpose. You can carry whatever tests you need.'

The students were given three springs (black long BL, black short BS, and silver S) and an elastic band, EB, a ruler, little objects with a mass similar to the small stones, scales, calibrated weights with a hanger including a needle, so lengths could be measured on the ruler, and a stay.

Springs could be used in order to measure mass if they behave in accordance with Hooke's law and, in fact, dynamometers are used by field zoologists. Two of them, BL and S, met this condition, while for the BS and the EB the relation among stretching and weight was not linear. Although it could be possible to use springs not behaving in a linear way, a non-linear calibrating was beyond the reach of the students. They were expected to decide which data were needed (measures of the stretching), to assemble the material and produce the data, to find the patterns (relations among stretching and weight) and to find an explanation for them.

– *The tracks of the thief*: the students are given an unknown sample, and told that an alien has stolen the video camera, leaving parts of its skin from which the microscope slide was prepared, and asked to identify it. The handout describes four suspects, two animal-like, two plant-like. The problem design is discussed in Jiménez, Díaz and Duschl (1998b). The samples of group F & G were plant epithelium and fish operculum; they were not expected to recognise them, just to identify them as plant tissue having two different cell types in the first case, and animal cells, all of a single type, in the second.

The teacher expectation was that they 'saw' the cells (decide which 'structures' correspond to whole cells or cell parts); matched them with animal cells studied in a previous session; found the patterns, for instance distinguish individual cells from groups; explain the patterns and identify the sample.

In summary it can be said that both tasks provided a problem-solving context in which the students were expected to cooperate in order to find a solution.

2.1 Data collection and analysis

The conversations of the students were audiotaped; the Biology students and some sessions of the Physics tasks were videotaped. The analysis focuses on the conversations and actions in small groups while solving the problems. The argumentative and epistemic operations were codified with the frames discussed.

3. EPISTEMIC OPERATIONS: FROM RAW DATA TO EXPLANATIONS

The epistemic operations are analysed using Duschl and Erduran (1996) model: the expected path was outlined for each task (as summarized in the methods section) and the performances of the groups were matched against it.

3.1 An explorer in the field: comparison of four small groups in the Springs task

In table 1 the steps taken by the four groups with and without teacher assistance are compared. The first row summarizes each step.

It is not easy to compare the four groups, for each had a particular development, and took steps forward and back, so we must be cautious about conclusions. On the whole, clear differences could be found among Groups 1 and 4 (discussed below), while Groups 2 and 3 were able to proceed from raw to hard data and began to find difficulties in identifying the patterns.

Steps	produce **raw data**	select **hard data**	identify **patterns**	**explain** patterns
	measure stretchings	decide intervals, discard useless measurements	relate stretching to mass: k	springs with constant k are appropriated
Group 1 first time	assisted	assisted difficulty: indirect measurement	assisted	assisted
Group 2 first time	unassisted	unassisted	part. assisted (when the pattern broke)	assisted
Group 3 not first time	unassisted	unassisted	part. assisted (Hooke's law not universal)	incomplete (did not request help)
Group 4 not first time	unassisted	unassisted	unassisted	assisted (discussed intervals)

Table 1. Progression in the Springs task and teacher assistance

Group 1 required assistance from the beginning, their first actions were erratic and seem guided by the assumption that all instruments provided must be used, without any idea about what to do with the data in order to solve the problem. The selection of hard data proved difficult for them:

Group 1, session 1

697 Teacher: Is it possible to measure mass with springs?

698 Juana: No.

699 Teacher: Why not?

702 Juana: Because... Because a mass is measured with... with... scales. I don't know! *(long pause)*

703 Juana: Because it is not normal.

One of the obstacles is that they do not understand that it is possible to measure mass indirectly. Pattern identification was also hindered by the lack of familiarity with the strategy of measuring a magnitude indirectly.

Group 4 was able to recall Hooke's law as a cognitive resource in the first session. They designed a methodology, measured stretching and calculated elastic constants. Then they identified patterns and began to relate deformability to capacity of use to determine mass. After consulting the teacher, deformability is related to k, and they refine the pattern identification. It is the only group mentioning intervals for which springs comply with Hooke's law.

It seems that the quality of the epistemic operations is higher in the two last steps: identifying patterns and explaining them and both proved difficult for all groups: only group 4 was able to identify patterns unassisted. Groups 2 and 3 were both able to proceed through the two first stages unassisted and although they took some steps in identifying patterns, both needed assistance for completing it. Group 3 was unable to provide a full explanation, but their strategies and use of resources were more sophisticated than those from group 2. Although it is clear that group 1 experienced more difficulties and group 4 had the best performance, it is not possible to attribute these differences to the previous experience of the last, as groups 2 and 3 had similar performances.

3.2 The tracks of the thief: constructing data in microscopic observation

Table 2 represents the progression of groups F & G. The first row summarizes each step. In a previous microscope session they had observed onion skin and cheek cells samples.

Steps	Produce **raw data**	Select **hard data**	Identify **patterns**	**Explain** patterns
	observe the slide and 'see' the cells	match sample with prototype of plant / animal	e.g. match sample with drawings in notebooks	Identify sample, e.g. as animal cells of a single type
Group F	unassisted (lines 26, 53: nucleus)	unassisted (lines 23, 115: green spot)	unassisted (lines 31-35: stomata)	unassisted (line 331: traits of phloem)
Group G	unassisted (lines 84, 109: epithelium)	unassisted (lines 129-131: cheek cells)	unassisted (lines 137, 153: one type)	unassisted (line 140: one cell type)

Table 2. Progression of the groups F and G in the microscope task

'Seeing' the cells may look trivial, but distinguishing structures is not. For instance, in the F sample there are stomata; the students describe them as doughnuts, but without recognising them, as seen below. An instance of selection of hard data in group G is the comparison, in line 129, of the sample with the cheek ('face') cells seen in a previous session:

Group F

113 Felix: Yes, it is a plant

114 Flavio: This doesn't have stomata

115 Fabri: but it has the green thing…

Group G

129 Gema: This looks like... like these cells from the face

The comparison is not immediate because cheek cells (for them prototype of animal cells) are loose, individualized, while the cells in the fish operculum are included in the extracelular matrix. The step from hard data to identifying patterns required several related operations as, in group G to distinguish whether they belonged to one or several cell types. Later they matched the sample with the appropriate 'suspect'. In summary, the students used epistemic operations, as appeal to analogies, a resource in knowledge construction, to explain new observations or data; or comparison with prototype, used in biology in identification of species, which requires the students to make use of the available cognitive resources (as knowledge of the prototype or assumption that all animal or plant cells are not exactly the same). They went through all the stages unassisted.

In summary, Duschl & Erduran (1996) model proved an useful tool for checking the progression of the students in epistemic operations. It has to be noted that, although we discussed the stages separately, they are not so clearly defined in the temporal sequence, but mixed, because students sometimes go forward and back from patterns to data or from explanations to patterns. Not all the steps presented the same difficulties for the students.

3.3 Argumentative operations: warranted arguments

For argumentative operations in each group, first we identified the sequence of arguments, considering a part of debate a different argument when it had a different claim; for instance, in the Biology task the first claims relate to the distinction animal/plant, while later they move to the type of tissue. Only substantive arguments, related to the knowledge domain, are considered for the analysis, not discussions about who should use the microscope. From each sequence, the focus is on the warranted arguments, considered of better quality than unjustified claims. In a laboratory context, the actions are as relevant as the dialogue, so their justifications are also of interest. Four types of justifications (in arguments and actions) have been found and placed in the following hierarchy:

- cognitive: justifications in terms of knowledge. For instance the reference to neutralisation as conceptual support to titration in task 3.
- pragmatic: justifications in terms of achieving a goal in solving the problem. For instance, coordination to measure times in the task about the motion of a ball on a rail.

– appeal to rules: in terms of established or perceived rules or steps. For instance, look for formulae to apply in order to reach a solution.
– dramaturgical: justifications under Habermas' category of performing a role. In the first Physics & Chemistry tasks the students from group 3 separated a small sample from the salt & sand mixture. Then they decided to separate the whole mixture (which was not necessary) only because other groups did it.

The progress would be shown both by the increase in number within a category and by the shift from one category to another as cognitive justifications are considered of better quality than pragmatical or appeal to rules and these three types, from the communicative domain, of better quality than the dramaturgical.

3.4 Physics and Chemistry tasks: progression along twelve sessions

The progression of Group 3 was analysed along the five tasks for explicit and implicit justifications. A justification was considered implicit when the claim or action followed immediately to an explanation (from the teacher or another student). Table 3 summarizes the types of justifications, in the cognitive category, the number of explicit justifications is included in parentheses. The expectation that a progression would be found was not fulfilled. Rather it seems that the justifications are context-bound and related to the cognitive demand of the tasks which was very different, being the two firsts quite simple, while the last three, particularly # 3, titration, proved very demanding. There were other differences: some tasks took only two or two and a half sessions while other needed three.

tasks / warrants	1 separate salt / sand	2 volume of recipients	3 titrate acid	4 motion	5 springs
Cognitive	55 (21)	40 (19)	99 (10)	64 (22)	74 (19)
Pragmatic	7	48	116	61	66
Rules	16	6	14	3	3
Dramaturgical	2	4	1	4	17
Not justified	13	11	54	33	48

Table 3. Explicit (in cognitive in brackets) and implicit warrants in-group 3

For tasks 1, 4 and 5, the higher numbers of justifications are cognitive, (in task 4 the pragmatics are close). For 2 and 3 the pragmatic, it may be related to the great number of manipulations and tests which both required; task 3 needed much assistance from the teacher; perhaps this continued assistance explains why so few cognitive claims and actions were explicitly justified. One of the contexts that promote justifications is the debate, for when all agree there is no need for warrants, or when the knowledge justifications had been explicitly stated a short time before, they are not repeated. Ogborn, Kress, Martins & McGillicuddy (1996) refer to these differences as semiotic tension, which for them is the motor of communication.

3.5 The tracks of the thief

On the whole session the students constructed or co-constructed respectively 18 arguments in group F, ten explicitly warranted, and 21 in group G, also ten warranted. From these warranted argument, in group F the first three deal with the issue of whether the sample belongs to a plant or to an animal, then they discussed the presence of nucleus and chlorophyll, discard animal tissue and establish that it is phloem. Four arguments are co-constructed and six individually constructed. In group G the first two arguments discuss if it is animal or plant and the presence of blood, and the debate proceeds back and forth between the identification of one or two cell types (some of them looked bigger). Also four arguments were co-constructed and the rest produced by individuals. In these twenty arguments all warrants are cognitive and have an empirical referent, relating evidence (observation) to biology knowledge. The sources for this knowledge were the information in the handout; other microscope sessions, or cell biology.

4. CONCLUSIONS AND EDUCATIONAL IMPLICATIONS

The purpose of this paper is not to show that the students learn more working with this approach nor to compare their performances with others, but to explore the processes of data construction and transformation, the arguments and justifications offered by the students to their peers while involved in problem solving in an experimental context. Although the use of open problems has been advocated, there is only a reduced number of studies about them, about how do students cope with problems when they have to decide the path to follow, and the difficulties that they encounter, partly related to a school culture which casts them in a role demanding just to follow instructions.

The analysis of the operations carried by the students and of their conversations shows an involvement in the practices of scientific culture, documented by their use of argumentative and epistemic operations, by their appeal to cognitive or procedural resources and how they justify their actions and claims when they have to convince the audience. Although most of the small groups studied were able to proceed through the first steps in data transformation, the operations of higher quality, pattern identification and explanation proved difficult and required assistance from the teacher in most cases.

About the quality in the argumentative operations, it can be said that the justifications for claims and actions belonged mainly to the cognitive and pragmatic categories, but no clear progression was found along a sequence of five open tasks. A complete argument is not just a claim; it makes appeal to warrants, and sometimes to backings. Such appeals, in addition to the data, seem to be context dependent. The exploration of the argumentative operations in two laboratory contexts in different content domains and particularly the analysis of the operations considered of higher quality, the warranted arguments, shows that the greater influence in them seems to be the context of the task and not even the previous experience with similar open problems. The role of initial knowledge, although it is not the focus of the paper, must be acknowledged. The differences in justification seem also related to the context of each task.

We believe that the design of open laboratory tasks could promote knowledge producing communities, and help students to work in a context closer to scientific practice, as their argumentative and epistemic operations show, but this approach is not devoid of problems as they encounter obstacles. Some are related to their lack of control of resources, other to the way they perceive laboratory tasks, lab instruments or even their role as students. When suggesting teachers to adopt this new approach, it is important to discuss these problems and how it is possible to cope with them, to offer a realistic image of lab work in a context of innovation, while communicating the exciting moments when we can observe students constructing their own meanings and designing their own paths to solve a problem.

ACKNOWLEDGEMENT

Work supported by the Spanish DGESIC, grant PB 98-0616.

REFERENCES

Duschl, R,.A. & Erduran, S. (1996). Modelling the growth of scientific knowledge. In: Weldorf, G., Osborne, J., Scott, P. *Research in Science Education in Europe* . London: Falmer Press.

Habermas, J. (1987). *Teoría de la acción comunicativa*. Madrid: Taurus.

Jiménez-Aleixandre, M., P., Díaz de Bustamante, J. & Duschl R.,A. (1998a). Scientific culture and school culture: epistemic and procedural components. Paper presented at the NARST annual meeting, San Diego, CA.

Jiménez-Aleixandre, M., P., Díaz de Bustamante, J. & Duschl R.A. (1998b) Plant, Animal or Thief? Solving problems under the microscope. In Bandiera, Caravita & Torracca *Research in Science Education in Europe*. Dordrecht, Kluwer.

Kelly, G.,J. & Crawford, T. (1997). An ethnographic investigation of the discourse processes of school science. *Science Education*. 81, 533-559.

Ogborn, J., Kress, G., Martins, I. & McGillicuddy, K. (1996). *Explaining Science in the Classroom*. Buckingham: Open University Press

Reigosa C. & Jiménez-Aleixandre M.,P. (2000). La cultura científica en la resolución de problemas en el laboratorio (Scientific culture when solving problems in the laboratory). *Enseñanza de las Ciencias*. 18: 275-284.

Toulmin, S. (1958). *The uses of argument*. Cambridge: Cambridge University Press.

MODELLING THE EVOLUTION OF TEACHING - LEARNING SEQUENCES: FROM DISCOVERY TO CONSTRUCTIVISM

Petros Kariotoglou, Dimitris Psillos, Vasilis Tselfes

School of Education, Aristotle University of Thessaloniki, Greece

ABSTRACT

In this study we study in retrospect three teaching learning sequences in the area of fluids developed over several years. We investigate the factors that affected the development of these sequences by analyzing the developers practice on the basis of Pickering's theory of scientific practice. We argue that, in science education, the researchers' practice towards meeting his/her/ objectives is influenced by the dominant approach to science teaching and learning but is also constrained by the material factor and the educational introduction.

In recent years, many research based teaching-learning sequences (TLSs) have been developed and studied on a wide range of topics (Anderson and Bach, 1996; Meheut, 1997; Psillos, 1998; Duit, 1999; Psillos and Meheut, 2001). TLSs are considered as a medium-scale curriculum development, a product of developmental research. By developmental research we mean the interlacing of research, development and application of a teaching-learning sequence on a specific topic, usually lasting a few weeks, in a cyclical evolutionary process enlightened by rich research data (Lijnse, 1995; Kattmann and Duit, 1996).

The nature and evolution of scientific knowledge, research on students' conceptions and reasoning, epistemological assumptions, learning beliefs, existing resources, current pedagogical theories and educational contexts are among the factors which are involved in the development of such TLSs. Yet, whereas the learning outcomes from the various TLSs are often published to a considerable extent, explicit and implicit assumptions and decisions that affect the design and development of TLSs are less widely discussed or even made clear by the researchers.

In our case, we have been involved in long-term research and development concerning the teaching and understanding of conceptual and procedural scientific knowledge involved in the area of fluids. Our work has focused on the development and investigation of three innovative sequences on fluids. In this context, rich data exist about teaching practices and learning results over a long period. In the present paper, we study these sequences in retrospect, in order to identify factors and assumptions that affected their development and evaluation in different educational settings in Greece.

D. Psillos et al. (eds.), Science Education Research in the Knowledge-Based Society, 259–268.

1. MODE OF INQUIRY

The existence of data about the sequences on fluids, which were developed and evolved by almost the same scientific staff, in a changing educational and scientific landscape over the past twenty (20) years, constitutes a significant advantage of the present paper. We consider these three sequences as three successive TLSs, namely TLS1, TLS2, TLS3. As data we take documents concerning the theoretical foundations of the TLSs within the context of Science Education appearring in the teacher guides and plans, the equivalent successive proposals on teaching/learning activities both for the teachers and the pupils (worksheets, plans). Besides we take into account the obtained learning outcomes from each TLS following their classroom application. In addition, our data include documents concerning the initial official curriculum/textbook that obtained during the initiation of the TLSs in the early '80s.

Our approach to the analysis of the data is an epistemological one and aims at revealing the practices that shaped the successive TLSs as scientific products in the domain of Science Education. In our case, following Pickering, we regard as scientific practice (Pickering, 1992; Pickering, 1995) a changeable "behavioural model" that unravels through time. Such a model is structured by the relations between the scientist's knowledge, handling beliefs and experiences as well as his social/institutional relationships, at least in the area of (science) education. We accept that the science education researcher's scientific practice on the one hand exposes the features of the TLSs, which either change or remain stable as time passes, and on the other reveals the relations between the stability or the variability of the factors that influence the science educators' scientific work running over the evolution of these particular TLSs.

More specifically, we assume that in our case the Science Education scientist (or didactician), through his practices, constructs and connects teaching / learning activities in order to produce a TLS. The various activities and the connections between them are constrained by a) educational and institutional, b) material and physical and c) scientific (Science Education) factors (Pickering, 1992; Pickering, 1995). In so doing the didactician is aiming at accomplishing his various objectives and overcoming the specific resistances his work runs into, through a process of accommodation.

As an example, we consider that the objective of a science education researcher in developing a TLS may be the learning of a specific topic by certain students as stipulated by the curriculum that obtains in a specific educational context. **He/she probably is approaching certain resistances to the TLS objectives in terms of the dominant teaching/learning theory (e.g. discovery, constructivism).** We consider though that the choice of the relevant teaching-learning activities does not depend only on the researcher's theoretical preferences. The objectives which have been set, the anticipated resistances and the envisioned adaptations under certain contextual and material constraints also affect the shaping of this TLS. In this way a TLS, being a product of scientific practice, is more of a changeable structure of interrelated teaching/learning activities than a final product. We envision the overall process to be illustrated in the schema that is presented in Figure 1.

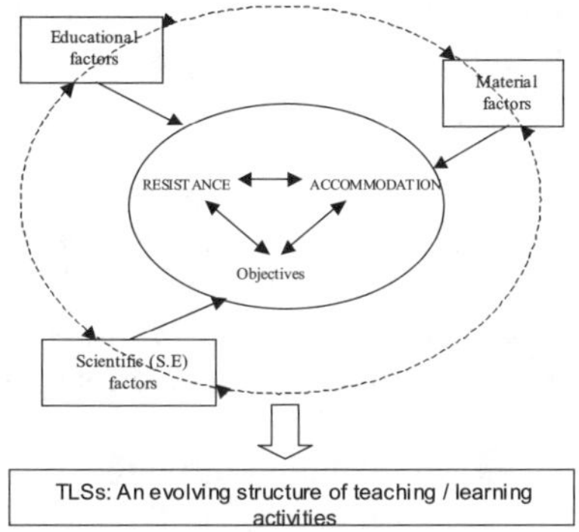

Figure 1. The dynamic that shapes scientific practice in the development of a TLS

Following the above rationale, the scientists' (science educators) practice may derive from the very analysis of the various features of each TLS as materialised in the relevant documents. Such an analysis is presented in the next sections. It was carried out, at an initial stage, by one of the writers (who did not participate in the construction of the sequences) and the results were validated afterwards, during the three (3) researchers' discussion.

2. DOCUMENTING THE EVOLUTION OF THE TEACHING-LEARNING SEQUENCES (TLS) ON FLUIDS

In this section, we describe the evolution of the three TLSs with regard to their content, learning objectives, relevant students' achievements and the dominant approach to science teaching and learning in which each TLS evolved. We also present the official Greek curriculum on fluids obtaining in the early '80s. We must note that these three TLS were small scale innovations not widely adopted changes.

The 1st TLS, developed in the early '80s, arose from the need to introduce students to labwork, since in Greece traditional teaching was based on lecturing and, possibly, on some demos. For that purpose, a 6-unit teaching sequence on Fluid Mechanics was developed on the basis of the official curriculum and the school textbook. The main objective of the TLS is student practice in experimental skills and some understanding of fluid concepts and laws in relation to everyday applications. TLS1 was applied to 13/14-year-old pupils in the 2nd form of Greek Gymnasium. For reasons related to educational context, the application of TLS1 activities follow some lecturing on the basic concepts/principles of fluid phenomena in line with the official curriculum. The units include labwork in small groups, using rotating laboratory exercises and structured worksheets) on themes such as the fundamental law of hydrostatics, Pascal's principle and factors affecting buoyancy.

TLS1 was evolved and applied at a time when discovery approaches to teaching and learning science were dominant. For this reason labwork was based on guided discovery approaches with regard to principles and laws. In addition, there are experiments that students are asked to interpret, aiming to connect physics with everyday life. For example a) compression/extension of the piston of a syringe containing first water and then air, in order to observe similarities and differences between liquids and gases, b) experiment with an egg floating in salted water which shows off the role of the density of the liquid in buoyancy, a case connected with everyday experiences, like brine c) experiment on shipping to point out floating condition cases where weight was a variable.

Table 1 presents the historical evolution of these TLSs, in connection with their content, learning objectives, and student achievements. At the beginning of this table, the content, objectives and perceived result of the official curriculum of that time are also presented as reference material.

PHASES PERIOD POPULATIO N	**CONTENT OF THE TEACHING SEQUENCE**	**LEARNING OBJECTIVES**	**RESULTS**
Official curriculum Early '80s, Gymnasium	Concepts, Laws/Principles and Phenomena of Fluids Mechanics	Conceptual learning of the basic issues of Fluids Mechanics	Limited conceptual improvement
TLS1 Early '80s, Gymnasium	Verification experiments of rules and principles Interpretation of experiments in order to connect Physics to everyday life.	Learning experimental skills Connection of Physics theory to practice	Positive attitudes to labwork Improvement of **experimental skills** Limited conceptual improvement
TLS2 Late '80s, Gymnasium	Fluids' content transposition (introduction of P as primary concept) Experiments on Fluids to facilitate conceptual knowledge	Learning conceptual knowledge e.g. differentiation between pressure and force Learning procedural knowledge	Sufficient learning of conceptual knowledge e.g. successful differentiation between pressure and force Insufficient learning of procedural knowledge
TLS3 '90s University	Fluids' content transposition (introduction of P as primary concept) Experiments on Fluids to facilitate both conceptual and procedural learning	Conceptual learning e.g. Differentiation between pressure and force Learning procedural knowledge	Improvement of conceptual learning, - Differentiation of Pressure / Force described in 3 distinctive steps Moderate **procedural learning** Difficulty in manipulating concepts as entities

Results from the application of have shown positive student attitudes towards the lab as well as substantial student familiarization with experimental skills. However, this was not the case on the conceptual level **(Kariotoglou et al, 1988)**

The 2nd TLS refers to the transformation of guided labwork into the constructivist context and the pursuit of mainly conceptual learning, driven by conceptual change. This transformation is based, on one hand, on the "students' conceptions" movement and on the other hand on content transposition. As far as "students' views are concerned, TLS2 takes advantage of their classification, regarding pressure, into three models: packed crowd, pressure – force and liquidness, in order to introduce and negotiate the pressure concept (see Kariotoglou and Psillos, 1993; Kariotoglou et al., 1995; for an extensive discussion of the models). To avoid reinforcement of the pressure – force model (non-differentiation between pressure and force), the pressure concept is introduced independently, qualitatively and experimentally, without connecting it with force. Such an approach arises from students' conceptions, which are reinforced by the introduction of the pressure concept in the traditional way (P=F / s). However, at the end of TLS2 the concepts of pressure and force are connected with each other following their assumed differentiation.

Table 1. Evolution of the content, objectives and learning achievements of the TLSs

In the process of differentiation we induce students to cognitive conflict. For example, they are exposed to the following critical experiment: predict and test experimentally a) the comparison of pressures in a wide / narrow vessel that contains water to the same depth, and b) the comparison of forces exerted in order to detach a narrow sucker and a wide one). Based on the model of non-differentiated concepts (pressure – force model), predictions about pressure and force are expected to be uniform, that is, students claim either that they are equal or that the bigger the size of the vessel / sucker the bigger the pressure or force exerted. However, the experimental data are different, that is, equal pressures and unequal forces were observed. Discussion on this issue leads to most students' differentiating the two concepts.

The results of TLS2 have shown success concerning conceptual learning and the differentiation of pressure and force concepts. As for the **experimental skills**, no significant improvement was observed since it was not the main objective. The experiments are mainly regarded as a means to conceptual learning in combination with guided discussion (Kariotoglou et al., 1993; Kariotoglou et al., 1995).

TLS3 is an extension of TLS2 that was applied to a different population, namely student primary teachers with a limited amount of conceptual and procedural knowledge in science. Empirical data point out that their conceptual knowledge had similar features to those of 13/14 year-old pupils , though richer in interpretations and terminology. Besides, the equivalent pursuit of learning experimental procedures and conceptual learning constitutes a fundamental change in the constructivist context. This fact bears a double justification: on the one hand, procedures are an inseparable part of science and, on the other, prospective teachers ought to become acquainted with this part of scientific knowledge as well. In that way, experimental activities become enriched and aim to facilitate not only conceptual learning but procedural learning as well. For instance, a) prediction, planning and realization of experiments aimed at testing variables affecting hydrostatic pressure, b) experimental investigation of Pascal's principle using a specially designed vessel fitted with three manometers and a pump to increase pressure (see also Table 1 and Table 2, below).

The results have shown success in the conceptual domain combined with a refined presumption of learning in three levels (Psillos and Kariotoglou 1999). The objective concerning procedural learning proved to be less successful, chiefly due to a difficulty in handling theoretical concepts like entities of the material world, with which students would be able to intervene in the experiments. For example, while pressure seems to be understood as a representation and differentiated from force, it cannot be handled as an entity and used to intervene in the experiment. That is, students cannot, when asked, intervene to increase or decrease pressure (see for details Kariotoglou, 1998) (see table 1).

3. ANALYSING THE SCIENTIFIC PRACTICE: RESISTANCE AND ACCOMMODATION

In Table 2, we present the official curriculum as well as all three TLS in terms of the factors we take into account in our analysis of scientific practice, namely educational context, the dominant scientific approach to teaching and learning and the material conditions. As mentioned before (see Figure 1), these factors are considered to be taken into account by a didactician in his/her attempt to promote the achievement of certain objectives by the targeted students an assumption which is verified by our analysis of the documents. We note that our analysis focus on the researcher's practice not policy makers' or teachers' practice. In this table we codify and present the proposed teaching/learning activities in order to show their relation with the envisioned factors and the relevant learning results. Table 2 reveals whether there is a resistance to achieving the set objectives in a TLS the interpretation of which will guide the accommodation of this TLS.

The 1st TLS was developed within a context. in which learning by reception of a scientific topic was a widespread practice, and not just within Greek educational tradition. This tradition is governed by the rationale of expository teaching and is reinforced by the limited material/technical school infrastructure: that is, absence of laboratory premises and a limited number of experimental set-ups. The basic resistance that appears within this practice has to do with the limited learning results in relation to scientific subject matter , and is located within the educational framework (compulsory education) whose objective is learning of scientific content

In the face of this sort of resistance, the options of the researcher for accommodation were either to modify the scientific content or to select an appropriate teaching approach. In the general education framework of the early 1980s, modification and/or abandonment of the relative content did not appear compatible with current views on science teaching. The second approach, however, that of discovery-oriented teaching, seemed more plausible. The students would discover knowledge and learn by doing. Adapted to the framework of the existing material restrictions (limited material/technical infrastructure), this option resulted in labwork performed on a rotation basis with a single set of instruments. On this basis, the didactical proposal of TLS1 was the addition of discovery based practical labwork to the lectures of the typical curriculum. This practice, regarded by the researchers as necessary and potentially effective as far as **the linking of theory with practice and the απόκτηση of laboratory skills is concerned**,was successfully realized within the contemporary educational context, although under material constraints. Despite the successful results of the application, the resistance towards conceptual learning persisted.

Table 2. Analysis of the scientific practice

Teaching Learning Sequence	EDUCATIONAL Factor	Scientific Factor (Science Education)	Material Factor	Learning Objective	Accommodation: Didactical Proposal in terms of Activities	Resistance
Official Curri-culum	General Education Context 9th form	Hypothesis of learning by reception	Limited number of standardized set-ups of simple demo experiments	Learning of scientific content	Lecture) + Demo experiments)	Limited Con-ceptual Learning
TLS1	General Education Context 9th form	Hypothesis of Discovery Learning	Standardized set-ups of simple demo experiments + Limited number of experimental set-ups with simple / everyday materials	Learning of scientific content + Linking theory and practice + Experimental skills	Lecture + student guided discovery labwork in a rotating lab form	Limited Con-ceptual learning
TLS2	General Education Context 9th form	Students' Pre-existing Conceptual Models + Hypothesis of Learning by Construc-ting Knowledge	Limited number of experimental set-ups both standardized and with simple / everyday materials	Learning of Specific content, transposed from scientific theories + Learning of scientific procedures	Familiarization experiments, Testing ideas by guided experiments Demos inducing conflict Discussion of applications	Limited learning of Proce-dural Know-ledge

TLS3	Primary School teachers' education context	Students' pre-existing conceptual models + Hypothesis of Learning by Construc-ting knowledge	Experimental set -ups with simple / everyday materials + Standardized experimental set - ups (corresponding to the multitude of students' practicing groups)	Learning of specific content transposed from scientific thories + Learning of scientific procedures + Experimental skills	As TLS2 +open investigation Guided discovery experiments	Diffi-culty in Mani-pulating concepts as entities of the real world

The new resistance, i.e. limited conceptual learning, that emerged at the end of the 1980s has to be seen in the context of the students' conceptions movement and the emergent constructivist approach to teaching and learning in science education. This new scientific framework affects the new accommodation. The 2nd TLS consequently exploits students' conceptions while at the same time transforming the didactical content in order to render it appropriate for the target population. We may detect here a substantial change in teaching aims, from the learning of the relevant scientific theory and models to the learning of a transposed scientific theory and models. Within this framework the objective of learning laboratory procedures remains, but is subordinated to the basic objective of conceptual learning. Based on this rationale, the learning of scientific methods was scarcely pursued and therefore was not achieved. This fact constituted the new resistance (see table 2, TLS 2).

In the TLS3 there was a change in a significant educational factor, with the attempt to apply the basic principles of the sequence to a new population, namely student primary school teachers. At the same time the resistance to procedural learning remained, since the new population required them for professional reasons. What was the outcome of this new accommodation? Since the need for conceptual learning remained, the role of students' ideas and the constructivist approach could not be ignored. But the requirement of scientific procedures restored the idea of guided labwork. The difference between this and the 2nd TLS lies in the fact that in this case experimental skills are pursued for themselves rather than as an adjunct to conceptual learning. Accommodation in the 3rd TLS led to a combination of constructivist teaching with guided labwork, a novel connection between the scientific and the material factor within the current educational context (see table 2, TLS 3).

The above description of the dialectical relation between resistances and accommodation in the evolution of the TLSs shows that material, scientific and educational factors have all played an important role in the changes that have taken place. These factors led to changes in both the objectives of the TLSs and the didactical proposal.

A global presentation of the preceding analysis is given in Table 2. The resistances and accommodations are shown in the last two columns of the table. The resistances are mainly related to learning of conceptual knowledge (Official Curriculum, 1st TLS) and of procedural knowledge (2nd and 3rd TLS). The accommodation that led to the development of the 1st TLS is related to the educational framework (content learning), the predominant scientific paradigm of the period (discovery) and the material factor (limited materials / rotating labwork). In the case of the 2nd TLS, we have a change in the scientific paradigm (constructivism), the same educational framework, and a material factor adapted to the new scientific paradigm: that is, demonstrations of critical experiments. Finally, in the 3rd TLS the scientific paradigm remains the same (constructivism) but is enriched (guided labwork) on account of the priorities of the new educational framework (university education). This new educational framework also affects the material factor (existence of sufficient material means). We observe that all three factors of our theory – that is, the scientific, the educational and the material – play a role in each separate accommodation. We further observe that in each case the accommodation is realised within the framework of the currently predominant scientific approach.

4. CONCLUDING REMARKS

This paper is a retrospective critical review of the evolution of a series of Teaching-Learning Sequences in the field of fluids with the objective of revealing the factors that affected the development and modification of these TLSs as they were applied at different educational levels. In our study we argue that the development of these Teaching-Learning Sequences is as scientific practice within science education. Accordingly we study such a practice by a method developed on the basis of Pickering's theory (1992, 1995) of scientific practices.

The analysis of the dialectical relation between resistances and accommodations with regard to the evolution of the TLSs that was carried out in the previous sections reveals that scientific and educational factors as well as material ones have played an important role in the changes that have taken place. The changes related both to the objectives and the relevant teaching proposal.. Our basic position is that prevailing approaches to science teaching and learning (expository, discovery-based or constructivist) constitute the scientific factor that influences the practice of the science education researchers and this is verified from the analysis of the data. Besides a privileged relation between learning results and scientific factor emerges as the researchers attempted to make up for the noted limitations in students' learning. Our study reveals that this relation is constrained by the educational and material factors which affect the specific materializations of the TLS. Yet in discussing TLSs in the public domain these factors are not usually taken into account.

We believe that the practices of science education researchers are becoming of an age so there may be enough historical documents whose study warrants investigation. In our case we believe that this review of the development practices applied to the sequences as the object of scientific investigation contributes from both the methodological and the conceptual point of view to the modelling of explicit and implicit assumptions and decisions that have a considerable effect on the design and development of a sequence and therefore on the relation of research to development in Science Education.

REFERENCES

Andersson B., & Bach F., (1996). Developing new teaching sequences in science: The example of "Gases and their properties". In: Welford, G., Osborne, J. & Scott, P. (eds), *Research in science education in Europe: current issues and themes* (pp. 7-21). London: The Falmer Press.

Duit, R., (1999). A model of educational reconstruction – A framework for research and development in Science education. In: Koumaras, P., Kariotoglou, P., Tselfes, V. & Psillos, D. (Eds.), *Proceedings of the 1st Panhellenic Conference on science education and Application of New Technologies in education* (pp. 30 - 34). Thessaloniki: Christodoulides (in Greek).

Kariotoglou, P., Koliopoulos, D. & Psillos, D. (1988). An approach to the application of experimental teaching of Physics in Gymnasium.. Contemporary Educatiom, 38, pp. 90 – 96 (in Greek).

Kariotoglou, P., (1998). Investigating aspects of an innovative experimental sequence: The case of fluids. Case Study final report, project "*Labwork in Science Education*", European Commission DG XII, SOE2-CT95-2001.

Kariotoglou, P. & Psillos, D. (1993). Pupils' Pressure Models and their implications for Instruction. *Research in Science and Technological Education*, 11(1), 95 - 108.

Kariotoglou, P., Koumaras, P. & Psillos, D., (1995). Différentiation conceptuelle: Un enseignement d' hydrostatique, fondé sur le développement et la contradiction des conceptions des élèves. *Didaskalia*, 7, 63- 90 (in French).

Kariotoglou, P., Koumaras, P. and Psillos, D. (1993). A constructivist approach for teaching fluid phenomena. *Physics Education*, 28, 164 - 169.

Kariotoglou, P., Tselfes, V., Evangellinos, D. & Psillos, D., (1999). An investigation on student teachers' laboratory practices during a familiarisation with phenomena phase of experimental teaching. *Paper presented to the 2nd International ESERA Conference, Kiel, Germany.*

Kattmann, U. & Duit, R. (1996). Educational Reconstruction-Bringing together Issues of Scientific Clarification and Students' Conceptions. *Paper presented at the First European Conference on Didactic of Biology (ERIDOB), November, Kiel, Germany.*

Linjse, P. L., (1995). Developmental research as a way to an empirically based "Didactical Structure" of Science. *Science Education,* 79(2), 189-199.

Méheut M., (1997). Designing a learning sequence about a pre-quantitative model of gases: the parts played by questions and by a computer-simulation. *International Journal of Science Education,* 19(6), 647-660.

Psillos D., Méheut M., (2001). Teaching-learning sequences as a means for linking research to development. In: Psillos D., Kariotoglou P., Tselfes V., Bisdikian G., Fassoulopoulos G., Hatzikraniotis E., Kallery M. (Eds), Proceedings of the Third International Conference on Science Education Research in the Knowledge-based Society (p.226-244). Thessaloniki, Greece.

Pickering, A., (1992). From science as knowledge to science as practice. In: A. Pickering (Ed) *Science as practice and culture,* Chicago: The University of Chicago Press.

Pickering, A., (1995). *The Mangle of Practice.* Chicago: The University of Chicago Press.

Psillos, D. (1998). Teaching introductory electricity. In: Tiberghien, A., Jossem, E. – L. & Barojas, J., (eds) *Connecting Research in Physics Education with Teacher Education.* http://www.physics.ohio-state.edu/~jossem/ ICPE/BOOKS.html.

Psillos, D. & Kariotoglou, P., (1999). Teaching Fluids: Intended knowledge and students' actual conceptual evolution. *International Journal of Science Education* (special issue), 21(1), 17 – 38.

NONLINEAR PHYSICS IN UPPER PHYSICS CLASSES: EDUCATIONAL RECONSTRUCTION AS A FRAME FOR DEVELOPMENT AND RESEARCH IN A STUDY OF TEACHING AND LEARNING BASIC IDEAS OF NONLINEARITY

Michael Komorek, Dimitrios Stavrou,
Reinders Duit

IPN -Leibniz Institute for Science Education at the University of Kiel, Germany

ABSTRACT

The model of educational reconstruction has proven a useful frame for developing instructional units, especially, for integrating empirical research on teaching and learning and instructional development. The model allows bringing issues of science content structure into balance with educational issues. Processes of educational reconstruction of a new topic, namely nonlinear physics and self-organization are presented. In cooperation with a group of experienced teachers we developed a teaching sequence and evaluated it in a real school setting. Our empirical data provide insight into the difficulties teachers experience in reconstructing a new scientific topic, how these teachers transform their reconstructed subject matter structure into an appropriate content structure for instruction, and into the students learning processes towards understanding nonlinear systems.

1. BACKGROUND, AIMS AND FRAMEWORK

Progress in teaching and learning science and in research in this field appears only to be possible if there is a balance between two traditions in science education research. The one can be called *science oriented* and the other *student oriented* (Duit, 2001). As a frame for analytical and empirical research processes we use a *model of educational reconstruction* (Fig. 1) which main aim it is to realize this balance. The model has been developed as a theoretical framework for studies focussing on whether it is worthwhile and possible to teach particular content areas of science (Kattmann et al., 1997). It comprises three closely interrelated components. The first component includes clarification of science matter structure and analysis of the educational significance of particular science content. These analyses constitute a basis for constructing the *content structure for instruction*. This process includes an analysis of the core and elementary concepts of the content in question and a reconstruction of these concepts and their relationships from educational perspectives.

D. Psillos et al. (eds.), Science Education Research in the Knowledge-Based Society, 269–276.

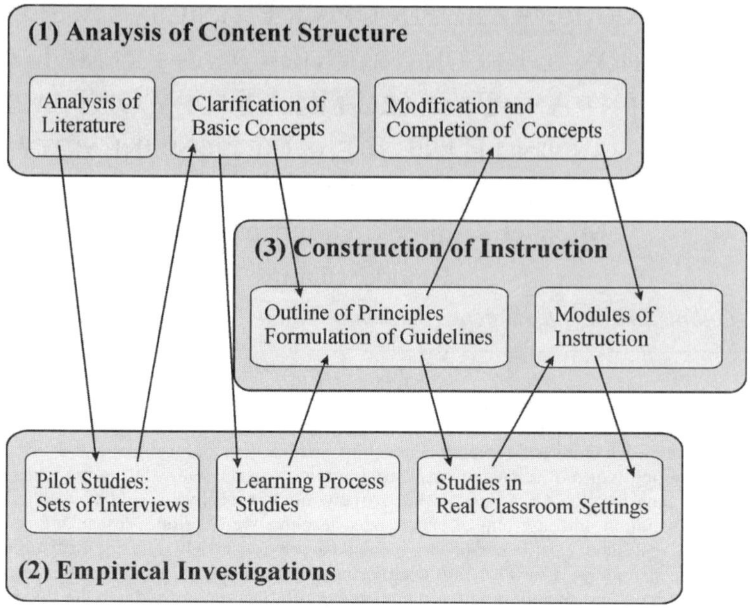

Figure 1. Process of Educational Reconstruction

Results from empirical investigations on students' perspectives and learning processes (component 2) play a significant part in designing a content structure for instruction that is accessible for students. Hence, it is necessary to include construction of instruction (component 3) from an early state. Educational reconstruction can be characterized as a cyclical process of theoretical reflection and conceptual analysis accompanied by small scale curriculum development and classroom research on teaching and learning processes (Fig. 1) (comp. Lijnse, 1995). Our research is based on a constructivist epistemological position. We believe that there is no such thing as the "true" content structure of a particular content area. What is commonly called "the science content structure" should be seen as a consensus of a particular science community. Every presentation of this consensus, including the presentations in the leading textbooks should be seen as an idiosyncratic reconstruction of the authors informed by the specific aims they explicitly or implicitly hold. In our research program the approach of students' perspectives and conceptions being idiosyncratic cognitive constructs is applied to scientific theories and concepts accordingly. Issues of the content's significance from the science perspective and from students' perspectives are intimately interconnected. It is furthermore a key assumption of ours that the curriculum developer's awareness of the students' point of view may substantially influence the reconstruction of a particular science content in that a different position improves the researchers' understanding of the referring science content (for examples of such feedback from mathematics education see Steffe & D'Ambrosio, 1996). Analysis of content structure is never solely influenced by the referring science content and by issues that stem from philosophy and the history of science but as well by educational concerns, such as the aims of teaching elementary features detected as the basic ideas of a science theory in question. We may add that the word "detected" as used in the previous sentence should be read from a constructivist perspective, i.e., any detection is after all, a construction of the analyser. Substantial feedback from investigating students' understanding to the reconstruction of content structure has been an integral and essential part of our projects.

We apply the model of educational reconstruction to the content area of nonlinear physics. Theories of nonlinear systems provide tools to study a large number of phenomena that could not be modelled theoretically before (Schuster, 1989: Agyris et al., 1994). Sometimes the new theoretical approach is interpreted as a new paradigm that provides a totally new view of natural phenomena to a similar extent as the quantum theory did. But the majority of physicists today is more cautious in this regard (comp. Kanitscheider, 1994). The reason for the enthusiasm that theories of nonlinear systems caused appears to be that it challenges the idea of deterministic predictability of natural events which is undoubtedly paradigmatic for traditional physics. This issue seems to be rather appealing to those who are very hesitant to believe in the idea of unrestricted predictability. Additionally the new theories deliver ideas how to describe and explain self-organizing structures in nature as an interplay of deterministic laws of nature and accidentally occurring small disturbances ("order in chaos"). For these reasons it appears to be no wonder that theories of nonlinear systems have become appealing for a number of other disciplines outside mathematics and physics (comp. Küppers, 1996) and have even gained considerable attention in a broader public. Figure 2 shows the core principles that rule nonlinear systems. The structure of concepts, which are, represented in that figure results from analysis of content structure of nonlinear physics.

In the last years we have designed learning process studies (Duit and Komorek, 1997; Komorek, 1999) and studies in real classroom settings on how students at the age of 15 to 16 are able to understand the principle of limited predictability of chaotic systems (Duit, Komorek, Wilbers, 1997; Duit, Roth, Komorek and Wilbers, 1998 and 2001). This principle is one of the core ideas of theories of nonlinear systems, which is worth to be taught. In the study presented here a group of teachers designed a teaching sequence on the two aspects of limited predictability and self-organizing natural structures (for the second aspect comp. Komorek, Duit, Naujack & Bücker, 2001). We want to know how teachers reconstruct a new scientific content area, how they cooperate (Huber, 1998), and how the results of this process reflect their professionalism as teachers. Moreover we want to study students' learning processes towards understanding the scientific concepts in question.

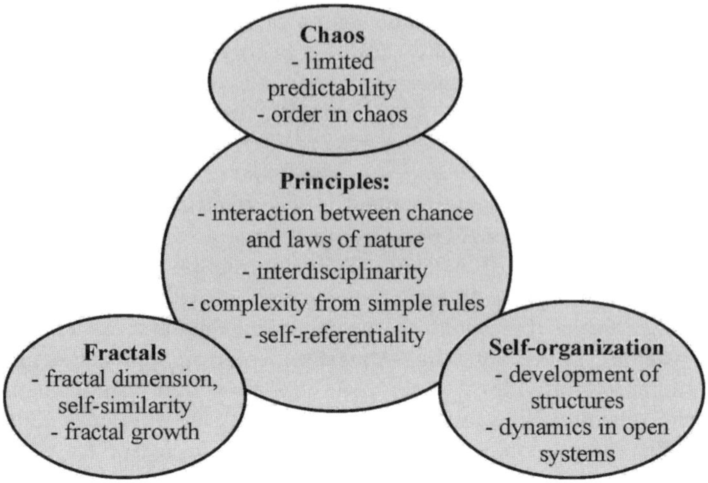

Figure 2. Results from analysis of content structure of nonlinear physics

2. METHODS AND SAMPLE

In cooperation with a group of six experienced physics teachers we designed an instructional sequence of 18 lessons. These teachers are experienced in teaching physics but had not taught nonlinear physics before. Because nonlinear physics was not even part of their education as teachers (about 15 to 20 years ago) in-service training on the new content was provided and information about results from empirical studies were given that tried to make nonlinear physics understandable to students. On the basis of this in-service training and selected materials the teachers developed an instructional unit, which contains the following aspects:

- Limited predictability of nonlinear chaotic systems (slightly different initial conditions and small disturbances restrict the predictability of dynamic instable systems)
- Simulation of nonlinear systems (like magnetic pendulum, double pendulum, or the solar system; students should use simulation software to test own hypotheses about dynamic instable systems)
- Simple experiments and analogs helping to understand nonlinear systems (experiences from former studies show that experiments or toys that students can easily decode are helpful in their learning processes towards the scientific views)
- Conceptions of the world (mechanistic vs. modern views) (students should learn about the history of scientific models of the world and how and why models were replaced by other ones; Laplace's demon for example is challenges by ideas of nonlinear physics)

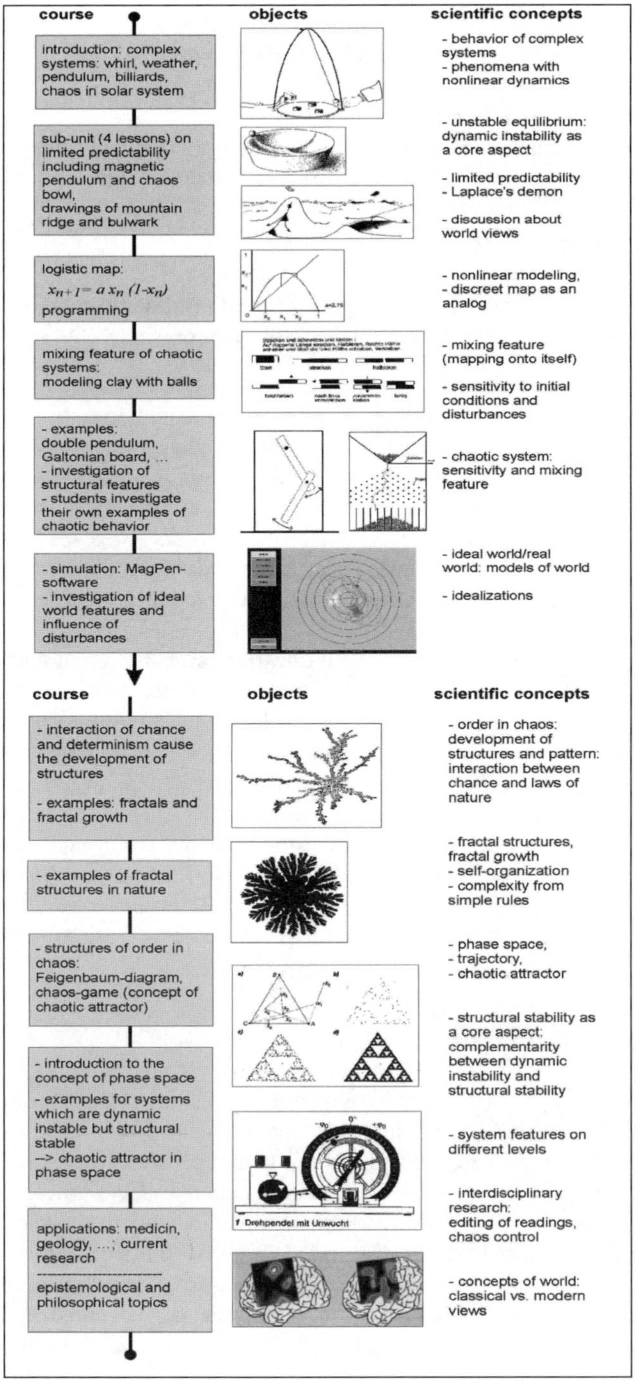

Figure 3. Unit on nonlinear physics

- from structural similarities to a "structural view"
  (one aim of the unit is that students should explain the behavior of systems by analyzing their structural features on a some kind abstract or global level)
- significance of the interplay of deterministic laws and chance
  (structures that nonlinear dynamics deal with are produced by an interaction between deterministic laws of nature and small influences from outside of the system that occur by chance)
- natural self-organizing structures (e.g. fractal structures)
- order from chaos (e.g. chaotic attractors)
  (there is a complementarity between dynamic instability of nonlinear systems and their structural stability in phase space; this stability is represented by chaotic attractors)

Figure 3 shows the structure of the course. The unit was tested four times in upper physics classes (grade 11 to grade 13). Two times there was "regular" teaching, which means two lessons a week over a period of nine weeks (grades 11 and 13, students at the average age of 16 and 18, respectively). In one case there was teaching over a period of three days immediately after each other, seven lessons a day (grade 11, average age 16). In the fourth case there was a kind of project oriented teaching in which the students were relatively free to choose tasks and learning processes by themselves (grade 13).

We gained data from several sources. There were questionnaires before and after the whole unit for all students, interviews with the teachers before the in-service training started and questionnaires after they had taught the unit. We videotaped all the lessons with two cameras and observed two groups of each class during the whole instructional sequence. Field notes were taken while teaching and we had small interviews (5 minutes) with different students directly after each lesson. After the unit had been taught we made interviews with four students from each class along a sequence of physical experiments that students already know from the unit plus new ones. We transcribed most of the video material and all the interviews. The analysis was based on categories that were defined on the basis of our research questions and were modified during the process of interpretation.

3. RESULTS AND IMPLICATIONS

The model of educational reconstruction has proven a useful frame for our research and developing activities so that the study can deliver results at several levels (fig. 4). First, about the question how teachers can reconstruct an area of modern physics that is new to them. We have got detailed data about the process of educational reconstruction of central parts of nonlinear physics. We have learned that the process of reconstruction had to be guided even when working with experienced teachers. In our group the teachers tried to construct a unit just before analysing the content structure and before discussing the educational aims of the teaching and learning sequence. Moreover interview and questionnaire data allow conclusions about teachers' conceptual change not only regarding scientific concepts but also regarding the conceptions their students have. Concerning basic scientific concepts like 'determinism', 'laws of nature', or 'predictability' teachers sometimes held similar 'wrong' - or better to say preinstructional - conceptions compared with their students. These results give valuable hints for developing future in-service training courses. In such a course some emphasis has to be given on the epistemological aspects of nonlinear physics.

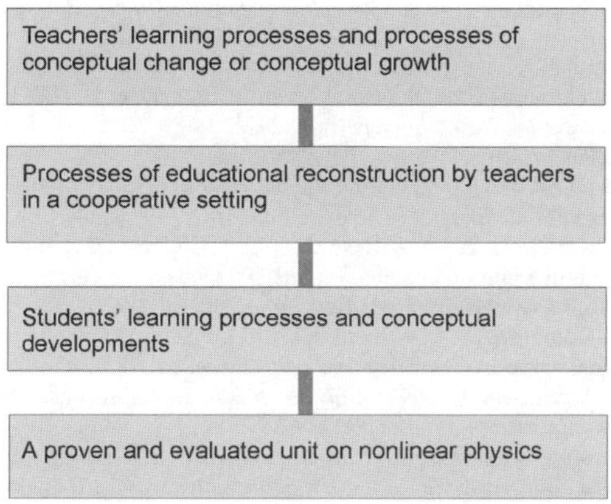

Figure 4. Levels of results

Second, the data offer an extensive insight into the learning processes of the students. To a certain extent students have learned about limited predictability and about the interplay of laws of nature and chance, which results in spatial and temporal structures of order. No discontinuous pathways from their preinstructional conceptions to the scientific concepts could be observed but a kind of differentiation of their conceptional 'clusters' and processes of meaning construction during the discussions within the groups. Apart from this they have learned about the process of science including the current frontier of science and they were able to draw conclusions about ideas of determinism. It can be described in detail how they learned and used the experiments and material.

And third, we could derive well-founded assertions from the first two runs about how to improve the unit developed and how to make it suitable for school reality. As a result there is a tested teaching and learning sequence for school practice, which turned out to be teachable to students. As a special feature of the unit 60 per cent of the time was designed as group work in which students could develop and prove their own hypotheses.

We think that the results of the study can support curriculum developers even in the current situation when the boom of nonlinear physics decreases - or better to say - when nonlinear physics has become part of 'normal' physics. School physics educators now should think about how to integrate this part of modern physics into the existing school curriculum. This seems to be worth because knowing about modern science is a precondition for developing students' competencies for the knowledge-based society.

REFERENCES

Argyris, J., Faust, G. & Haase, M. (1994). *Die Erforschung des Chaos* [*Researching Chaos*]. Wiesbaden: Vieweg.

Duit, R. (2001). Educational Reconstruction: Science Subject Matter and Educational Issues in Harmony - Research and Development Intimately Linked. In: Psillos, D. et al. (Eds.), *Science Education Research in the Knowledge Based Society* (pp. 227-229). Proceedings of the third international conference of ESERA.

Duit, R. & Komorek, M. (1997). Understanding the basic ideas of chaos-theory in a study of limited predictability. *International Journal of Science Education*, 19, S. 247-264.

Duit, R., Roth, W.-M., Komorek, M., & Wilbers, J. (1998). Conceptual change cum discourse analysis to understand cognition in a unit on chaotic systems: Towards an integrative perspective on learning in science. *International Journal of Science Education* 20(9), 1059-1073.

Duit, R., Roth, W.-M., Komorek, M. & Wilbers, J. (2001). Fostering conceptual change by analogies - between Scylla and Charybdis. *Learning and Instruction*, 11, 283-303.

Duit, R., Komorek, M. & Wilbers, J. (1997). Studies on educational reconstruction of chaos theory. *Research in Science Education* 3, Issue 27, 339-357.

Huber, G., L. (1998). *Kooperation von Lehrkräften: Kooperatives Lernen kooperativ lernen [Cooperation of teachers: to learn cooperative learning in a cooperative way]*. Modulares Material zum BLK-Programm »Effizienzsteigerung im naturwissenschaftlichen Unterricht«, http://blk.mat.uni-bayreuth.de/blk/blk/material/ipn.html

Kanitscheider, B. (1994). Philosophische Reflexionen über Chaos und Ordnung [Philosophical ideas about chaos and order]. In: H.-O. Peitgen, H. Jürgens & D. Saupe. Chaos - Bausteine der Ordnung. Stuttgart: Klett-Cotta, S. 1-33.

Kattmann, U., Duit, R., Gropengießer, H., & Komorek, M. (1997). Das Modell der Didaktischen Rekonstruktion - Ein theoretischer Rahmen für naturwissenschaftsdidaktische Forschung und Entwicklung [The model of educational reconstruction - A theoretical frame for science education research and development]. *Zeitschrift für Didaktik der Naturwissenschaften ZfDN* 3, Issue 3, 3-18.

Komorek, M. (1999). Lernprozessstudie zum deterministischen Chaos [Learning process study in the field of deterministic chaos]. *Zeitschrift für Didaktik der Naturwissenschaften ZfDN* 3(5), 3-22.

Komorek, M., Duit, R., Bücker, N. & Naujack, B. (2001). Learning process studies in the field of fractals. In: Behrendt, H., Dahncke, H., Duit, R., Gräber, W., Komorek, M., Kross, A. & Reiska, P. (Eds.) *Research in Science Education - Past, Present, and Future* (pp. 95-100). Dordrecht (The Netherlands): Kluwer Academic Publishers.

Küppers, G. (Hrsg.), (1996). *Chaos und Ordnung - Formen der Selbstorganisation in Natur und Gesellschaft [Chos and order - Kinds of self-organization in nature and society]*. Stuttgart: Reclam.

Lijnse, P. (1995). "Developmental research" as a way to an empirically based "didactical structure" of science. *Science Education, 79,* 189-199.

Schuster, H., G. (1989). *Deterministic Chaos*. Weinheim: Physik Verlag.

Steffe, L., & D'Ambrosio, B. (1996). Using teaching experiments to understand students' mathematics. In: Treagust, D., Duit, R. & Fraser, B. (Eds.): *Improving teaching and learning in science and mathematics* (pp. 65-76). New York: Teacher College Press,

PROMOTING UNDERSTANDING THROUGH REPRESENTATIONAL REDESCRIPTION: AN EXPLORATION REFERRING TO YOUNG PUPILS' IDEAS ABOUT GRAVITY.

Terry Russell, Linda McGuigan

University of Liverpool, U.K.

ABSTRACT

This paper describes an attempt to operationalise some aspects of the cognitive developmental hypothesis of Representational Redescription, (Karmiloff-Smith, 1992, 1994 and 1997). It is illustrated by a study that examined processes and outcomes of teaching and learning about gravitational force with pupils in the 4-11 age range. The researchers collaborated with a group of teachers in an attempt to apply to classroom practice some implications of the theoretical underpinning upon which the R-R view of knowledge development is based. Some insights into how the transformation of pupils' internally stored representations might be provoked in the classroom are described. Our attempts at educational application do not claim to test the R-R hypothesis, but in the educational context of learning science ideas, some difficulties in applying the theoretical formulation are discussed, and some implications for ways forward are considered.

1. BACKGROUND, AIMS AND FRAMEWORK

The authors have reported the outcomes of constructivist research addressing a number of micro-domains within science knowledge construction (see e.g. Russell et al. 1998). We have described the characteristics of an approach that reconciles assumptions about pupils' individual knowledge construction with the constraints of classroom ecology. We have maintained an applied theoretical orientation towards working collaboratively with teachers to promote pupils' conceptual change.

The work reported here constituted an attempt to advance our applied theoretical understanding of conceptual change, the context being pupils' understanding of gravitational force. Attention to this area is warranted by deficits in pupil performance identified in our national curriculum assessment activities, (Russell and McGuigan, 2001). We draw on Karmiloff-Smith's (op.cit.) hypothesis applied to the conceptual change process. Karmiloff-Smith, like Fodor (1983), adopts a domain-specific theory of cognitive functioning. While for Fodor 'modules' are innate, encapsulated entities, Karmiloff-Smith envisages developmental plasticity in the formation of neural structures. Another contrast is in the relationship between domain and central processing, where Karmiloff-Smith posits the importance of language in transforming domain-specific information into a form accessible to conscious processing:

D. Psillos et al. (eds.), Science Education Research in the Knowledge-Based Society, 277–284.

'A fundamental aspect of human development is the hypothesised process by which information that is *in* a cognitive system becomes progressively explicit knowledge *to* that system. I call this the "representational redescription" hypothesis.' (K-S, 1994, p.694.)

It is suggested that, during development, internal representations that are initially stored implicitly (and which are not automatically available to conscious access) may become increasingly explicit and thus accessible to the learner in a meta-cognitive sense. Representations in domain-specific form are subjected to meta-cognitive 'explicitation' usually mediated by language-based central processes. This formulation informed our framework for the examination of pupils' knowledge construction about gravitational force. Of specific interest were:

1: the ways in which teachers succeeded in helping children to become aware of their own representations by making them explicit to themselves (*and their peers*);
2: the variety of representations (*including those expressed through different sensory modalities*) relating to gravity which teachers discovered and invented to encourage pupils to encode relevant experiences;
3: the changes in children's representations, amongst individuals and across the target 4-11 age range;
4: the impact of all these practices on the efficacy of learning outcomes i.e. frequencies of movement towards conventional scientific understandings of gravity.

The points in italics flag our practical extrapolations beyond what our reading of R-R strictly warrants. Firstly, R-R describes intra-psychological processes, while teachers attempt to exploit the group-work efficiencies of inter-psychological interactions. Point 1 confirms that we would be inclined to see explicitation as something capable of being enhanced by social encouragement (given a classroom environment that does not threaten ridicule for public errors). Also, point 2 acknowledges that because the relationships between 'modules', 'domains' and 'modality' are not specified, inventive liberties were taken in attempting to extrapolate to classroom ecologies. While Karmiloff-Smith is content to speak of the 'domain of science', it is evident that ideas about gravity may be experienced by pupils kinaesthetically, visually (using 3-D models or 2-D drawings), and so on – through different sense modalities. We may thus have introduced our own view of development in understanding as involving the re-coding of representations stored in different modalities. Point 3 admits to a significant non-attendance to an important element of Karmiloff-Smith's formulation, namely, her rejection of Piaget's domain-general notion of intellectual development and advocation of a phase model of change. The nature of our intervention precluded the monitoring of fine-grain processes of micro-developmental change in individual children. Finally, point 4 reflects our interest in conceptual change in school populations. Significant change in the desirable direction would not confirm the R-R hypothesis, but it would encourage further investigation of its potential utility.

2. RESEARCH ACTIVITIES

The study took place between 1998 and 2000, the research sample comprising a group of 246 children (4-11 years) and their ten teachers. A group of 214 children (4-11 years) drawn from one local primary school was used as a control. (A control group distributed through a range of schools would have helped to confound teacher effects, but this was not logistically viable.) At the onset of the enquiry, all children in the 'research' and 'control' groups responded to the same concept probes: pencil and paper responses following a directly observable stimulus activity. A stratified sub-sample of six children in each of the research classes was interviewed individually. A period of intervention followed the initial explicitation of ideas. Teachers were encouraged to select those activities considered most appropriate to their class. Teachers encouraged R-R by helping learners make their representations explicit. No attempt was made to influence the activities of the control pupils; it was assumed they would be exposed to the normal gamut of national curriculum science during this period.

Teachers kept 'field notes' and university researchers made some classroom observations and opportunistic video recording. Following intervention, both the research and control groups responded to the same concept probes as earlier. The interview sub-sample was re-interviewed by researchers.

3. ENCOURAGING THE EXPLICITATION OF REPRESENTATIONS

'Explicitation' within the R-R model is a central process by which inherent representations are brought to conscious awareness. Intuitive and previously implicit ways of knowing in the form of information in the system (i.e. in the mind/brain) are made explicit to the system via a quintessentially human, meta-cognitive process. An important consideration for education is the role of the teacher in this process. Teachers were encouraged to help children to access and articulate internally stored representations of which they might have been unaware previously. In the classroom (unlike the developmental psychologist's clinical interview) pupils attempt to make their representations known to their peers as well as to themselves. Empathising with others' struggles to express their tentative representations might have been, in itself, facilitative of learning.

To elicit children's representations about gravitational force, a range of stimulus material was used, drawing on kinaesthetic, haptic, visual and linguistic modes. Our assumption is that the different modes afford a range of representational opportunities for expressing different properties of an object or event. It is arguable that different attributes of a belief gain salience in different representational forms and furthermore, different ways of encountering a concept promote different understandings of that concept. Encouraging a range of representational forms reveals different ways in which a concept is understood by an individual - to the individual, to peers and to the teacher.

4. CONCEPT PROBES

Concept probes were devised by the teachers and researchers to provoke initial exposure of pupils' ideas. The probes could be presented in individual, group or whole class settings. Three principles underpinned their development:

- They should be accessible to children and engage their attention;
- They should exploit a variety of response modes;
- They should be conceptually focused but invite open-ended responses.

Each probe was introduced by a practical activity that orientated children towards an aspect of gravity: the media in which the force might operate; ideas about magnitude and direction; effects on moving and stationary objects. Teachers encouraged pupils' elaborated expression of ideas. Children annotated their initial responses (with teachers' support if needed). This process could be termed 'diagnostic' in that it alerted teachers to the range of pupil understandings and promoted children's meta-cognitive awareness of their own representations.

5. INTERVENTION ACTIVITIES

Classroom intervention aimed to maximise opportunities for children to experience new contexts which might provoke reflection upon their understanding of gravitational force, and to make those understandings explicit. The representational possibilities of different modalities were anticipated by considering linguistic, sensory/kinaesthetic, spatial, drawing and modelling opportunities, as well as formal notations such as arrow drawing to represent the direction and magnitude of forces. Analogies and argumentation to justify ideas was encouraged, as well as reference to supporting or competing interpretations of representations. Opportunities for children to use their imagination to explain how a hypothetical event might occur were also presented. Discussion provided opportunities to consider links between representations expressed in different modes.

Intervention also encouraged children to draw on sensory phenomena including whole-body experiences such as weight bearing, gesturing, demonstrating and handling real objects. In some instances these were provided in the classroom. For instance, children were invited to place bags of sugar on different parts of their body, then to describe the weight sensation. Some children were asked to jump up in the air or to lift parts of their body to encourage the sensation of the effect of gravity on their bodies. Direct experiences were sometimes combined with encouraging children to reflect on sensory experiences that were inaccessible in the classroom, e.g., describing the feeling of moving on a swing.

Children were provided with globes and asked to show and describe how (and where) people were located on Earth. In some classes, a polystyrene sphere with a hole through the centre provided the stimulus for children to represent their ideas about the imagined movement of a ball through the centre of the Earth (Nussbaum and Sharoni-Dagan, 1983). Children were asked to draw people holding a helium balloon to show its orientation at different locations around the Earth, using arrow notation to draw the direction of gravitational force.

Teachers were involved in the development and selection of intervention activities for their own class. Some resonance was anticipated with an approach based on the familiar classroom maxim, 'Let's look at it another way'. Each intervention activity specified the modes that might be encouraged, indicating where children might be encouraged to model, draw or write their ideas. Teachers' professional experience guided the selection of age-appropriate activities and recording mechanisms. Following each activity, a period of reflection was suggested during which the teacher sought to actively provoke R-R by encouraging individual and shared pupil interpretations of one another's representations. Links between representations were considered; the strengths and limitations afforded by different representations were discussed.

6. QUANTITATIVE OUTCOMES

Responses to the concept probes from research and control groups pre- and post-intervention provided data about changes in quality and frequency of explicit representations. Qualitative response categories were established which were then used to quantify pupil responses. While performance against 125 variables was analysed, this paper does not report quantitative outcomes in full. Our intention is to give an impression of measured outcomes to check whether, in general terms, our approach was productive of shifts in understanding. Table 1 summarises pre- and post-intervention references to gravity in explaining why a plumb line hangs as it does.

Table 1. Explanation of why a plumb line hangs as drawn: Reference to gravity (percentages)

	KS1		Lower KS2		Upper KS2		All	
	Research group n=68	Control group n=81	Research group n=66	Control group n=67	Research group n=112	Control group n=66	Research group n=246	Control group n=214
Pre-intervention	6	2	32	16	39	47	28	20
Post-intervention	35	-	63	22	76	56	65	25

The frequency of references to gravity in the research group following intervention activities was greater than that in the control group (see Table 1). While the increase in gravity references in the research group averaged 37%, that in the control group averaged 5%. The increase in the use of 'gravity' as an explanatory concept was found to be statistically significant in both the control and the research group. However, when the *improvement* of each group was compared (as defined by those cases in which 'gravity non-referrers' in the pre-test became 'gravity referrers' in the post-test) the performance of the research group was significantly higher than that of the controls. (Mann-Whitney, $p < 0.001$). A more detailed discussion is not possible within the space constraints of this paper, but the claim is made that the data warrant further consideration of attempts to explore the utility of applying the

R-R hypothesis in classroom contexts. Equally important to the development of our understanding as researchers were some qualitative aspects of the study, which are introduced below.

7. QUALITATIVE OUTCOMES

Figure 1. Representation of gravitational force during act of jumping

Figure 1 illustrates a child's recoding of the kinaesthetic experience of jumping in the air as a visual representation of her understanding of gravitational force. We submit that the different modes used in this representation afford different meaning-making opportunities. The drawn figures show a linear temporal order and also illustrate the simultaneous spatial relationship of parts to the whole. The use of the arrow notation enables the child to show magnitude and direction of the force while the written narrative provides additional insights into the child's appreciation of the link between elevation from the ground and the magnitude of gravitational force thought to be operating.

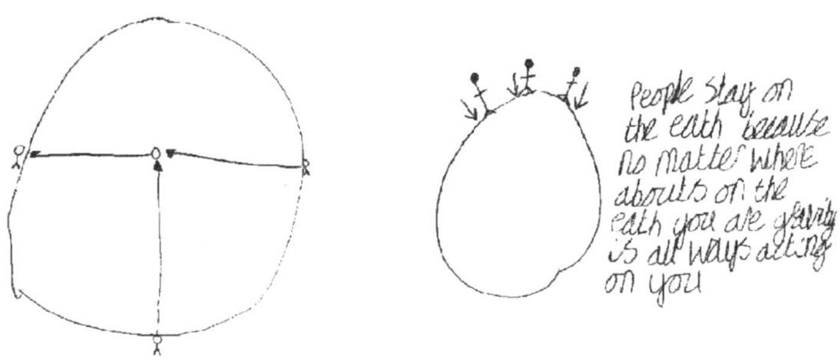

Fig. 2a. *Fig. 2b.*

Figure 2. Representation of people located on the surface of the earth

Figures 2a and 2b are examples of drawings with arrow notation and writing following 3-D activities using a globe and models of people. Figure 2a reveals a discrepancy between the child's 3-D and 2-D representations in that the people have been drawn with head and sides of bodies touching the Earth while her model showed people correctly orientated towards the Earth's surface. The arrows indicating the direction in which gravitational force is believed to be acting are also inconsistent. The circle at the Earth's centre was often referred to by children as the point towards which gravity acted, and was frequently confused with the notion of the core – a separate entity rather than a point in a continuous mass. The child's annotation of Figure 2b refers to people staying on the Earth 'no matter whereabouts', yet shows them clustered around the North Pole. We can assume the two representations, writing and drawing, are consistent in this child's view, and that 'no matter whereabouts (– but it can only be at the top!)' may be a more accurate formulation of this child's belief. Perhaps by triangulating representations, we can identify beliefs to which pupils are committed.

8. CONCLUSIONS AND IMPLICATIONS

A first important point to establish - a prerequisite to taking our research enterprise forward - is that teachers were comfortable with the language of 're-describing representations'. A view of pupils as active constructors of knowledge rather than passive receivers is fundamental to our exploration of the R-R model. Primary teachers (in the UK, at least) tend to be familiar with 'encoding' and 'decoding' in acts of writing or reading and in this study were content to base their transactions on the notion of pupils making representations (rather than science 'facts') explicit.

Karmiloff-Smith's formulation is based on assumptions about individual psychological change, though it does not exclude socio-cultural inputs. Reflecting on R-R in a classroom context draws attention to the possibility of useful reusable external representations in the form of conventional science models, descriptions analogies or empirical experiences that may be culturally promoted.

It proved not too difficult to move to a social stance in which teachers made the process of 'explicitation' overt to pupils as a tool for developing ideas. Thus 'explicitation' proved capable of being used as a device to guide teachers' and pupils' actions in an interventionist and facilitative manner. This quality of teacher-pupil interaction has close similarities to that prescribed by the advocates of formative assessment who point to the significant educational gains associated with its use.

More precision in what constitutes a 'domain' will be helpful. While Karmiloff-Smith is clear that domains do not imply modules in the Fodorian sense, it is tempting to think of sensory inputs whose neurological localisation is relatively well-established having some relationship to a modular or domain-specific hypothesis. A 'microdomain' such as understanding of gravitational force may offer several modalities through which understanding might be apprehended. An unravelling of the taxonomies of domains, modules and modalities would be instructive.

Another feature of R-R in a social context is the possibility of multiple and competing representations. Encouragement to articulate the evidence influencing the commitment to one representational form in preference to another adds another dimension to explicitation.

REFERENCES

Fodor, J., (1983). *The Modularity of Mind: An essay on faculty psychology*. Cambridge, Massachusetts: MIT Press.

Karmiloff-Smith, A. (1992). *Beyond Modularity. A developmental perspective on cognitive science*. MIT Press

Karmiloff-Smith, A., (1994). Précis of beyond modularity. *Brain and behavioral sciences* 17, 693-745.

Karmiloff-Smith, A., (1997). Continuing Commentary on Précis of beyond modularity. *Brain and behavioral sciences,* 20, 351-369

Nussbaum, J & Sharoni-Dagan, N., (1983). Changes in second grade children's preconceptions about the Earth as a cosmic body resulting from a short series of audio-tutorial lessons. *Science Education*, 67, 99-114.

Russell, T; McGuigan, L. & Hughes A., (1998). *Forces, Science Processes and Concept Exploration research report*. Liverpool University Press.

Russell, T. & McGuigan, L. (2001). Making formative use of a national summative assessment regime. In: Behrendt H. et al. (eds.). *Research in Science Education – Past, present and future* (pp.71-76). Kluwer Academic publishers.

DIFFERENT TYPES OF CLASSROOM DEBATES ON BIOTECHNOLOGY. ARE THESE SIMPLY AN EXERCISE IN RHETORIC OR DO THEY ENCOURAGE A WELL- FOUNDED CRITICAL ATTITUDE ?

Laurence Simonneaux

Ecole Nationale de Formation Agronomique, France

ABSTRACT

The objectives discussed in this paper were to test research methods used to analyse student debates, to report on them, and to compare students' arguments in role-playing and debating situations on the topic of animal transgenesis. Analyses were made from different perspectives, from microscopic analysis of debate episodes to macroscopic analysis of discussion dynamics, thus in turn identifying episodes for more detailed close-up study. During a formal debate, as opposed to role-playing, students speak for longer periods and develop more complex arguments based on valid data. When role-playing, moreover students used more ironical and provocative rhetorical techniques.

1. BACKGROUND, AIMS AND FRAMEWORK

Biotechnology applications raise questions in several areas, including ethics, sociology, economy, ecology, regulation and politics. Debates often lead to dilemmas because of uncertainty as to the risks involved. Our aim is to help students to contribute to debates so that they understand that decision-making can be complex when important social issues are involved, for example, economic, ethical and other issues.

Debates generate difficulties for students, since they find themselves in a situation of potential inter or intra-subjective conflict. Moreover students in France are not used to debating in school activities. Debates generate difficulties for teachers too, in that they have to learn how to chair them, instead of being 'one of *those who know*'.

Some authors hold that argument is an integral part of present-day communication, originating in classic rhetoric (Breton, 1996). Others deal with the subject from a linguistic point of view (Adam, 1992). Others again address argument at several different levels in science teaching: as a means of improving the understanding of concepts, promoting a better grasp of epistemology in science, developing investigative skills (particularly in practical work), or improving the quality of decision-making on socio-scientific subjects (Geddis, 1991, Solomon, 1992, Ratcliffe, 1996, Driver and Newton, 1997, Jiménez Aleixandre *et al.*, 2000a,

D. Psillos et al. (eds.), Science Education Research in the Knowledge-Based Society, 285–293.

2000b, Duschl and Ellenbogen, 1999, Mortimer and Machado, 1999, Osborne, 1999).

In analysing different means of persuasion, it is possible to distinguish, in theory at least, between manipulation, propaganda, seduction, argumentation and demonstration. Actually, situations relying purely on seduction, argumentation or demonstration are very rare, a fact which accounts for the development of rhetoric. In France, rhetoric ceased to be taught in schools and universities at the end of the 19th century, and even became discredited as a corpus of knowledge. This seems to be linked with a conflict between a culture - based on evidence, drawing on advances in science and positivist theory, and a culture based on argumentation (Breton, 1996). Interest in rhetoric came to the fore again in the 1960s, with a growing awareness of the impact of media technology. In 1967, Blanché, in his *Introduction à la logique contemporaine* stated that 'evidence is no longer a guarantee of soundness'. Advances in science are driven by debate and disagreement: there is therefore a need to understand the role of controversy in modern science, and its rhetorical functions in handling uncertainty (Osborne, 1999). The objective of the present work is to encourage students to develop a critical approach to scientific statements on the repercussions of biotechnology applications.

Studies on argumentation in science education also draw on research in linguistics. Working from a theoretical typology of text sequences, Adam (1992) defines several prototypes: narrative, descriptive, argumentative and explicative sequences, and dialogue. The following diagram shows how Adam (1992) defines his prototype argument sequence. The pattern of the argument rests on linkages that are made between '*data*' and a '*conclusion*'. These linkages may be implicitly or explicitly supported (by the '*guarantor*' and the '*material*') or opposed (by '*refutation*' or '*exception*').

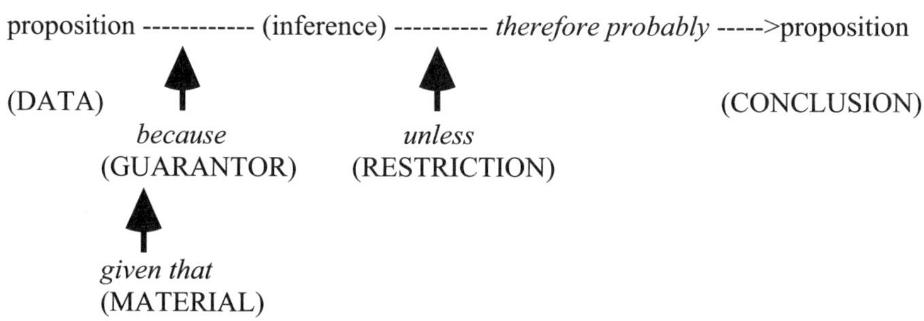

Figure 1. Adam's pattern of argument

Conscientious readers or interlocutors will be attentive to linguistic clues, which lead them to expect some kind of argument. From the point of view of the cognitive treatment of a text, it seems that the understanding of schematic prototype representations that subjects develop over time has an impact on the way they store the information they have considered while coming to grips with a discourse, and on the way they look for blocks of information through a process of anticipation (Adam, 1992).

Analysing an argumentative text is a new subject in the teaching of French as a first language in upper secondary school. The importance attached to it has to do with its ultimate social purpose, which is related in particular to students' perceptions of citizenship. However this topic is rarely, if ever, taken up in science teaching. Moreover, in French teaching, what is taught is determined by a standard division of descriptive reference theories, and does not touch on the question of values – although this is crucial to the study of argument. Furthermore the main media used in these activities is textual and not oral.

In science education studies, many authors refer to Toulmin's layout of arguments (Toulmin, 1958, Pontecorvo and Girardet, 1993, Kelly et al., 1998, Jiménez Aleixandre et al., 2000a, 2000b, Diaz de Bustamante and Jiménez Aleixandre, 2000, Sonora Luna et al., 2000). This layout is very similar to Adam's pattern of argument. A representation of this layout is shown below. Toulmin characterised the components of the argument as follows: "Data" are the facts the proponent of the argument explicitly appeals to as a foundation for the claim. The "claim" is the conclusion based on the merits 'established' by the argument. The "warrant" consists of the rules, principles or inference licence that demonstrate that the step from the data to the claim is a legitimate one. The strength of the warrant may be indicated by modal "qualifiers". The "rebuttal" indicates the circumstances for which the general authority of the warrant is merited. The "backing" establishes the general conditions, which give authority to the warrants.

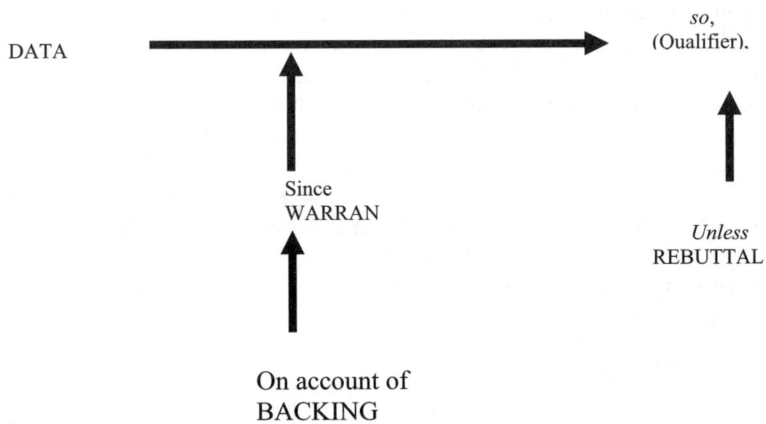

Figure 2. Toulmin's pattern of argument

The objective addressed in this paper is to compare students' arguments in role-playing and debating situations.

2. METHODS AND SAMPLE

A role-playing scenario on animal transgenesis has been developed around the production in a seaside village of giant transgenic salmon expressing a foreign gene for the growth hormone (Simonneaux, 1997). The fish are dubbed 'Sumotori salmon' after the Japanese wrestlers. The local population is very concerned about the project and the mayor has thus decided to organise a public debate. The role-playing was then compared with a conventional debate on the same topic. Our study was conducted in two classes at the Agricultural Lycée in Auch (Gers *département* in South-West France), with students in their 2nd year of upper secondary vocational education geared to scientific subjects. All the dialogues were recorded on audio and video tape and transcribed in full. Although a formal debate, the debating situation (17 students) was organised in exactly the same way as the role-playing (18 students).

In the role-playing scenario, the teacher played the role of mayor. She introduced the scenario and was responsible for the timing. At the end the teacher (the mayor) asked the students to vote.

During the debating situation, the teacher explained the issue being debated by making an analogy with the citizens' conference that took place in France on the subject of biotechnology applications, and set out the aims of the debate, i.e., to reach a decision through argument, explaining that the first task was to discuss the economic, political, ecological and human health aspects of the issue. The teacher remained neutral, leaving the students to take up the various points spontaneously. If one or more aspects were neglected, the teacher asked for their opinion on them. Next, the teacher prompted a debate on the various topics addressed in the role playing scenario, unless they had already been taken up spontaneously by the students. For instance, an increase in livestock productivity, increasing overproduction, technology transfers to the developing world, famine reduction, disruption of the ecosystem, risks to human health, labelling and consumer reactions, patenting living organisms and company monopolies. As the debate proceeded, the teacher supplied the information that was available in the role-playing scenario. The analysis of students' opinions on the Sumotori fish farm is developed in Simonneaux (2000). In this paper, we analyse the discursive method used by students during the debating situations.

The experiment protocol is shown in table 1.

Table 1. Experiment protocol

Role playing	Debate
Pre-test opinions	ditto
Introduction to the history of domestication and chronology of the transfer of the growth hormone gene	ditto
Presentation of the role playing	Presentation of the debate
Written essay by students on the installation of a Sumotori fish farm (opinions to be supported)	ditto
Role playing performed	Debate
Post-test opinions	ditto
Written essay by students on the installation of a Sumotori fish farm (opinions to be supported)	ditto
Students explain under what circumstance(s) they might change their opinions.	ditto

Different methods for analysing argument have been proposed for scientific education. None of them appear to be suitable as is. This paper describes the way in which we have adapted them and the particularities of our approach. Our intention was not to suggest a statistically reliable method.

We were not trying to identify universal characteristics of the way students argue about biotechnology, but rather to compare the effect of different teaching situations (role-playing, debates) on the quality of students' discursive and debating skills.

We see our method as an example that would need to be adapted for specific contexts in further studies.

As noted by Dumas-Carré and Weil-Barais (2000) debate transcripts may be used 'for analysing from different perspectives'. Microscopic analysis of episodes during which a 'subject' is being debated and macroscopic analysis which allows for a description of discussion dynamics and for choosing episodes for microscopic analysis.

We did this in several stages.

- Stage 1 involved writing a synthesis of discussion dynamics by identifying major debate topics.
- Stage 2 involved microscopic analysis of argument episodes.
- Stage 3 concerned the teacher's role.

The argumentation in the two debating situations was compared gradually during the three stages. A certain number of indicators were quantified: the duration of discussion, the number of interventions, the number of interventions per minute, the number of times the teacher intervened, the number of 'message units'. The higher the numbers of interventions per minute, the less detailed were the arguments. The messages units are units of linguistic meaning that reveal the ways actors construct their actions in a conversation.

It is worth describing the methodology used for the second stage in detail, as it turned out to be the most difficult stage. We first used the Toulmin model, completed successively by Resnick *et al.* (1993), Kelly *et al.* (1998), Sonora Luna *et al.* (2000), Pontecorvo & Girardet (1993). Resnick *et al.* added 'the challenge' to the argument components defined by Toulmin, Kelly *et al.* added 'empirical data' and 'hypothetical data', Sonora Luna *et al.* added 'concession', Pontecorvo & Girardet added 'opposition'. As we gradually developed our analysis of the argument episodes, we changed and completed our conceptual argument system in order to take into account the rhetorical techniques used. We perceive debate as being a socio-discursive interactive event. Within this conceptual framework, language is considered as an activity. We use the concept of "*a linguistic action which integrates and links up the representations of the production context and theme content* parameters" Bronckart, 1996. In order to be able to analyse the complexity of debates we used different argument components borrowed from the authors we have just referred to as well as terms we ourselves had introduced: 'declaration' (equivalent of Toulmin's 'claim'), 'question', 'rejected question or declaration' (equivalent to the 'opposition' of Sonora Luna *et al*), 'critical question' (equivalent to Resnick *et al's* 'challenge'), reformulated question', 'restarting', 'answer', 'hypothesis,' 'objection', 'agreement', 'avoidance', 'revealing gaps or uncertainty', 'value judgement', etc. Each linguistic action can support another action. For instance, a declaration can support an opposition, an agreement, a question,... From rhetorical procedures we selected the following components, 'provocation', 'suspicion', 'promise', 'irony'.

We identified the spheres of knowledge or social references on which the students had based their arguments and evaluated the validity of the arguments used. We analysed the argumentation strategies by making a distinction between simple arguments based on only one bit of evidence from multiple strategies consisting of several encapsulated bits of evidence or linear evidence.

3. RESULTS

The summaries of discussion dynamics reveal recurring themes, which correspond to the topics, debated in the classroom and which were controversial. Recurring themes are discussions that students have spontaneously repeated several times. For the role-playing the topics are biodiversity, consumption, labelling, human health and economic implications. For the debate they were economic, ecological and political repercussions in a context of First World/Third World relationships.

The following table compares the different criteria used to analyse students' discussions

Table 2. Comparison of different criteria

	Debate	Role playing
Duration in minutes	94	55
Number of interventions	101	175
Number of interventions per minute	1. 07	3.18
Number of times the teacher intervened	35	16
Number of student message units	242	296
Number of arguments	36	35
Number of multiple strategies	35	8
Number of invalid arguments (with respect to scientific disciplines or to information provided)	2	8
Number of rhetorical procedures	2	24

During the debate, interventions were longer and more complex (the number of interventions per minute was 1.07 as opposed to 3.18 in the role playing). The arguments were more developed and based on valid data. In the role-playing students spoke for shorter periods of time, arguments were simple and sometimes based on invalid data (students incorrectly interpreted the information with which they had been provided in the description of their role). During the debate the students used 35 multiple argument strategies, i.e. including several encapsulated or linear justifications whereas in the role-playing, the students only used 8 multiple strategies.

In the role-playing students tended to use more rhetorical procedures including irony and provocation (24 rhetorical procedures during 55 minutes of role playing as opposed to 2 rhetorical procedures for 94 minutes of debate). They were acting a role as best they could. Their arguments were superficial and were not based on their own point of view but on the description of the role they were playing. They do not necessarily agree with the opinions of the individual they were playing.

There is some discussion on the existence of a prototype argumentation text pattern as defined by Toulmin and Adam. This pattern, if one does exist, could be an argumentation teaching aid. Toulmin and Adam's patterns have been criticized (Brassart, 1987; Golder, 1996) because among other reasons, they do not integrate dialogue. Our analysis supports this criticism; we have not been able to identify a canonic typical argument. Looking at dialogue has led us to identify various types of linguistic actions that take place in exchanges (question or rejected declaration, critical question, reformulated question, reminder, answer, objection, agreement, avoidance, revealing gaps or uncertainty), rhetorical procedures as well as axiological (value judgements) or prescriptive statements. Argumentative discourse cannot be reduced to a series or organisation of linguistic components which are the same in all situations (Golder,1996).

4. CONCLUSION AND IMPLICATIONS

This work showed how difficult it is to analyse classroom debating situations. A methodology had gradually to be adapted to deal with the complexity. Analyses from different perspectives are complementary. Macroscopic analysis highlights discussion dynamics and the topics debated, in particular recurring themes. Dynamics and topics are affected by the teacher's interventions and by the kind of debated implemented (role-playing or formal debate). The quantified criteria show that in formal debates students' arguments are more structured and valid. A formal debate appears to encourage well-founded reasoning. Role-playing is mixed with rhetorical expressions. But of course, each debating situation is different depending on the context, the topic, interlocutors… It is a socio-discursive interactive event.

Microscopic analysis best reveals the complexity of discussion. Evidence for this is the significant number of argumentation elements which we identified. Of course debates do not unveil canonical argumentative approaches, as there is some interference from the communication situation itself. Analysis of a debate in action requires drawing on at least two theoretical disciplines, linguistics and communication.

The study of classroom debates on controversial social issues should be continued in order to refine analysis of argument in action and to evaluate the effect of training on argumentation. How to train students to develop argument on controversial knowledge? Teaching them a pattern of argument or helping them to analyse controversial discourses. In the latter case, analysis could be focused on the social and physical situation (who is speaking? what are the stakes?…) and on argumentation itself (kind of argument, validity, strength, justification,…).

REFERENCES

Adam, J.,M., (1992). *Les textes: types et prototypes*. Paris: Nathan - Université.

Blanché, R., (1967). *Introduction à la logique contemporaine*. Paris :PUF.

Brassart, D.,G., (1987). *Le développement des capacités discursives chez l'enfant de 8 à 12 ans : le discours argumentatif (étude didactique).* Thèse pour le Doctorat de Sciences Humaines. Strasbourg.

Breton, P., (1996). *L'argumentation dans la communication*. Paris : Ed. La Découverte.

Bronckart, J.-P., (1996). *Activité langagière, textes et discours. Pour un interactionisme socio-discursif.* Paris : Delachaux & Niestlé.

Diaz de Bustamante, J. & Jiménez Aleixandre, M.,P., (2000). Communication in the laboratory sessions and sequences of arguments. *Paper presented in the 3rd ERIDOB Conference*, September 2000, Santiago de Compostella.

Driver, R. & Newton, P., (1997). Establishing the norms of scientific argumentation in classrooms. *Paper presented in the ESERA Conference,* 2-6 September 1997, Rome.

Dumas Carré, A. & Weil-Barais, A., (2000). Analyse du travail de jeunes élèves au cours d'activités scientifiques; dynamique des échanges et conceptualisations des élèves. *Paper presented in Journées d'étude franco-quebecoises – Didactique des disciplines: Recherches sur les pratiques effectives*, 5-6 October, 2000, Toulouse.

Duschl, R.,A. & Ellenbogen, K., (1999). Middle school science students' dialogic argumentation. *Paper presented in the ESERA Conference*, Kiel.

Geddis, A., (1991). *Understanding Scientific Reasoning*. (3rd ed.) Fort Worth, TX: Holt, Rinehart and Winston.

Golder, C., (1996). *Le développement des discours argumentatifs*. Paris : Delachaux et Niestlé.

Jiménez Aleixandre, M., P., Pereiro Munoz, C. & Aznar Cuadrado, V., (2000a). *Promoting reasoning and argument about environmental issues, Research in Didaktik of Biology* (pp. 215-230). Göteborg: IPD.

Jiménez Aleixandre, M.,P., Bugallo Rodriguez, A., Duschl, R.,A., (2000b). " Doing the lesson " or " Doing science ": Argument in High School Genetics. *Science Education*, 84, 757-792.

Kelly, G., J., Druker, S. & Chen, C., (1998). Students' reasoning about electricity: combining performance assessment with argumentation analysis. *International Journal of Science Education*, 20, 849-871.

Mortimer, E., F. & Machado, A.,H., (1999). Mediational tools and discourse interactions in science classrooms. *Paper presented in the ESERA Conference*, Kiel.

Osborne, J ., (1999). Promoting rhetoric and argument in the science classroom. *Paper presented in the ESERA Conference*, Kiel.

Pontecorvo, C. & Girardet, H., (1993). Arguing and reasoning in understanding historical topics. *Cognition and Instruction*, 11 (3&4), 365-395.

Ratcliffe, M., (1996). Adolescent decision-making, by individual and groups, about science-related societal issues. In: Welford., G., Osborne, J., Scott, P., (Eds.), *Research in science education in Europe: current issues and themes*. London : Falmer.

Resnick, L.B., Salmon, M., Zeitz, C.M., Wathen, S.H. & Holowchak, N. (1993) Reasoning in conversation. Cognition and Instruction, 11 (3&4), 347-364.

Simonneaux, L., (1997) *Sumotoris in Unit 11 Transgenic animals.* On http://www.eibe.org

Simonneaux, L., (2000). Comparison of the impact of a role playing and of a classical discussion on students' argumentation concerning an issue in animal transgenesis. *Paper presented in the 3rd ERIDOB Conference*, September 2000, Santiago de Compostella.

Solomon, J., (1992). The classroom discussion of science-based social issues presented on television: knowledge, attitudes and values. *International Journal of Science Education*, 14 (4), 431-444.

Sonora Luna, F., Garcia-Rodeja Gayoso, I. & Branas Pérez, M., P., (2000). Discourse analysis: pupils' discussions of soil science. *Paper presented in the 3rd ERIDOB Conference*, September 2000, Santiago de Compostella.

Toulmin, S., (1958). *The uses of argument*. Cambridge: Cambridge University Press.

PART IV

Science Education and Information and Communication Technologies

WISE RESEARCH – PROMOTING INTERNATIONAL COLLABORATION

Marcia C. Linn

Graduate School of Education, University of California at Berkeley

ABSTRACT

International collaboration in the design of science instruction enables us to study the impact of technology and inquiry under systematically varied conditions. Projects in the Web-based Inquiry Science Environment (WISE) engage students and their teachers in complex science investigations. By studying WISE in Europe and the United States, we have the chance to exploit naturally occurring differences in professional development and educational philosophy and gain insight into what works. This paper describes the WISE knowledge integration framework and discusses how teachers and students gain understanding of inquiry and technology as they use multiple WISE projects. It concludes with 5 research questions fruitful for international collaboration.

1. INTRODUCTION

New technology-enhanced systems for providing inquiry science projects offer viable opportunities for international research programs. This paper describes the philosophy and features of the Web-based Inquiry Science Environment (WISE). Directions for international collaboration come from examples of research on the WISE *Cycles of Malaria* project in the United States (Slotta, in press) and Norway (see Jorde, this volume).

2. WISE PERSPECTIVE ON KNOWLEDGE INTEGRATION

New understanding of how people learn (Bransford, Brown, & Cocking, 1999; Driver et al., 1996), better ideas about how technology can enhance instruction (Brown and Campione, 1994; diSessa, 2000; Reiser et al., 2001; Scardamalia and Bereiter, 1992; Vanderbilt, 1997; White and Frederickson, 1998), coupled with innovative new technologies can enhance inquiry science instruction. WISE builds on an established perspective on learning called knowledge integration (Linn, Davis, & Bell, in press) and a set of design principles that reflect this perspective, called the scaffolded knowledge integration framework (Linn & Hsi, 2000).

The WISE multidisciplinary research program continues a tradition that has roots in the Computer as Learning Partner (CLP) (Linn & Hsi, 2000), the Knowledge Integration Environment (KIE) (Linn, 2000), the Hypermedia Case Studies in

D. Psillos et al. (eds.), Science Education Research in the Knowledge-Based Society, 297–307.

Computer Science (HCS) (Linn & Clancy, 1992) and the Synthesis Coalition (Agogino & Linn, 1992). The WISE research partnership brings together experts in analysis of student learning (both educators and information technologists), experts in delivery of media (text, video, simulations and tele-presence), classroom teachers, and experts in pedagogy who contribute to each other's professional development.

The knowledge integration view of learning and instruction stems from two central findings. First, learners hold multiple conflicting ideas about virtually any phenomena (diSessa, 2000; Eylon & Linn, 1988; Linn & Hsi, 2000). These conflicting views may be connected to context (e.g. objects in motion come to rest on the playground but remain in motion in science class), discipline (heat in chemistry is mathematical, in physics it is energy), or experience. Second, learners invest intellectual energy in sorting out, linking, connecting, critiquing, reconsidering, prioritizing, selecting, and organizing their ideas (Piaget, 1957; Vygotsky, 1962).

The knowledge integration perspective suggests that instruction will succeed when students have the opportunity to add new stimulating ideas to their mix of views, and when instruction motivates them to make sense of their diverse ideas (e.g., by developing norms for distinguishing among ideas, by creating a coherent set of generative ideas, and by using these ideas to interpret new situations). The process of knowledge integration is interpretive, cultural, and deliberate. Knowledge integration is *interpretive* in the sense that learners make sense of new ideas in light of their existing views of the phenomena or situation (Piaget, 1957; Eylon & Linn, 1988). Knowledge integration is *cultural* in the sense that norms, standards, and expectations from diverse individuals govern the process (Lave and Wenger, 1991; Vygotsky, 1962). Knowledge integration is *deliberate* in the sense that learners have responsibility for guiding their own intellectual effort (White & Frederiksen, 1998; White, 1988).

The framework has emerged from research showing how to support students as they conduct projects or carry out case studies (Linn, 1995; Slotta & Linn, 2000; Linn et al. 1999). Too often, students flounder and fail when attempting complex projects (Feldman et. al, 2000). The scaffolded knowledge integration framework synthesizes research to give designers a head start on creating research based instruction.

Students debate three different approaches for controlling Malaria worldwide: (a) developing of an effective pesticide that targets the anopheles mosquito, (b) developing a vaccine to prevent malaria, and (c) creating social programs that reduce exposure to mosquitoes (e.g., through distribution of bed nets or community clean-ups). Students explore evidence relating to each control method, and debate alternative approaches. WISE teachers build on their own teaching practices, participate in professional development, and customize the curriculum (for details on this research see Slotta, in press).

2.1.1 Make science accessible

Courses should offer projects that connect to student interest and abilities. Projects should represent content at an appropriate level of analysis so that students can restructure, rethink, compare, critique, and develop their own cohesive ideas. In the *Cycles of Malaria* project, biomagnification represents the metabolic processes that concentrate toxins in one organism and not another and connects to student knowledge. To make the Malaria debate accessible to students, WISE introduces Kofi, an African child suffering from Malaria. Additionally, teachers can customize the project by making connections to more personally relevant diseases in North America, (e.g., HIV, Sickle Cell Anemia, or West Nile Virus).

2.1.2 *Make thinking visible*

Both technologies and classroom teachers should make scientific visualizations and scientific arguments more visible to students using modeling environments, reasoning tools, and embedded assessments as well as encouraging students to design innovations (diSessa, 2000). For example, WISE represents the inquiry and problem solving processes through Web-based maps that lead students through inquiry steps (see Figure 1). To make thinking visible, the Malaria project includes animations and video of the mosquito and parasite lifecycles, as well as maps showing the worldwide incidence of Malaria.

Figure 1: WISE Malaria Project—inquiry map on left

Powerful examples called *pivotal cases* come from careful study of student reasoning (Linn, in press-a). Pivotal cases enable students to make critical comparisons within a curriculum activity, encouraging them to rethink, reorganize, and reconceptualize their view of a scientific phenomenon. For malaria, pivotal cases compare countries choosing to use DDT with those discontinuing use. Students look at both infant mortality rates and bird population size. They use the pivotal case to interpret the arguments in Rachel Carson's *Silent Spring* and in the legislation calling for a global ban of DDT.

2.1.3 Help students learn from each other

Courses should facilitate the use of collaborative tools, such as online discussions, peer reviews, and collaborative design to help students take advantage of their classmates' ideas (Hoadley & Linn, 2000; Hatano & Inagaki, 1988). These discussions also allow students to jointly create standards and criteria for arguments. In addition, online discussions can provide equitable opportunities for participation in the discourse of science. In classroom research studies, over 90% of the students participated in the online discussions. In contrast, class discussion only involves about 15% of the class and far more of the participants are male than female (Hsi, 1997).

Designing discussions to cross-geographic boundaries offers special advantages for international collaboration. The WISE Malaria project enables Norwegian and American students to debate solutions to the worldwide Malaria threat (Jorde, this volume).

2.1.4 Foster lifelong learning

Courses should engage students in reflecting on their progress, evaluating their plans, critiquing solutions, and diagnosing their weaknesses to help them become lifelong learners. WISE projects ask students to design solutions, debate different approaches, interpret experiments, or critique alternatives. Reflection notes and hints as shown in Figure 1 prompt students for reflection. To ensure success, notes need regular testing and redesign (Davis & Linn, 2000).

Research suggests that critical evaluation of new material and reflection on connections among ideas can help students become lifelong learners. In WISE students criticize Internet materials and sort out promising as well as less helpful ideas. For example, in the Malaria project, students compare different arguments for uses of DDT, reflect on personal travel decisions, and critically evaluate Internet evidence that supports competing views about dealing with Malaria. Students who perform the activity become more successful at critiquing new evidence, make connections to environmental stewardship, and return to their teachers year after year with new news articles about Malaria treatment and issues.

3. WISE RESEARCH PROGRAMS

3.1.1 Customizing WISE

The WISE learning environment and others like it enable research partnerships to study compelling questions, gather detailed and precise information about student trajectories and teacher customizations, and compare findings across countries. By researching how teachers customize these projects in varied contexts we can begin to understand how to support teachers as they adapt instruction to their own situations. International investigations have the advantage of expanding the range of contexts and curriculum frameworks and can result in more robust findings.

Slotta (in press) has studied how two middle school teachers, each teaching 180 students in 6 classes, use the WISE *Cycles of Malaria*. Both teachers participated in WISE mentored professional development. These teachers differed in how they customized the project to their curricula and teaching styles. Students in both classes benefited from the same pivotal cases, knowledge integration activities, and assessments. All students completed WISE pretests, posttests, and embedded assessments. The teachers participated in interveiews and were observed in their classrooms.

3.1.2 WISE mentored professional development

The WISE mentored professional development follows the scaffolded knowledge integration framework. A mentor respected by the teachers and familiar with WISE introduced WISE and orchestrated the implementation.

The mentor made inquiry science *accessible* by modelling WISE teaching in each participant's classroom. The mentor modelled WISE teaching for the first class and then turned the instruction over to the teacher but remained for support.

The mentor made the process of inquiry teaching *visible* with pivotal cases. For example, to help teachers learn how to listen to students, the mentor suggested listening with the purpose of identifying obstacles or conundrums that concern all the students. The mentor told the teachers to stop the class and discuss the problem after three or four groups appeared to face the same dilemma. To help teachers elicit information from students, the mentor recommended asking specific questions such as "Why is this evidence less connected to the argument than the evidence you used?" rather than giving general encouragement such as "That's great." (Linn, in press-b)

To help students *learn from each other*, the mentor encouraged visits to each other's classes. The mentor held frequent informal meetings with the teachers to help them solve problems. To encourage the teachers to become *lifelong learners*, the mentor helped the teachers customize the project based on student success.

The teachers interpreted this mentoring differently. One teacher emphasized monitoring student progress and circulated around the room regularly. This teacher spent an average of only about twenty seconds with each group and rarely asked for elaborations or identified obstacles.

The other teacher focused on the questions identified by the mentor—even writing down the actual wording and referring to the notes. This teacher often spent 5 minutes tutoring one student group and averaged over 2 minutes per group. For the tutoring teacher, the WISE inquiry map was particularly valuable because other students continued to work while the teacher spent time with individuals. This teacher rarely interrupted the whole class to share insights.

3.1.3 WISE Student learning

The teachers reported excellent insight into their inquiry teaching as reflected in interviews. During the first year, the tutoring teacher reported, "I first tried to touch bases with every single group, like, walk around and listen to the conversation, and read their responses, which I usually don't do." The teacher worried, however, that some groups were neglected because they could not be visited during each class. In interactions with the students, the tutoring teacher saw the role of the teacher as helping students to make connections. This teacher remarked, “I’m hoping they see the interaction of all the different sciences, you know—within one subject they were studying, like, malaria—it’s not just cells, but the whole interaction of the people, the insects, the life cycle, the studies, how we’re going to solve it—so they saw a complete picture of how science works outside the classroom.” The teacher was aware that the students had trouble getting the big picture and saw the instructor as asking questions so that the big picture would emerge, as well as “Questioning what they wrote down—what did they mean? Things like that,” to insure that students made valid connections.

The mentoring teacher remarked that the role of the instructor was “just wandering around to make sure that all the students are focused.” This teacher noted, like the other teacher, “They weren’t getting the big in-depth thing, so you had to stop and redirect them and say why it is important.” The teacher experimented with explaining things versus having the students read the material and discuss it with their partners, saying, “I think they understood it better reading it versus listening to me drone on.” The monitoring teacher reported that students had difficulty supporting a point of view they disagreed with but concluded that those who defended a position they didn’t agree with wrote more coherent and stronger arguments than when they chose their own position. This experiment, conducted by the teacher, reveals a benefit to inquiry teaching with technology like WISE. Because teachers typically repeated the same lesson for five or six classes over one or two days, they can quickly test new practices. Of course, such informal experiments could be misleading, since many factors contribute to successful inquiry. Nevertheless, instilling in teachers the belief that practices can be tested and refined over a few days of teaching, strengthens professional development

Students taught by these distinct methods made substantial and similar gains from pretest to posttest in understanding of the topic of malaria (Slotta, in press). However, students with the tutoring teacher were more likely to construct complex arguments effectively. Whereas, students with the monitoring teacher were more likely to generate effective explanations of class experiences. These differences, while significant, were relatively small. Overall, each teacher was successful in using the WISE materials to improve student understanding.

3.1.4 Teaching WISE a second time

After the first teaching of WISE Cycles of Malaria, both teachers reflected on their experiences and decided to teach the project to their next class. The mentor kept in contact with the teachers but did not visit regularly. The teachers reported using more complex inquiry teaching strategies. The embedded notes written by the new students were more sophisticated and made more connections (Slotta, in press). Both the students — who were doing their second or third WISE project — and the teachers had more expertise in the second year and the progress of the students reflects this. Nevertheless, the teachers remained distinct; the monitoring teacher circulated and the tutoring teacher talked more with the students. In the second year, students made more progress than they did the first time the teachers used WISE. As in year one, students of the tutoring teacher generated more sophisticated and integrated explanations.

During the first project year, both teachers complained of exhaustion after teaching with WISE, pointing out that not only did they have to concentrate on new subject matter, but they also had to institute new teaching practices. By the second year, they had found ways to reduce the exhaustion while maintaining their personal contributions. Interviews suggested that the two teachers were both experimenting with new inquiry strategies – clear evidence that the mentoring had set them on a path of lifelong learning.

The tutoring teacher had added a portfolio requirement to the science course, and reported that many of the students picked their WISE projects for the portfolio. The teacher remarked, "They felt like real scientists looking at evidence out there. They felt the connection between what we do in school and what's currently happening in the world." This teacher also remarked that students were developing analytic skills and that in the second year they seemed more adept at analyzing information than they had in the first year, possibly because they were now conducting their second WISE project, having also used WISE in sixth grade. This teacher, in the second year, focused more on helping students connect ideas, saying, "It's difficult to show the whole picture" when using a textbook, but WISE "is the perfect vehicle."

The tutoring teacher experimented with on-line rather than face-to-face interactions. The teacher said, "Last year I used to try to get to every single computer and talk to each group… I had them all print out their finals so I could take my time grading them. This year, I decided to do something different. I decided to use the online comments, so I started typing comments to them." As a result, the interactive teacher appeared to be quite non-interactive, typing comments at a computer in the back of the room while the students were working. While observers found this frustrating, assuming that the teacher was not engaged with the students, the teacher felt differently. Having taught the unit before, the teacher reported that it was now possible to ask more informative and specific questions using the online comments. Furthermore, the teacher reported, "I like making comments to the kids on the computer, and they can also come up to me as they're working. I'm talking to them, which is excellent." This teacher still depends heavily on the WISE software to guide students. This teacher has moved from emulating the model of the mentor to experimenting with new practices.

By the second year, both teachers reported a deeper understanding of their students' ideas. The monitoring teacher remarked, "You can go there and kneel down between them and get to know the kids better. It became much more of a personal relationship between me and the kids. I was able to clean up some misconceptions they had at the same time, by pointing them back where to go." The teacher also noted: "They picked up writing skills. They realized that the way they were reading and interpreting the question wasn't necessarily on target, and they also learned from the first time around that if there was anything that wasn't understood, they have to communicate more openly with their partner and with me."

Comments from the monitoring teacher also suggest a changing role in the classroom: "The first time around I was closely monitoring what was going on, and then I had the tendency to go 'hands off' and let them do their own thing. After reading the results of the first project that they did, I realized that I needed to be more involved and to monitor more closely what was going on, and take a sampling of the answers that each group was coming up with." The teacher also reported that spending more time analyzing the project helped with instruction: "The first time around I hadn't gotten into the project myself, and read through it, seeing it more often, becoming more familiar with it—seeing what the kids were actually going to be reading, I was better prepared. I was able to prepare them going into the project by giving them a little more background on what they were going to be doing and seeing." This teacher also reported integrating the WISE material more closely into the curriculum, remarking, "I just made a reference to the malaria project this morning, because now we're studying anthropoids."

Both teachers credited the mentor with modelling classroom interactions. They also realized that experimenting with varied approaches and studying the unit more intensively had improved both the WISE experience and student learning outcomes.

3.1.5 Teaching WISE in Norway

Teaching WISE in Norway required customization to connect to the existing curriculum (Jorde, this volume). Since Norwegian students spend more time on environmental stewardship, they could handle more sophisticated ideas about control of malaria. The Norwegians created visualizations that have subsequently been tested with American students. Classes in the two countries have held joint discussions of the proposed international ban on DDT. Comparisons of professional development needs for the two countries reveal differences. Norwegian teachers have more preparation time and wanted to learn more about customizing the curriculum.

4. CONCLUSIONS

These comparisons between teachers in the same school and between the same curriculum in different countries illustrate how research programs that employ learning environments and emphasize customization can advance understanding of learning and instruction. These studies reveal the benefit of an instructional framework like knowledge integration to guide research and to frame conclusions that inform future work. These studies show that knowledge integration processes are robust for variations in teaching practice but require that teachers recognize the importance of deep, coherent understanding of the scientific phenomena.

Although these environments work differently depending on the cultural context, tools can be exchanged and improvements shared, problems jointly attacked and issues identified. Teachers in distinct contexts can emphasize environmental stewardship. Discussions between Norwegians and Americans suggest that students can learn from each other about stewardship issues and inform each other's thinking.

These illustrations suggest a broad range of potential research questions that might be addressed using a learning environment like WISE and an instructional framework that emphasizes knowledge integration.

First, questions concerning making science accessible lend themselves to cross-national comparisons. Studying the customizations that teachers and students introduce in making projects like Malaria accessible can shed light on the overall nature of personal relevance in science, and help us design more and more activities that promote lifelong science learning by enabling students to revisit, rethink, and reconceptualize their ideas not just during science class but throughout their lives.

Second, to make thinking visible, research clarify which visualizations succeed and why. Attempts to exchange visualizations across cultures and even within the same school district have frequently been difficult. Quite often students don't recognize the meaning or interpretation of a visualization that makes sense in another context. Often visualizations are more difficult to learn than the subject matter itself.

Third, to help students learn from others cross-national discussions can help but these discussions need careful design. Sometimes translation of comments or vocabulary can help to foster communication. Worldwide environmental problems require that we learn how to communicate across national boundaries about complex issues in science, technology and society.

Fourth, promoting lifelong science learning has particular relevance to international collaboration, because curriculum emphases and course sequences differ dramatically from country to country. Studies looking at the factors that lead to persistence and understanding of complex science could help everyone. Designing and studying a topic using a learning environment like WISE allows comparison of paradigm trajectories of students in varying cultures who study the same topic.

A fifth area for research that this study emphasizes concerns variations in professional development and professional practice. Differences between Norway and the United States lead to very distinct opportunities for preparation and reflection on teaching. Comparing these opportunities provides insight into the challenges of teaching when demands are high and resources and preparation time is low or non-existent. We can learn from these comparisons and we encourage continued research along these lines.

The opportunity to conduct research using some similar and some distinct materials, curriculum, innovations, technological tools and professional development models has yielded research findings that offer promise for defining new research programs and for promoting international collaboration. Please join us at our website to continue this discussion (http://wise.berkeley.edu).

This material is based upon research supported by the National Science Foundation under grants 9873180, 9805420, 0087832, and 9720384. Any opinions, findings, and conclusions or recommendations expressed in this publication are those of the authors and do not necessarily reflect the views of the National Science Foundation. This paper was partially prepared while Linn was a Fellow at the Center for Advanced Study in the Behavioral Sciences with support from the Spencer Foundation. Thanks go to Web-based Inquiry Science Environment (WISE) and the Science Controversies Online: Partnerships in Education (SCOPE) partnerships. Preparation of this manuscript was made possible with help from L. Safley, S. Hsu, C. Kinnison, and D. Crowell.

REFERENCES

Agogino, A. M. & Linn, M.C. (1992 May-June). Retaining female engineering students: Will early design experiences help? NSF Directions. M. Wilson, National Science Foundation. 5: 8-9.

Bransford, J. D., Brown, A. L., & Cocking, R. R. (Eds.). (1999). How people learn: Brain, Mind, Experience, and School. Washington, DC: National Research Council.

Brown, A. L. & Campione, J.C. (1994). Guided discovery in a community of learners. Classroom lessons: Integrating cognitive theory and classroom practice. K. McGilly. Cambridge, MA, MIT Press/Bradford Books: 229-270.

Davis, E. & Linn, M. C. (2000). Scaffolding students' knowledge integration: Prompts for reflection in KIE. *International Journal of Science Education* [Special Issue], 22, 819-837.

diSessa, A. A. (2000). *Changing minds: Computers, learning and literacy*. Cambridge, MA: MIT Press.

Driver, R., Leach, J., Millar, R., & Scott, P. (1996). *Young People's Images of Science*. Buckingham, UK: Open University Press.

Eylon, B. S. & Linn, M.C. (1988). Learning and instruction: An examination of four research perspectives in science education. *Review of Educational Research*. 58(3): 251–301.

Feldman, A., Konold, C., & Coulter, B. (2000). *Network Science, A Decade Later: The Internet and Classroom Learning*. Mahwah, NJ: Lawrence Erlbaum Associates.

Hatano, G. and K. Inagaki (1988). Constrained person analogy in young children's biological inference. American Educational Research Association, New Orleans.

Hoadley, C. M. & Linn, M. C. (2000). Teaching science through on-line, peer discussions: SpeakEasy in the Knowledge Integration Environment. *International Journal of Science Education* [Special Issue], 22, 839-857.

Hsi, S. (1997). *Facilitating knowledge integration in science through electronic discussion: The Multimedia Forum Kiosk*. Unpublished doctoral dissertation, University of California, Berkeley, CA.

Lave, J., & Wenger, E. (1990). *Situated Learning: Legitimate Peripheral Participation*. Cambridge, UK: Cambridge University Press.

Linn, M. C. (1995). Designing computer learning environments for engineering and computer science: The Scaffolded Knowledge Integration framework. Journal of Science Education and Technology, 4 (2), 103-126.

Linn, M. C. (2000). Designing the Knowledge Integration Environment: The partnership inquiry process. *International Journal of Science Education* [Special Issue], 22, 781-796.

Linn, M. C. (in press-a). WISE design for lifelong learning-Pivotal Cases. In P. Gärdenfors & P. Johannsson (Eds.), *Cognition, Education and Communication Technology*. Mahwah, NJ: Lawrence Erlbaum Associates.

Linn, M. C. (in press-b) WISE Teachers—Using Technology and Inquiry for Science Instruction. In R. Floden and J. Bell Eds., *Meaningful Learning*. Battle Creek, Michigan.

Linn, M. C., & Clancy, M. J. (1992). The case for case studies of programming problems. *Communications of the ACM*, 35(3), 121-132.

Linn, M. C., Davis, E. & Bell, P. (in press) *Internet Environments for Science Education*. Mahwah, NJ: Lawrence Erlbaum Associates.

Linn, M. C., & Hsi, S. (2000). *Computers, Teachers, Peers: Science Learning Partners*. Mahwah, NJ: Lawrence Erlbaum Associates.

Linn, M. C., Shear, L., Bell, P., & Slotta, J. D. (1999). Organizing principles for science education partnerships: Case studies of students' learning about 'Rats in Space' and 'Deformed Frogs'. *Educational Technology Research & Development*, 47(2), 61-84.

Piaget, J. (1957). *Logic and psychology*. New York,, Basic Books.

Reiser, B. J., Tabak, I., Sandoval, W. A., Smith, B. K., Steinmuller, F., & Leone, A. J. (2001). BGuILE: Strategic and conceptual scaffolds for scientific inquiry in biology classrooms. In S. M. Carver & D. Klahr (Eds.), *Cognition and instruction: Twenty five years of progress* (pp. 263-305). Mahway, N.J.: Erlbaum Associates.

Scardamalia, M. and C. Bereiter (1992). A knowledge building architecture for computer supported learning. In E. De Corte, M. C. Linn, H. Mandl and L. Verschaffel, *Computer-based learning environments and problem solving*.. Berlin, Springer-Verlag.

Slotta, J. D. (in press). The Web-based Inquiry Science Environment (WISE): Scaffolding knowledge integration in the science classroom, In M.C. Linn, E. Davis and P. Bell (Eds). *Internet Environments for Science Education*. Mahwah, NJ: Lawrence Erlbaum Associates.

Slotta, J., & Linn, M. C. (2000). How do students make sense of Internet resources in the science classroom? In M. J. Jacobson & R. Kozma (Eds.), *Learning the Sciences of the 21st Century*. Hillsdale, NJ: Lawrence Erlbaum & Associates.

Vanderbilt, C. a. T. G. a. (1997). *The Jasper Project: Lessons in curriculum, instruction, assessment, and professional development*, Lawrence Erlbaum Associates.

Vygotsky, L. S. (1962). *Thought and language* (E. Hanfmann and G Vaker, Trans.). Cambridge, MA: MIT Press.

White, B. Y. & Frederiksen, J.R. (1998). "Inquiry, modeling, and metacognition: Making science accessible to all students." *Cognition and Instruction* 16(1), 3-118.

White, R. T. (1988). *Learning science*. Oxford, England, Basil Blackwell, Inc.

RESEARCH ABOUT THE USE OF INFORMATION TECHNOLOGY IN SCIENCE EDUCATION

Hans Niedderer [1], Florian Sander [1], Fred Goldberg [3], Valerie Otero [3], Doris Jorde [4], Jim Slotta [5], Alex Strømme [4], Hans E. Fischer [2], Lorenz Hucke [2], Andrée Tiberghien [6] and Jacques Vince [6]

[1] *University of Bremen, Germany;* [2] *University of Dortmund, Germany;*
[3] *San Diego State University, USA;* [4] *University of Oslo, Norway;*
[5] *University of California, Berkeley, USA;* [6] *University of Lyon 2, France*

ABSTRACT

In this paper, we investigate some aspects of effectiveness in two different kinds of learning environments with use of information technology. The first part shows research results related to a web-based integrated science environment (WISE) and its effects on changing the role of the teacher, the curriculum, and the student. In the second part, three different approaches for scaffolding conceptual development in physics by using interactive computer models are analysed. Research results about different aspects of their effectiveness are presented. Both parts together show empirical results in relation to different promising approaches of using information technology in science education.

1. INTRODUCTION

Information technology can be used in science education in many different ways. From a social constructivist perspective, those forms bear the most potential for learning in science education, which provide help for social learning in groups combined with interactive work with the computer. The computer then can have several supportive functions: It can supply information, for instance from the Web, it can supply new representations of models and theories, or it can stimulate social interactions about representations on the computer screen. All this is interactive, it can be manipulated by students, thus making thinking visible. In this way, the computer becomes a part of an integrated learning environment to foster social and active learning in science, which can be called "scaffolded knowledge integration" (Linn 1998, 275). This paper does not look for learning processes in detail as in other papers of the same authors (Fischer 1994; Petri & Niedderer 1998). Instead, this paper describes certain features of ICT use in the science classroom and research results about their effectiveness. More studies of that kind will be published elsewhere (Psillos & Niedderer 2002, chapter 5: Labwork Based on Integrated Use of New Information Technology).

D. Psillos et al. (eds.), Science Education Research in the Knowledge-Based Society, 309–321.

2. ICT[1] IN SCIENCE EDUCATION – CHANGING ROLES OF CURRICULUM, TEACHER, AND STUDENT

Web-based Inquiry Science Environment (WISE) - refining our technology and curriculum. The role of teachers and teaching is quickly taking a different pathway as phrases like, "Children responsible for their own learning", "Information, communication and technology" "teaching across the curriculum" and "Project directed teaching" enter the science curriculum (e.g. the national curriculum of Norway in 1997). The WISE learning environment is developed to help scaffold students as they perform innovative science inquiry projects. In our view of *Scaffolded Knowledge Integration,* inquiry activities should engage students in the intentional process of diagnosing problems, critiquing experiments, distinguishing alternatives, planning investigations, researching conjectures, searching for information, constructing models, debating with peers, and forming coherent arguments (see also the paper of Linn *in this book*). WISE provides an Internet-based platform for middle and high school science activities where students work collaboratively on inquiry projects, making use of "Evidence" from the Web. Typical projects engage students in *designing* solutions to problems (e.g., building a desert house that is warm at night and cool during the day), *debating* contemporary science controversies (e.g., should we have wolves in Norway?), or *critiquing* scientific claims found in web sites (e.g., arguments for life on Mars). Figure 1 displays the WISE interface where students navigate through activity steps in the left-hand frame of their Web browser, called the "Inquiry Map." All of these materials can be seen, and activities explored at our project Web site: http://wise.berkeley.edu or at the Norwegian site: http://viten.no

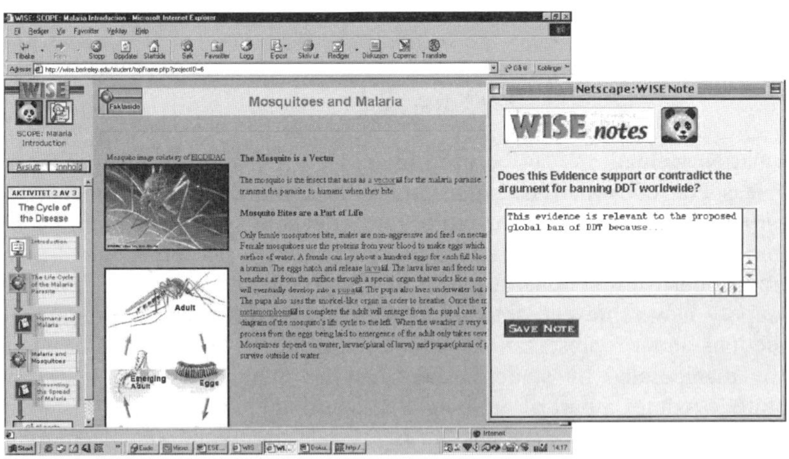

Figure 1: Example of a WISE screen page and the WISE note taking tool

The changing role of the teacher and the curriculum. In an empirical study we followed two teachers in the US (Sandra and Gilbert) and one teacher in Norway

[1] Information and Communication Technology

(Marit) as they taught the *cycles of Malaria* project first in year one, then in year two. We compared teaching strategies of the three teachers as they integrated the WISE unit into the curriculum, looking for differences in teaching styles. We measured student performance in WISE by analysing pre and post test gains as well as looking at embedded assessments in the form of note taking. Finally, teachers were interviewed after each year's implementation.

In the first year run of *Cycles of Malaria* we started to see the emergence of different teaching styles when using WISE materials. Sandra had a pattern of engaging deeply with individual students for several minutes at a time, while Gilbert and Marit showed a different pattern of cycling through the classroom, stopping at each group to see if they were progressing. Whereas all of the three teachers seemed to be letting the technology run their classes, Marit and Gilbert relied more on the technology to lead the way. In interviews with all of the teachers they expressed concern that they were not familiar enough with the curriculum materials and did not completely understand how the net-based program would work in the science classroom.

In the second run of the curriculum unit (year two of *Cycles of Malaria)*, we started to notice differences in how teachers were preparing their students. Marit planned the Malaria unit as a part of her yearlong syllabus, including chapters on micro-organisms and disease and sickness before using *Cycles of Malaria* as a capstone project to integrate information. She was more careful to ask students to support the answers they provided on their notes and was more successful in talking to groups of students about their progression throughout the unit by looking with them at their completed on-line work. Marit was able to customise the malaria unit to fit into her pedagogical style of teaching as well as into her content goals for her science syllabus. She commented that she was much more comfortable teaching the unit the second time and expressed a sense of ownership in materials as compared to the first year when she was both sceptical and anxious to using this new form in her science teaching. When asked about using WISE, Marit's students commented that they wanted more WISE projects, thus supporting Marit's observation that this is a positive way to introduce variation in teaching science.

Pre-post student learning gains. There were dramatic changes in pre-post gains demonstrated by all three teachers in year two. Using an example from Marit's students we asked about the way malaria could be spread. The pre and post test comparisons demonstrate gains in all classes after completion of the unit.

Student comments. It was not unexpected that students' logs would reflect positive attitudes about using ICT in science teaching. Our experience is that students like most things connected to computers and with variation in teaching methods. A common comment in the student logs was "give us more WISE", especially more about malaria and disease. On the negative side, we have learned that students at this age do not like reading lots of text on the computer and particularly in Norway, they did not all like looking at English language sites.

3. DIFFERENT SOLUTIONS FOR MODELLING PHYSICS

Experiments, models and simulations. Modelling in a general sense is the most fundamental activity in science and science education. An important aspect of learning in science is to connect the two worlds: the world of theories and models and the world of objects and events (Tiberghien 1994). The general task of computer-based modelling is to help students develop a relation between the two worlds. Scaffolding in science education therefore refers not only to the development of particular concepts, but also to the development of an understanding of the process of building explanatory models. In physics teaching, experiments play a crucial role and provide opportunities for students to interact with objects and events in the physical world. In inquiry-based physics courses students are often expected to move directly from these interactions with phenomena in the world to the development of conceptual models that can explain those phenomena. This is a very difficult task for many students (Kuhn, 1993). Researchers have found that very often students do not link theory to their actions during experiments. "To many students, a 'lab' means manipulating equipment but not manipulating ideas" (Lunetta 1998). The integration of experimenting and modelling is an important step towards conceptual development, which can be fostered by appropriate computer tools (Schecker 1998). Efforts in computer-enhanced instructional design have led to the development of computer tools that can help students form connections between their experiments and their evolving conceptual models.

Research described in this paper features three different computer tools for fostering model building in physics learning:

- A CPU[2] static electricity and magnetism simulator (Otero, Goldberg)
- SimulaSON - macro and micro models of vibration/sound (Vince, Tiberghien)
- The model building system STELLA (Hucke, Sander, Fischer, Niedderer)

Common features of all three tools are that a model is represented graphically on the computer screen, that the students can manipulate it and that it is related to specific laboratory experiments. In all three learning environments, the interrelation between real experiments and computer representations of models is crucial.

Static Electricity and Magnetism Simulator (Otero, Goldberg). A unique aspect of the CPU materials (Goldberg 1997) is the way that the computer simulator is embedded in the activity documents (Otero et al., 1999). Students are expected to move back and forth between laboratory observations, computer simulators, and the activity documents. The CPU Static Electricity simulator (Goldberg et. al. 2000a, 2000b; Hickman et al. 1999) shows what would happen in a hands-on experiment and overlays representations of a corresponding conceptual model. This property of dual representation was designed to help students to make connections between their hands-on experiments and their evolving conceptual models of observed phenomena. The CPU static electricity simulators colour the surface of objects red (R) or blue (B) indicating the charge.

Figure 2 shows three snapshots while working with the Static Electricity and Magnetism simulator: (a) Two neutral insulators are near each other. (b) After

[2] Project "Constructing Physics Understanding"

rubbing together, the rubbed surfaces are coloured red (R) or blue (B). A neutral conductor with a conducting indicating flag sits nearby. (c) An R-charged insulator brought near the conductor causes the nearby surface of the conductor to be coloured oppositely (B), and the far surface to be coloured the same (R) as the charged insulator. This models the phenomenon of polarisation.

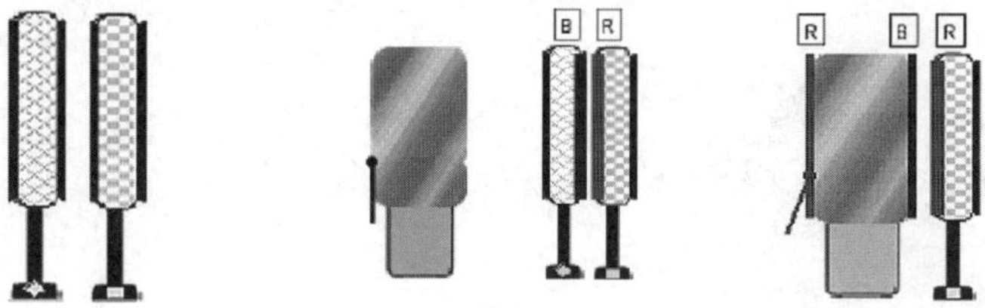

(a) (b) (c)

Figure 2: Snapshots from the Static Electricity and Magnetism simulator.

This model-like representation with colours does not depict charge as discrete units and does not use the convention of positive and negative in the visual representations; it does not impose a specific physical description of charge. The students are expected to develop concepts of charges on their own. The model-like results made available by the computer simulator allow students to make "observations" that they could not make with laboratory apparatus. The static electricity simulators provide a variety of tools and set-ups which allow students to make observations that are relevant to concepts in static electricity such as charge transfer by touching, charge polarisation, and charging by induction.

SimulaSON - microscopic models of vibration and sound (Vince, Tiberghien). SimulaSON simultaneously presents the simulated objects and the "observable" events including audio. It includes a microscopic model of vibration and propagation and aims at helping students understand concepts such as vibration, propagation, frequency or amplitude. The user acts on a graphic representation of the amplitude and the frequency of the movement of the vibrating line and simultaneously can hear the sound. This helps to see the relation between frequency and high/low pitches or between amplitude and loud/faint sound (Vince 2000b).

The design of this programme is based on hypotheses about learning along with an analysis of the knowledge to be taught carried out at a fine grain level (Vince & Tiberghien 2000). It is stated that the construction of conceptual understanding is mainly related to the relations established between these different representations.

On the same screen page, depending on the tasks, four main windows can appear (Figure 3): (1) the control window at the centre of the screen represents "the state of a sound" in terms of frequency and amplitude along two orthogonal axes. The user can act on it and modify either the frequency or the amplitude. (2) The vibration window on the left side with a "vibrating line" (in black) between two red lines representing the amplitude. The black line vibrates according to the values given by the user in the window in the middle of the screen. (3) An upper window with a microscopic representation of a particulate model, where one particle and/or a set of particles can be coloured. This particulate representation of a sound tube vibrates like the "black" line of the second window. (4) The oscillo window is on the right side. Its oscilloscope picture is related to two captors, which may be introduced in the microscopic window.

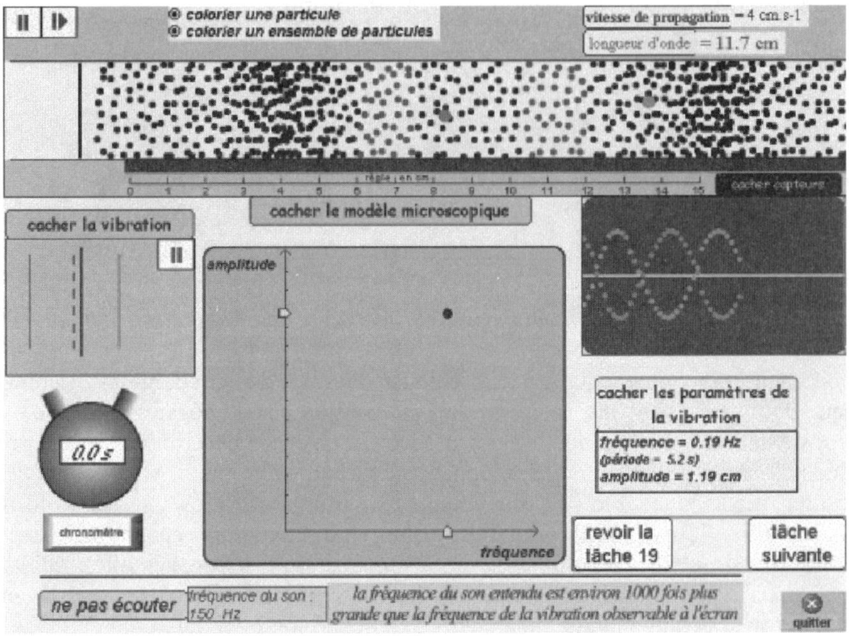

Figure 3: Example of SimulaSON *screen (experimental version).*

Moreover, there is a possibility to hear a sound with the amplitude and frequency chosen in the control window by clicking on the listen button. These different representations aim to help students to construct an understanding of vibration and propagation of sound at different levels: a macroscopic level of vibration and a microscopic particulate level. This understanding requires establishing relations between the behaviours of a single particle, of a set of particles, and the propagation of the vibration with its characteristics (wavelength and velocity).

Applications of model building system STELLA in quantitative mechanics (Hucke, Sander, Fischer, Niedderer). The basic idea of modelling with a model building system software like STELLA is to give students a comfortable computer tool to develop their own theoretical model for an experiment in the form of a concept map, which in addition is powerful enough to carry out calculations. These calculations (graphs) can be compared with results from measurement. Computer tools are used both for theory (STELLA for modelling) and experiment (any MBL device for data collection and processing). We illustrate the integrated use of computer tools for measuring and modelling with a case from a lab about spring oscillations. This started with measuring the free movement of a bob hanging on a spring (Figure 4).

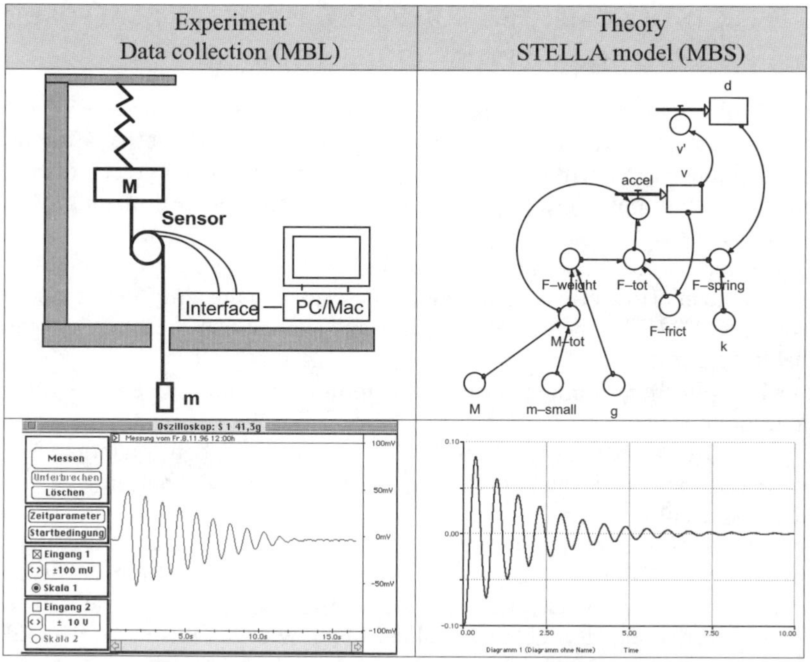

Figure 4: Experiment, computer model and results from both

Afterwards, two students develop their own computer model in many small steps, starting from an empty STELLA desktop. Students can manipulate their own theoretical approach in the following ways:

- They can put elements like concepts or variables on the screen and give them any name they want.
- They can select and specifically define relations between concepts and variables, such as defining the rate of change or more algebraic relations.

Let us keep in mind that the aim is to foster a qualitative understanding of physics phenomena. Students are expected to learn how to use basic physics concepts (like force or momentum) in relation to a specific experiment. The three most important steps in this development are

- to develop the acceleration->velocity->distance sequence of variables and rates of change (kinematics part of the model),
- to apply the force-mass-acceleration relation (Newton 2) as a power tool for all force and motion problems, and
- to introduce the specific forces for the spring problem.

The students performed the following lab activities: With computer measurement (MBL) the students got a graph, which due to friction had decreasing amplitude. Then they built a model with STELLA (MBS). Finally, they got a good agreement between measurement results and the simulation graph from their model (Figure 4).

4. EFFECTS OF COMPUTER MODELLING TOOLS FOR SCAFFOLDING CONCEPTUAL UNDERSTANDING

Research questions and methods. Use of such tools in the classroom has led to interesting research questions in science education about the effects of different kinds of computer modelling tools on students' communication processes, sense making and conceptual development. The three studies deal mainly with two research questions:

- How effective are the computer modelling tools in fostering conceptual development? How effective are they with respect to the aim "to link theory to experiment"? How much do experiments and simulations contribute to sense making processes?
- In what way are the computer modelling tools scaffolding students' learning of conceptual models?

Mainly video data were used. They have been analysed both quantitatively with certain categories, and qualitatively in case studies for analysing scaffolding processes in detail.

Effectiveness of modelling activities. In the studies of Hucke (2000), Sander (2000) and Buty (2000) a method for analysing video tapes from lab work with integrated use of computer tools was developed to investigate the relationship between labwork contexts and the verbal use of knowledge (Niedderer et al. 2002).

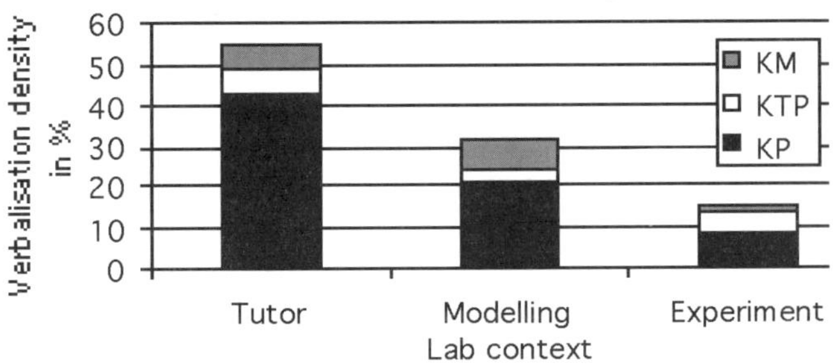

Figure 5: Density of verbalising theoretical knowledge in three types of lab contexts

From Figure 5 we can see that working in experimental contexts contributes little to talking about theoretical knowledge[3] but use up a lot of lab time. On the other hand, during model building phases we see a rather high density of verbalising theoretical knowledge (about 30 %). This is the second highest density behind talking with the tutor. This strongly supports the assumption that the use of model building software (STELLA) in the lab contributes to the objective "to link theory to practice". Hucke (2000) found KP densities in model building contexts up to 75 %. The high

[3] Categories for verbalising theoretical knowledge: KP: Physics knowledge; KTP: a combination of physics and technological knowledge; KM. Mathematics knowledge.

verbalisation density in the context 'tutor' (55 %) shows the important role of the tutor to link theory and experiment.[4]

With a similar research question and method, Otero (2001) determined the percentage of time the groups were engaged in sense-making (model-based reasoning or discussions) and compared this result to conceptual development of students within one group throughout the electrostatics unit. The percent of activity time spent in sense-making by a group throughout the unit is shown in Figure 6.

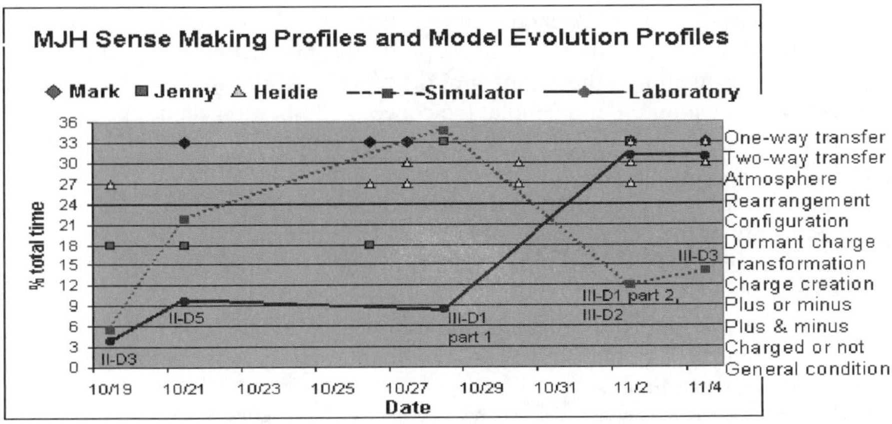

Figure 6: Percentage of sense-making model based discussions and individual student's conceptual models

The upper curve (dotted line) represents the sense-making in simulator experiments and the lower curve (solid line) represents sense-making in laboratory experiments. This data was superimposed with the models inferred for three students in a group throughout the unit (see Figure 6). The student-generated models for charging insulators by rubbing are listed on the right side of the graph (one-way transfer was the target model for instruction). These superimposed graphs suggest that the simulator mediated sense-making discussions when students' models were very different from one another. (Notice that sense making based on the simulator is much greater during the first half of the unit compared to the second half. It was during the first half that students' models were very different, see Figure 6). After students' models for charging insulators by rubbing began to converge to a one-way transfer model (around 11/2), students began engaging in more sense-making discussions directly from their laboratory experiments; they no longer relied on the simulator and colouring-scheme model to mediate their scientific discussions. It was not expected that sense-making with respect to simulator experiments would decrease over time since new features within the simulator became available and the content became increasingly complex. This data (and similar data for another group)

[4] Note that only students' contributions are counted in the knowledge verbalisation densities, not the contributions of the tutor talking himself.

suggests that the simulator served to scaffold the students' learning about the process of constructing explanatory models.

Scaffolding conceptual development with SimulaSON. This software includes a series of 25 tasks corresponding to three sessions of about 1h 30. To carry out these tasks the students have to use a text and schemas giving the physics model and the meaning of the representation of the different windows. Some of the tasks need real experiments. To study the role of SimulaSON software in conceptual understanding, a case study has been carried out. The collected data are an automatic trace and the videos of one dyad during 3 sessions (entirely transcribed). The analysis of students' productions was oriented in terms of steps of the problem solving, the type of students' re-construction of the meaning of the representations in relation to concepts and perception of sound. For example, the concept of vibration should be developed and related to the other concepts of frequency and amplitude, and with objects and events. The main results particularly relevant to simulation software emphasise (1) the importance of the description of the representations for themselves (familiarity); (2) The locality of the correct answers in a rich environment; (3) The crucial role of the questions which should help the students to understand the environment (Vince 2000a).

Concerning the first type of results, the following example shows that describing the representations for themselves is important and is not isolated from the conceptual understanding.

Example 1: During the second task, the dyad takes 6 minutes to solve the following question: If the frequency increases how does the back and forth movement evolve? They give the right answer. The next similar question is: If the amplitude increases how does the back and forth movement evolve? Instead of giving a direct answer using the similar previous question, the dyad takes more than 1 minute to read the question, then tries to modify the amplitude, and describes the symbolic representation (after 2 minutes). Looking to the representation given on the left of the screen page (see Figure 3) representing a black line moving back and forth between two red lines, and acting on the control window to modify the value of the amplitude, the dyad starts a dialogue showing that they construct a meaning to the amplitude. Student 1 says "*the amplitude is the space you see*", and in the same time they relate the variation of the value of amplitude (with a constant frequency) and the behaviour of the vibration. This same student acts on the control window to modify the amplitude and says: "the *more it [amplitude] increases ... the less it* [the red line] *goes fast. No? Ah no because it becomes smaller* [the interval between the two red lines]". The other student agrees and interprets the representation of the vibration: "*the amplitude is the red lines and the back and forth movement is the black line* ". This example also shows how the students' formulation is not straightforward, and time is necessary to construct the relevant language.

Concerning the second type of results, the following example shows both the local construction of knowledge and in the same time how this local construction can contribute to conceptual understanding.

Example 2: During the tasks 10 to 13, the students still relate velocity and amplitude of a sound, which is not right, even after measurements and calculation of the correct value of the sound velocity. In a following task (Task 19), asking how the signal given by a captor[5], when the air at the level of this captor is compressed, evolves, they give a right answer. This answer comes from a relevant interpretation relating two models of the vibration. The representation at the particulate level allows interpreting the behaviour of the particles in the air and an oscillogramme given from a captor indicates the evolution of the vibration at a given point of space. After a teacher's explanation student 2 says "*each time that the air is compressed* [in the microscopic window] *and, well, it makes a curve above* [on the oscillo window, the point tracing the curve is at the maximum]". This animated particulate representation allows the student to construct a behaviour of the vibration and to relate it to the behaviour of the membrane of the captor and

[5] in the real experiment a microphone

the shape of the curve given in the oscilloscope. The main aspects of the conceptual understanding are involved in such a situation even if the intermediary physical quantity, the electrical potential, is omitted.

This case study shows that computer modelling tool can scaffold students' conceptual development under certain conditions. The representation of models should be understood, then specific activities should be devoted to allow students to construct this understanding. Another aspect is the importance to foster the relations between the concepts, their diverse representations and the objects and events that can be interpreted. The analysis shows that the role of the questions associated to modelling tools are crucial factors in the construction of conceptual understanding. More generally, the design of modelling software programme requires a fine grain analysis of the knowledge involved in order to manage meaningful representations and to offer relevant action on them.

Scaffolding conceptual development using STELLA with force and motion. The goal of integrating modelling and measuring tools is to enable students to change freely from theory to experiment and vice versa so that the two activities mutually profit from each other (Schecker 1998, Niedderer & Schecker 1997). Our analyses of lab reports and videotapes show that these intended active interrelations are not found very often. Nevertheless, these are important activities in scaffolding concept development. Below we describe five types of those interrelations found in the data.

- Identify relevant quantities: in the course of modelling students reflect upon which quantities have to be measured in the experiment for the model to work.
- Correct model parameters: students change experimental parameters that are covered in the model, sometimes affording new measurements.
- Explore the system: the model is run in a simulation and the experiment is carried out in order to observe the behaviour of the system; the findings are compared.
- Change basic model structures: the model (conceptual layer and equations layer) is reconsidered and adapted, e.g. in order to include new experimental influences (e.g. friction).
- Develop new experimental ideas: students extend the experimental setting or apply new measurement techniques that were not given in the labguide; this also results in model structure changes.

Example: Students' discussion during the model building process centred on the acting forces. They held two different points of view: A "Newtonian" point of view, due to which forces reduce velocity, and a more "Aristotelian" point of view due to which force is proportional to velocity, and friction is a hindrance but no force. They especially discussed the lower point of the movement. An intermediate model already showed the total force as the sum of spring force and weight force. The spring force related to the distance. No friction force was considered. This resulted in an oscillation with no damping. They compared the model's prognosis with the graph from measurement (see Figure 4). Although not explicitly demanded in the labguide, the students engaged in a longer session of modelling activities in order to get a decreasing graph like the one measured. They started to discuss friction forces, especially air friction. They needed some time until they found that the friction force is always in the opposite direction of velocity. In more content specific detail, during model building with STELLA, the students talked about the following concepts of physics related to their specific experiment: (1) Weight force and spring force. (2) Acceleration at different points of the oscillation. (3) Balance between different forces. (4) Relation between position of the moving object and the force of the spring. (5) Directions of forces and signs of their values in the equations. (6) Friction conceptualised as a force in addition to the other acting forces;

its direction with respect to movement. (7) Magnitude of the friction force: Is it constant or air friction related to velocity?

The example shows that students are able to actively build their own computer model. During this process, a lot of theory is used and developed, especially in situations, where measurement graph and simulation graph are essentially different.

5. CONCLUSIONS

Two different types of using information technology for scaffolded knowledge integration in science teaching have been investigated. The approach of a Web-based Inquiry Science Environment (WISE) can be seen as a broad frame for integrating several forms of information technology into a social constructivist classroom. This has effects on the role of the teacher, the curriculum, and the role of the students. Specific results are:

- The WISE environment can be localised into science classrooms allowing teachers to personalise the way they adapt the materials into their teaching.
- Students enjoy learning science via ICT; they demonstrate learning gains.

A second type of using information technology in physics education has the main aim to scaffold conceptual development by using different computer modelling tools. Specific results were:

- During modelling phases of the teaching and learning process students spend more time in sense making activities or talking about physics than they do in most other phases of instruction.
- Students develop a deeper understanding of concepts and models by applying them step by step to specific experimental situations.

Results of these studies suggest that manipulating representational objects on the computer screen facilitate communication among students and can serve as a cognitive tool to scaffold students' development of the target instructional model. Through the collaborative construction of ideas, groups of students can come to a consensus on models that are best supported by evidence from both laboratory and simulator experiments. Ultimately, students learn to construct theoretical models for a new observable phenomenon. Conceptual tools such as CPU Simulators, SimulaSON, and STELLA can greatly facilitate this process.

REFERENCES

Buty C. (2000). *Etude d'un apprentissage dans une séquence d'enseignement en optique géométrique à l'aide d'une modélisation informatique*. Thèse de Sciences de l'éducation, Université Lumière Lyon II.

Fischer, H.,E. (1994). Physiklernen: Eine Herausforderung für Unterrichtsforschung. In Nachtigall, D. (Ed.): *Didaktik und Naturwissenschaften*, Band 3. Frankfurt: Lang.

Goldberg, F. (1997). Constructing physics understanding in a computer-supported learning environment. In: Rigden, J. (Ed.) *Proceedings of the International Conference on Undergraduate Physics Education* Volume II. American Institute of Physics.

Goldberg, F. et al. (2000a). *CPU (Constructing Physics Understanding): Simulation Software. Set of 20+ Java simulations on CD*. The Learning Team, Armonk, NY.

Goldberg, F. et al. (2000b). *CPU (Constructing Physics Understanding): Curriculum Units. Set of 7 full CPU curriculum units and teacher guides on CD*. The Learning Team, Armonk, NY. See also http://cpuproject.sdsu.edu/CPU/main.html

Hickman, P., Morse, R.A., Otero, V., Johnson, A. & Goldberg, F. (1999). *Static Electricity and Magnetism. In Constructing Physics Understanding: Curriculum Units and Simulation Software* (two separate CDs). The Learning Team, New York.

Hucke, L. (2000). Handlungsregulation und Wissenserwerb in traditionellen und computergestützten Experimenten des physikalischen Praktikums. . In H. Niedderer, H. Fischler (Hrsg.): *Studien zum Physiklernen*, Bd. 8. Berlin: Logos.

Kuhn, D. (1993). Connecting Scientific and Informal Reasoning, *Merrill-Palmer Quarterly*, 39 (1), pp. 74-103.

Linn, M. (1998). The impact of technology on science instruction: historical trends and current opportunities. In: Fraser, B.J. & Tobin, K. (Eds.): *International Handbook of Science Education*. Dordrecht: Kluwer

Lunetta, V., N. (1998). The School Science Laboratory: Historical Perspectives and Contexts for Contemporary Teaching. In: Tobin, K. & Fraser, B. (Eds.*), International Handbook of Science Education*. Dordrecht: Kluwer.

Niedderer, H., Buty, C., Haller, K., Hucke, L., Sander, F., Fischer, H.E., v. Aufschnaiter, S., Tiberghien, A. (2002). Talking Physics in Labwork Contexts - A Category Based Analysis of Videotapes. In Dimitris Psillos and Hans Niedderer, Eds. (2002). *Teaching and learning in the science laboratory*. Kluwer Academic Publishers (forthcoming)

Niedderer, H., Schecker, H. (1997). Laboratory tasks with MBL and MBS in introductory physics classes for prospective high school teachers. In. E. F. Redish, J. S. Rigden (Eds.). The changing role of physics departments in modern universities, Proceedings of the International Conference on Undergraduate Physics Education (ICUPE). New York: American Institute of Physics; http://didaktik.physik.uni-bremen.de/niedderer

Otero, V., Johnson, A. & Goldberg, F. (1999). How Does the Computer Facilitate the Development of Physics Knowledge Among Prospective Elementary Teachers. Journal of Education 181 (2), 57-90.

Otero, V. (2001). The Process of Learning about Static Electricity and the Role of the Computer Simulator. Unpublished doctoral dissertation. San Diego State University.

Petri, J., Niedderer, H. (1998). A learning pathway in high-school level quantum atomic physics. Int. J. Sci. Educ., Vol. 20, No. 9, 1075-1088

Psillos, D., Niedderer, H., Eds. (2002). *Teaching and learning in the science laboratory*. Kluwer Academic Publishers (forthcoming)

Sander, F. (2000). Verbindung von Theorie und Experiment im physikalischen Praktikum - Eine empirische Untersuchung zum handlungsbezogenen Vorverständnis und dem Einsatz grafikorientierter Modellbildung im Praktikum. Dissertation Universität Bremen. In: H. Niedderer, H. Fischler (Hrsg.): *Studien zum Physiklernen*, Bd. 13. Berlin: Logos

Schecker, H. (1998). Integration of Experimenting and Modeling by Educational Technology: Examples from Nuclear Physics. In: Fraser, B.J. & Tobin, K. (Eds.): *International Handbook of Science Education*. Dordrecht: Kluwer

STELLA software. High Performance Systems http://www.hps-inc.com/

Tiberghien, A. (1994) Modelling as a basis for analysing teaching-learning situations. Learning and Instruction, 4(1), 71-87.

Vince, J. (2000a). *Approches phénoménologique et linguistique des connaissances des élèves de seconde sur le son. Contribution à l'élaboration et l'analyse d'un enseignement et au développement d'un logiciel de simulation*. Thèse, Université Lumière Lyon 2.

Vince, J. (2000b). *SimulaSON* http://sir.univ-lyon2.fr/GRIC-COAST/Home/jvince/index.html

Vince, J.; Tiberghien, A. (2000) "Simuler pour modéliser.Le cas du son." *Sciences et techniques éducatives*, vol7, n°2, 333-366

PHYSICS LEARNING AND MICROCOMPUTER BASED LABORATORY (MBL)
LEARNING EFFECTS OF USING MBL AS A TECHNOLOGICAL AND AS A COGNITIVE TOOL*

Jonte Bernhard

Linköping University, Campus Norrköping, S-60174 Norrköping, Sweden

ABSTRACT

Different cases of physics instructor's implementations of Microcomputer Based Laboratory (MBL) in physics teaching have been studied. When implemented as a technological tool only poor learning results were observed while when MBL were used as both a technological and cognitive tool good learning results were observed. New technology thus does not necessarily lead to better learning. When developing and implementing computer aided learning we must focus as much on the cognitive aspects as on the technological aspects. Also we must focus on instructor's conceptions of teaching and learning since this affects their understanding of curricular reforms and lead to transformations of original developers educational intentions.

1. BACKGROUND, AIMS AND FRAMEWORK

Microcomputer Based Laboratory (MBL) was introduced into physics teaching more than one decade ago (Tinker 1996). An attachment of a sensor to a computer creates a very powerful system for collection, analysis and display of experimental data. Today several systems, specially developed for schools and undergraduate courses, are commercially available for different computer platforms. In a Microcomputer Based Laboratory (MBL-lab) students do *real* experiments, not simulated ones, using different sensors (force, motion, temperature, light, sound, EKG...) connected to a computer via an interface.

According to several studies (for example Tinker 1996, Thornton 1987, 1989 and 1997a, Thornton and Sokoloff 1998, Hake 1997, Laws 1997, Bernhard 2000a and 2000b, Hamne and Bernhard 2001) MBL is very effective in fostering a good functional understanding among different groups of students. According to a presentation by Euler and Müller (1999) at the ESERA-conference in Kiel 1999 MBL is the only method in physics teaching using computers with a proven positive learning effect. Bernhard (2001) has proven that MBL can give long-lived conceptual understanding. One of the main educational advantages of using MBL is the real-time display of experimental results and graphs thus facilitating direct connection between the real experiment and the abstract representation. This makes it possible to develop new types of lab experiments designed to facilitate better student learning

D. Psillos et al. (eds.), Science Education Research in the Knowledge-Based Society, 323–331.

and to use labs to foster a better conceptual understanding. Because data are quickly taken and displayed a predict – observe – explain cycle (POE) is easily implemented.

The question thus arises: Is the reported good learning effects of MBL due to inherent properties of the MBL-technology or is the educational implementation crucial?

2. METHODS AND SAMPLES

In this study different implementations of MBL-labs in introductory university level physics (mechanics) courses at a smaller Swedish university (Högskolan Dalarna) were studied. The main student body in the courses were either engineering students (Case II) or pre-service teachers (Case I, III, IV and V) preparing to be certified for mathematics and science teaching in Swedish grade 4 – 9. Between 20 – 40 students were registered in the courses. Different instructors had different educational "views" which led to different ways of implementing the MBL-labs. The cases are summarised in table 1.

Table 1. The student body, approaches and instructors of the different cases discussed in this paper. The instructors A – E were either senior lecturers or experienced lecturers. Instructor C is the author of this paper

Case	Course	Year	Registered # of students	Main student body	Approach in labs		Instructors	
					Kinematics, Newton 1st and 2nd laws	Impulse, collision, Newton 3rd law	Lecture	Lab
I	Pre-service	95/96	40	Pre-service Science Teachers (grade 4-9)	MBL – conceptual	No lab	A	B
II	Mechanics I	97/98	40	Engineering	MBL – conceptual	MBL – conceptual	C	C, D, E
III	Pre-service	98/99	31	Pre-service Science Teachers (grade 4-9)	MBL- formula verification	MBL- formula verification	B	E
IV	Pre-service	99/00	25	Pre-service Science Teachers (grade 4-9)	MBL- formula verification	MBL – conceptual	B	B
V	Pre-service	00/01	20	Pre-service Science Teachers (grade 4-9)	MBL- formula verification	MBL – conceptual	B	B

Case I and II: In Case I (Pre-service Teachers 95/96) and in Case II (Engineering students, Mechanics I) MBL-labs were implemented as "conceptual" labs. The educational approach taken in both cases were inspired by, but not identical to, the approach used in *RealTime Physics* (Sokoloff *et al* 1998, Thornton 1997a and references therein) and in Case II also by the "New Mechanics" paper by Laws (1997). The approach is described in some detail in Bernhard (2000a) and in more detail in a forthcoming paper (Bernhard, to be published).

- In both cases MBL were used as conceptual labs and among other things the labs emphasised concepts and connections between different concepts and the active engagement of students.
- Co-operation among students was encouraged.
- Students' preconceptions were addressed by using a POE-cycle (Predict – Observe – Explain). In implementing this cycle the rapid display of the results by the computer in graphical form is thought of being of crucial value.
- Each lab-group of 2-3 students was asked to submit a written report from each lab. This reinforces and strengthens student understanding, since they have to describe the lab in their own words.
- The instructions for the labs in both Case I and II were written in Swedish by the author.
- MBL was used both as a technological tool (measurement, processing and display of experimental data) and as a cognitive tool (sense making).

Case III: Pre-service Teachers 98/99.

- MBL-technology was used in the labs.
- The original labs were transformed into formula verification labs with instructions written by instructor E (se table 1).
- No POE-cycle was used.
- MBL was mainly used as a technological tool.

Case IV and V: Pre-service Teachers 99/00 and 00/01.

- MBL-technology was used in the labs.
- The "impulse and collision" lab from Case III was changed back into a conceptual lab using a further improved instruction written by the author.
- The labs dealing with Newton's 1st and 2nd laws were not changed from Case III (formula verification) but the students were orally asked by the instructor to do a POE-cycle.
- MBL was used as a technological tool and partly as a cognitive tool.

The conceptual tests Force Concept Inventory (FCI, Hestenes *et al* 1992) and Force and Motion Conceptual Evaluation (FMCE, Thornton and Sokoloff 1998) were used as evaluation instruments (pre- and post-test). The results of testing using the FMCE-tests are displayed in figure 1–6. Between 80 – 90 % of total number of registered students participated in the conceptual tests.

3. RESULTS

When implemented (Case I and II) as conceptual labs MBL achieve good learning results as displayed in figure 1 and 2.

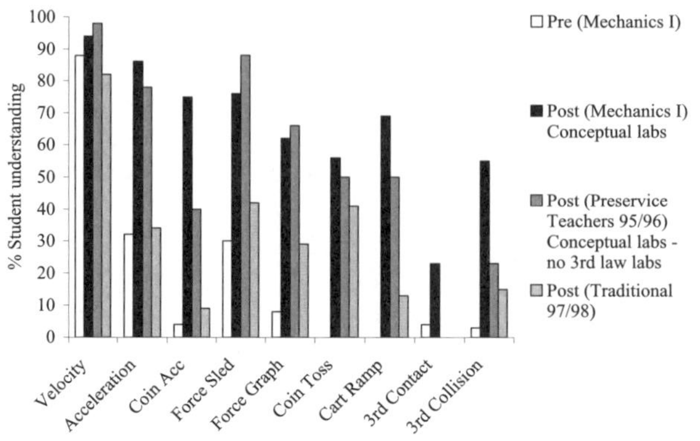

Figure 1. Student understanding in different conceptual areas of mechanics after having studied at a university level course in mechanics as measured by the FMCE-test (Thornton and Sokoloff 1998). The pre-instructional understanding of the students in the Mechanics I course and the post-instructional understanding of students in a traditionally taught course in mechanics (engineering students) is included as a comparison.

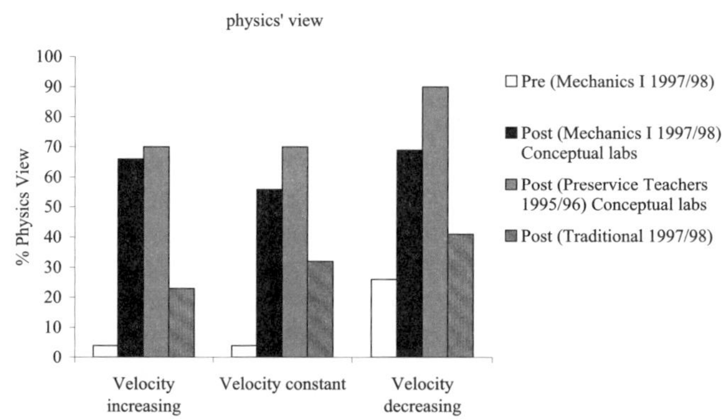

Figure 2. Fraction of students holding a "Physics' view". Student's views are assigned from FMCE-data using a method developed by R Thornton (1997b). Pre = pre-test = testing done prior to instruction. Post = post-test = testing done after instruction

In Case II both male and female students also achieved the same normalised gain [Normalised gain (Hake 1997) is defined as gain divided by the maximum possible gain] measured by the FCI-test.

However figure 3 show that when MBL were used for formula verification labs (Case III) significantly poorer learning results were obtained compared to Case II. The difference between the conceptual lab approach and the formula verification approach is very apparent when Cases III, IV and V are investigated in more detail. In Case III all labs were formula verification but in Cases IV and V the lab dealing with Newton 3rd law were conceptual. In figure 4 the conceptual understanding of the students in the areas where formula verification labs were used are presented and compared to Case II which used conceptual labs. As can be seen the results for the cases using a non-conceptual approach are quite similar except for "coin toss" and "cart ramp". However for Newton 3rd law understanding, as measured by the FMCE-test, there is a significant difference between Case III and Cases IV and V. In Cases III, IV and V identical lab-equipment using MBL-technology were used in both versions of collision lab. However the educational approach chosen was different. Cases IV and V used a conceptual active engagement approach and Case III a non-conceptual formula verification approach. As can be seen in figure 5 there is a large learning difference. The formula verification approach even led to negative learning in the case of contact forces.

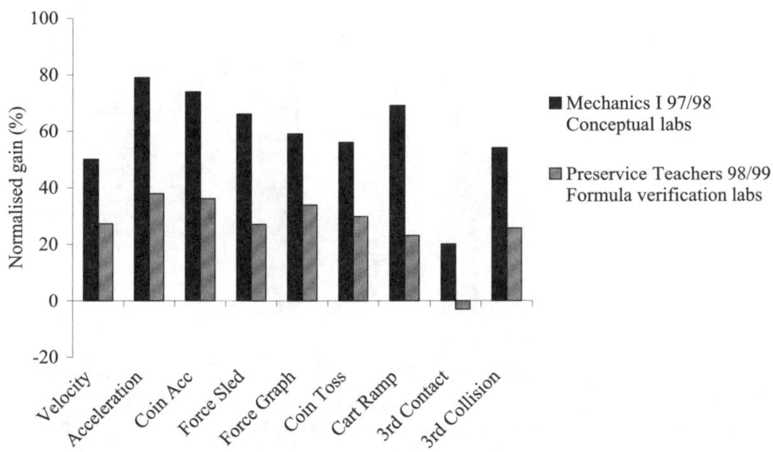

Figure 3. Comparison of normalised gains (Hake 1997) of two courses using the same MBL-technology in labs but different educational approaches. Note the negative learning gain for Newton 3rd law contact forces

A detailed analysis of data has shown that the formula verification approach has been especially disadvantageous for female and for poorly prepared students. Figure 6 shows for example that for Case III that the fraction of students believing in a force-follows-velocity view didn't change very much as a result of teaching.

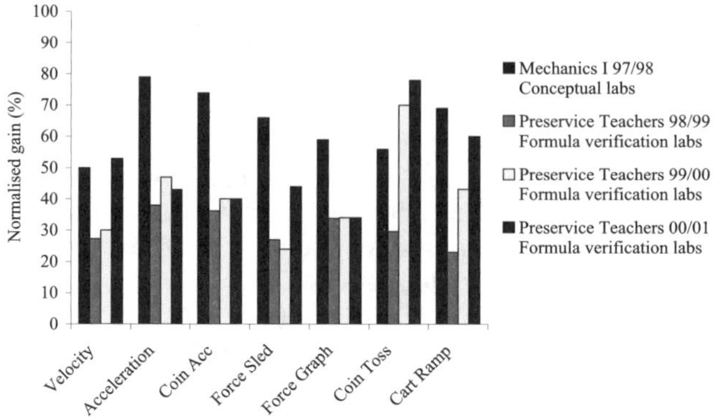

Figure 4. Fraction of students holding a "Physics' view" and a "Force-follows-velocity view" for Case III. Student's views are assigned from FMCE-data using a method developed by R Thornton (1997b)

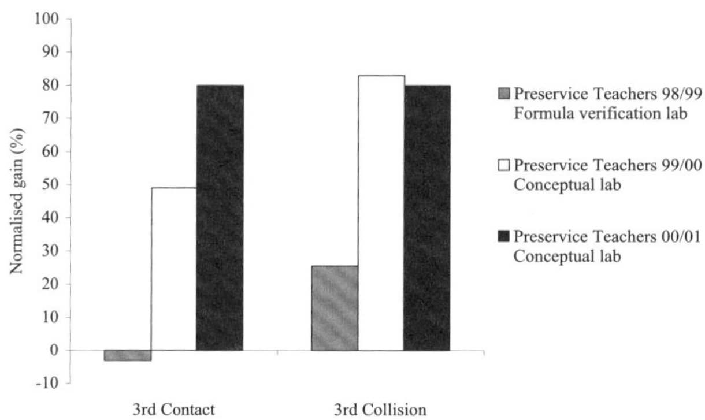

Figure 5. Comparison of learning in the domain of Newton's 3rd as measured by the FMCE-test

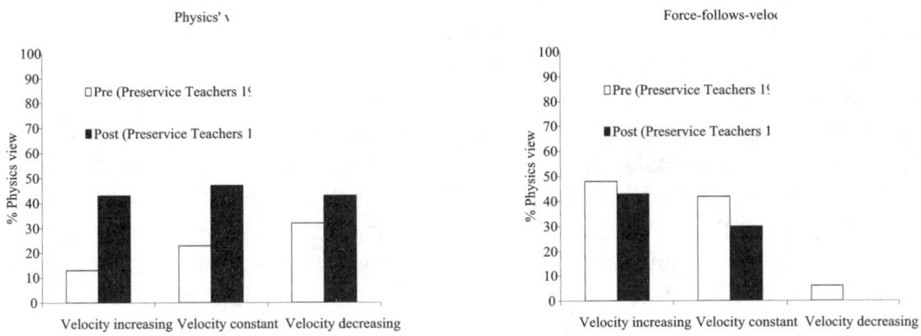

Figure 6. Fraction of students holding a "Physics' view" and a "Force-follows-velocity view" for Case III. Student's views are assigned from FMCE-data using a method developed by R Thornton (1997b)

4. DISCUSSION, CONCLUSIONS AND IMPLICATIONS

This study shows that there is a significant difference in learning when MBL is used for implementing conceptual labs (MBL used as a technological and a cognitive tool) and for implementing formula verification labs (MBL used as a technological tool). Could however the effects reported be due to either different student backgrounds or different lab instructors? Data suggests that the answer is no to this question. In figure 1 and 2 one can se that engineering students and pre-service teachers obtained very similar learning results. Instructor B was the lab-instructor both in Case I and in Cases IV and V with different learning results for kinematics and Newton 1st and 2nd law. The same instructor also taught both the labs in Cases IV and V using non-conceptual and conceptual approaches. Instructor E participated in the teaching of Cases II and III. Thus the results correlates to the educational approach but not to the teacher or to the type of student body. The results are also in line with a large body of literature (se for example Hake 1997 and McDermott 1997) that the active engagement of students is essential in obtaining a good functional understanding.

It should be noted that instructor E, when he rewrote the lab-instructions for Case III, thought he improved the labs. Case III is not invented as a control group by a person who didn't actually believe in the approach chosen. Thus the question arises what constitutes a "good" lab. In short I would propose that according to a "formula verification lab view" a "good" lab proves the "theory" with a good accuracy or measures a physical quantity like *g* with a good accuracy. Better accuracy – better lab. If a law is proven with a high accuracy it's tacitly assumed that the students now must accept the law and incorporate it in their understanding.

However in a "conceptual lab view" a "good" lab is a lab, which helps students in their meaning making and helps them to build concepts and understand concepts and connections between them. Better "understanding" – better lab.

The result of this study shows that MBL doesn't automatically give good learning results. MBL-technology, and other forms of computer aided learning, can not be implemented as only a technology. The educational implementation is of crucial importance and hence there is no definite answer to the common question if computers help to achieve "better" learning. Another conclusion is that we, when developing new curricula, must be as aware of instructor's conceptions of teaching and learning as we in teaching and curriculum design must be aware of student's preconceptions. Physics instructors perceive the intentions of a curricular reform or the advantages of an educational tool through their own understanding and conceptions ("misconceptions"?) of teaching and learning. In active engagement curricula using MBL-tools these tools are intended to be used as booth a technological and a cognitive tool. However, many instructors, perceive the MBL-tools as only a technological tool [see also similar results by Sassi (2001)]. Tinker (1996, page 3) points out: "*It is not usually advantageous to simply replace a traditional lab with an equivalent one using MBL. This kind of 'substitution' policy is easiest for schools to implement, but the result of such a substitution is often a simple lab made more difficult and expensive by the inclusion of computers with no educational gain. The MBL context adds capacity and flexibility that, to be exploited requires the lab to be reconceptualized, giving students more opportunity to explore and learn through investigations. This, in turn, often requires a change in teaching style that takes time and institutional commitment*".

To achieve positive learning effects, when using computer aided learning, we must focus as much on the cognitive aspects as on the technological aspects.

* Supported in part by the Swedish National Agency for Higher Education, Council for Renewal of Higher Education. Part of this work was done while the author was employed at Högskolan Dalarna.

REFERENCES

Bernhard J. (2000a). *Teaching engineering mechanics courses using active engagement methods.* Proc. of PTEE2000, Budapest, June, 2000.

Bernhard J.(2000b). *Improving engineering physics teaching - learning from physics education research.* Proc. of PTEE 2000, Budapest, June, 2000.

Bernhard J. (2001). Does active engagement curricula give long-lived conceptual understanding? In Pinto, R. & Surinach, S. (eds), *Physics Teacher Education Beyond 2000* (pp 749 – 752). Elsevier, Paris.

Euler M. & Müller A. (1999). Physics learning and the computer: A review, with a taste of meta-analysis. In Komorek, M. et al.: *Proceedings Second International Conference of the European Science Education Research Association.*

Hake, R. R. (1997). Interactive-engagement vs traditional methods: A six-thousand-student survey of mechanics test data for introductory physics courses. *Am J Physics*, 66, 64–74.

Hamne, P. & Bernhard J. (2001). Educating pre-service teachers using hands-on and microcomputer based labs as tools for concept substitution. In Pinto, R. & Surinach, S. (eds): *Physics Teacher Education Beyond 2000* (pp 663 – 666). Elsevier, Paris,.

Hestenes, D. Wells, M. & Swackhamer, G. (1992). Force Concept Inventory. *The Phys Teacher*, 30, 141–158. (Translated into Swedish by J Bernhard, 1997).

Laws, P. L. (1997). A New order for Mechanics. In: Wilson, J. (ed.), *Proc. Conf. on Intro Physics Course* (pp. 125–136). Wiley, New York.
McDermott, L. C. (1997). How research can guide us in improving the introductory course. In: Wilson, J. (ed.), *Proc. Conf. on Intro Physics Course* (pp 33–45). Wiley, New York,.
Sassi, E. (2001). Lab-work in physics education and informatic tools: advantages and problems. In: Pinto, R. & Surinach, S. (eds), *Physics Teacher Education Beyond 2000* (pp 57 – 64). Elsevier, Paris.
Sokoloff, D. R. Thornton, R. K. & Laws, P. W. (1998). *RealTime Physics.* Wiley, New York.
Thornton, R. K. (1987). Tools for scientific thinking - microcomputer-based laboratories for teaching physics. *Phys Ed*, 22, 230–238.
Thornton, R. K. (1989), Tools for scientific thinking, Learning physical concepts with real-time laboratory measurements tools. In: Redish (ed.) *Proc Conf Computers in Physics Instruction* (pp 177-189). Addison Wesley, Reading,.
Thornton, R. K. (1997a). Learning Physics Concepts in the Introductory Course: Microcomputer-based Labs and Interactive Lecture Demonstrations. In: Wilson J (ed.), *Proc Conf on Intro Physics Course* (pp. 69–86). Wiley, New York,.
Thornton, R. K. (1997b). Conceptual Dynamics: Following Changing Student Views of Force and Motion. In: Redish & Rigden (eds), *AIP Conference Proceedings* 399 (pp. 913 – 934). New York, American Institute of Physics.
Thornton, R. K. & Sokoloff, D. R. (1998). Assessing student learning of Newton's laws, The Force and Motion Conceptual Evaluation and the evaluation of active learning laboratory and lecture curricula, *Am J Physics*, 66, 338–352. (Translated into Swedish by J Bernhard, 1998).
Tinker, R. F. (ed.),(1996). Microcomputer-Based Labs: Educational Research and Standards. *NATO ASI Series F,* 156, 1996, Springer, Berlin.

PHENOMENOGRAPHICAL APPROACH TO DESIGN FOR A HYPERTEXT TEACHER'S GUIDE TO MBL

Kalle Juuti, Jari Lavonen, Veijo Meisalo

Department of Teacher Education, University of Helsinki

ABSTRACT

We have researched teachers' needs for a MBL guide through Web-based MBL guide design and development process. In-service teachers (N=23) were asked to write their ideas about MBL, three student teachers were interviewed and five student teachers and a training school teacher were asked to evaluate a designed prototype. We used a phenomenographical approach on data analysis. Teachers need an MBL guide that offers guidance for different pedagogical approaches. They wish for sample experiments that are not only recipes, but well stated chemically, physically and, especially, pedagogically. Furthermore, a guide has to introduce novel – computer aided – ways to do practical work.

1. MBL AND USE OF MBL

Plenty of research effort has been placed on developing user-friendly Microcomputer Based Laboratory (MBL) hardware and software, as well as research into the pedagogical uses of the MBL. Over the last two decades, the pedagogical use of a microcomputer-based laboratory (MBL or data-logging in the UK) has been one of the most important research areas in science education research (Tinker 1996). There are many research-based descriptions of user-friendly hardware based on the "plug-and-play" idea and user-friendly and clear software, with easy set-up and data processing (Tinker 1996; Adams *et al.* 1997, Czerniak *et al.* 1999, and Newton, 2000). In the software, there should be, for example, user-friendly set-up options and data processing tools. There is also research-based evidence that an MBL could improve science learning when meaningfully integrated into the curriculum (e.g. Sassi & Balzano, 1996; Sassi, 2001). However, it is well known that teachers need support when using an MBL in the school laboratory, but little is known regarding the type of guide materials teachers need.

Presently, MBLs are only occasionally used in schools. Linn (1996) argues that when teachers start using the MBL they need to become familiar with conducting laboratories, managing computer software, understanding interfaces, dealing with students' unequal prior experience with technology and creating a curriculum. Newton (2000) interviewed five teachers to ascertain teachers' rationales concerning data logging, obstacles to the implementation of data logging methods, strategies for overcoming these obstacles and developing learning objectives. Based on his research, he claims that teachers first need up-to-date hardware and software, technical support to ensure that the equipment is serviceable and to be skilled in the

D. Psillos et al. (eds.), Science Education Research in the Knowledge-Based Society, 333–341.

technical use of the equipment. Then teachers are able to foster the next level of professional awareness of the MBL: an understanding of how data logging can change classroom management needs, understanding of a fuller range of the benefits of data logging approaches, and developing teaching approaches, which exploit the technology.

Although much research has been completed on MBL tools, their user-friendliness and how they affect learning, it is obvious that more research and development, as well as design efforts, are needed to comprehend how guidebooks and tools could be made more user- and curriculum-friendly emphasising that an MBL package is for science education not for computer science or technology. It is obvious that a curriculum-friendly MBL guide allows different approaches for different teachers, for example, approaches emphasising constructivist learning, co-operative or collaborative learning, or even based on a 'chalk and talk' approach. Moreover, a curriculum-friendly guide helps teachers to use an MBL within different domains of subject knowledge in science.

Equipment producers offer equipment specific supporting guides. These guides may have been integrated into software. Different ICT instructions materials are also offered to teachers. According to Driscoll and Dick (1999), these instruction materials are not systematically designed and are not integrated into the curricula.

The research question in this study is: What kind of MBL guides do teachers need? The development task is to design a guide fulfilling the identified needs.

Through developmental research design, we acquire information about teachers' needs for an MBL guide. According to Khan's (1997) review, Web-based instructional publishing provides rich learning environments in a global, democratic, and interactive manner. We, especially, respect the Web's practical possibilities such as easy updating, delivering and hypertext.

2. AIMS AND FRAMEWORK OF THE STUDY

The aim of this study is to develop a versatile hypertext teacher's guide to facilitate the use of the MBL and to integrate it into the curriculum. Development is essential for this study. Clements and Battista (2000) introduced a model showing how the design of instructional computer programs could be integrated into curriculum research. Their model has nine phases. The first three focus on initial designing, the next four phases focus on evaluation, the eighth focuses on combining design and evaluation, and the final phase is publishing. The international standard for human-centred design processes for interactive systems (ISO 13407) emphasises the importance of understanding and specifying the context of use as well as specifying the user and organisational requirements. The method of carrying out a developmental study depends on the phase and properties of the design process. Usually, a quasi-experimental design is used for the evaluation of the educational innovations, and qualitative methods are used for defining the needs of users. (Richey and Nelson, 1996)

Figure 1 describes our interpretation of the progress of the developmental research based on the framework summarised above. The *Results* chapter includes a detailed description of the realisation of assessing the initial needs, defining objects, designing, and evaluating an MBL guide, as well as a description of teachers' needs for an MBL guide through this process.

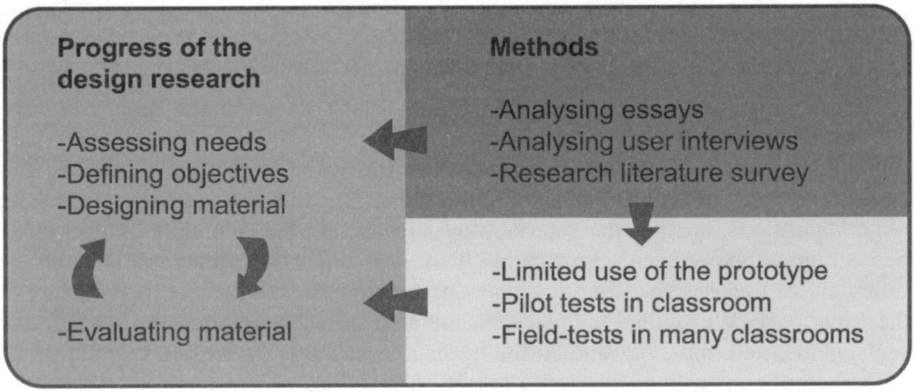

Figure 1. Structure of the development study process.

3. METHODS AND SAMPLES

We decided to collect the ideas of chemistry and physics teachers, as well as three student teachers, about using an MBL in science education. Furthermore, we collected the experiences of a training school teacher and five student teachers experiences with regard to a prototype of the MBL guide we designed. By triangulation in data gathering, described below, we attempted to attain proper solutions to our research question and to design an appropriate MBL guide.

For the initial needs assessment, we asked 23 voluntary secondary school chemistry and physics teachers during an MBL in-service training course to write down: why and for what purpose they use or do not use the MBL and what sort of support material (an MBL guide or WWW-pages) do these experienced teachers need. To ensure that we obtain the views of young and inexperienced teachers regarding the MBL and practical work, we interviewed three student teachers. We asked the student teachers for their experiences during their teaching, their ideas on the experimental nature of physics, the purpose of practical work in science teaching and learning, the role of the MBL in science education, as well as their opinions on the existing Web material for supporting teaching. The needs assessment comprised 30 transcript standard pages.

A training school teacher and five student teachers at the teacher training school tested the prototype of the MBL guide (limited testing and pilot testing) in order to evaluate the guide we had designed. Furthermore, this evaluation gave us information about the needs and usability of the MBL guide. The training school teacher was interviewed twice. Both times we asked her to describe the MBL guide, describe her MBL guide using, and describe disadvantage of the MBL guide.

Furthermore, we asked a general question about the role of practical work in physics. Before the first interview, she wrote notes on the printed version of the prototype of the MBL guide. We also asked five student teachers to plan a teaching sequence (a two-hour lesson or workshop) and write an essay: What in the MBL guide helped you plan a teaching sequence in which the MBL is utilised and what would you still need? The training school teacher's interviews comprised 36, and the student teachers' essays comprised 16 transcript standard pages.

Analysing the interviews and written feedback, we followed the principles of qualitative methodology. (Lincoln & Guba, 1985) In particular, we took into consideration the requirements of the phenomenographical methodology. By asking 'what' and 'how' questions, we obtained information about how teachers and student teachers perceive and experience the MBL and MBL guide. By asking 'what' questions, we got the interviewees to describe the reference of his or her perception or experience. By asking 'how' questions, we were able to get the interviewees to describe different ways of perceiving or experiencing the structure of the experience. We studied the teachers' and student teachers' ideas about MBL and MBL guide, and, thus, we studied the needs and usability of the MBL guide in the second order perspective. (Marton, 1981; Marton & Pang, 1999).

4. RESULTS

The results are structured, based on the progression of the development study (Figure 1). First, we describe initial needs assessment then objectives for designing the guide and the designed prototype. Last, we describe the prototype's evaluation.

4.1 Initial needs assessment

When analysing the teachers' and student teachers' written and spoken comments on the MBL through inductive classifying, we obtained five referential aspects of the teachers ideas of the MBL and practical work, and the structural aspects are also presented. In the following, direct quotations are printed in *Italics*.

1) The MBL offers something new for teaching and learning physics.
   a. Perspicuity of the graphical presentation *"it's easy to present the research results to the whole class."*
   b. Quickness of the graphical presentation *"Quick and clear graphical presentation" "Processing of the research results is easy, because the computer does the calculations for the pupils."*
   c. Phenomena are easy to investigate *"Repeating the measurement is easy."*
2) Computers as such are important
   a. Computers are the equipment of the future *"I think it is fashionable ' to have an MBL. One feels like a 'fossil' if one does not keep up-to-date."*
   b. Pupils have to be familiar with computers *"You can learn to use computers by using one."*
   c. Computers motivate the pupils *"We made a lot of fancy measurements and had a lot of action with the MBL"*

3) There are no educational reasons for using an MBL

   a. Teachers do not know when they should use an MBL *"I don't know what to do and how"*
   b. Traditional tools are clear *"It seems that the MBL sometimes gives too 'cold' data " I need tips for appropriate demos of phenomena that can be realised using MBL tools."*
   c. The technology may play too important a role *"the phenomenon may 'disappear' If there were too complicated experiment design and equipment"*

4) There are practical difficulties when using the MBL

   a. *"There were problems because first we used Scotch tape but it* [photo gate] *'saw' through. We couldn't practice that demonstration before and we had 15 anxious minutes before teaching."*
   b. The preparation of the experiments takes too long *"It takes time just to learn how to use the systems. There is just so much to learn before you can begin to use the MBL. Sometimes, I just spend the whole lesson trying to work out what caused the error because I'm not familiar with the equipment."*
   c. The MBL is not to hand *"MBL tools are not in the chemistry classroom but in the physics classroom"*

5) Introducing meanings for concepts through qualitative and quantitative experiments

   a. Qualitative experiments are important *"The main point is to connect the concepts to the pupils' own experiences."*
   b. Quantitative experiments are important *"They were very keen to measure and they asked what the resistance is. It's important to do practical things"*
   c. Teachers do not seem to notice the gap between qualitative and quantitative concepts. *"Well, it seems that, for example, yesterday the practical work stopped short of the goal, so we need to repeat the experiments tomorrow" "Today, one group was interested in the phenomenon we had studied last week. They started to investigate the causes and consequences."* When teaching seemed to start a quantitative level of physical concepts, pupils have difficulty understanding the meaning of the concepts. It seems that studying causes and consequences in the qualitative level first, and then studying quantitative measuring, helps pupils understand the meaning of the concepts.

4.2 Objectives

We defined the objectives for the MBL guide based on the classification of the initial needs. The MBL guide emphasises novel properties, such as graphs and an interpretation suggestion in the guide. Furthermore, fast and weak (e.g. decreasing temperature of a liquid due to evaporation) phenomena are introduced in sample experiments. One motivation of the teachers to use the MBL was that computers, as such, are important. Thus, the guide should introduce versatile uses of the computer and describe teaching methods when using a computer. Teachers needed pedagogical reasons for using the MBL. The guide introduces a variety of phenomena and it should be explained why MBL tools are appropriate for every phenomenon. The guide should be designed in such a way that it helps with the practical use of the MBL in schools. The guide offers easy set up, interactive support and detailed descriptions, how phenomena can be demonstrated in the classroom. The guide includes plenty of qualitative and quantitative experiments and places special attention on the transfer from the qualitative level of phenomena, to quantitative level of models and theories in the concept formation process.

4.3 Prototype of the mbl guide[1]

After defining the objectives, we started to redesign and edit the manuscripts in hypertext form. The MBL guide includes a hierarchical index that emphasises the hierarchical levels of the physical concepts. The teacher can find the article by clicking the interesting experiment topic on the index.

The hierarchical index helps teachers notice the hierarchical levels of physical concepts (qualitative – quantitative). The vertical direction is for the development of a conceptual structure and the horizontal direction is for specific examples of the phenomena. Currently, there are now 137 experiments in the index. The horizontal structure is: a phenomenon area (e.g. magnetism), a subcategory of the phenomenon area (e.g. magnetic field; electric wire as a magnet), and a concrete example (e.g. magnetic field of a bar magnet; magnetic field of a bobbin).

We follow the physics teaching principles that Arons (1997) introduces and which, for example, Sassi and Balzano (1996) specify for kinematics. Furthermore, we took into consideration the research into pupils' ideas (e.g. Driver, Guesne, and Tiberghien, 1985) An article includes background information about the phenomena - how the phenomenon is seen in the environment - idealisation tips about phenomena - how phenomena can be studied in school laboratory - with photos, video clips and graphs using real measurements, as well as interpretation and further study suggestions. Terms, such as “classifying” and “preconceptions” lead through hyperlinks to the hypertext teaching methods guide. Moreover, terms, like “calibration” and “curve fitting” lead through the hyperlinks to the operating manual for the MBL tool.

There are currently 54 hyperlinks to the hypertext teaching methods guide, 328 hyperlinks to the MBL tool operating manual, 292 photos or figures, 16 video clips, 119 hyperlinks to the MBL tool with ready-made set-ups and a template measurement file, and 168 graphs based on real measurements. The hypertext format enables the teacher to avoid repeating texts. Therefore, terms such as, “classifying” and “preconceptions” could lead to the hypertext teaching methods guide. In this guide, we could concentrate on the appropriate use of the MBL and help the teacher improve the practical work in school science

[1] Prototype is available online: http://www.malux.edu.helsinki.fi/malu/kirjasto/mbl/

4.4 Evaluation

The teacher in the training school carried out a limited test of the designed MBL guide. She tested the material in the mechanics and thermodynamics upper secondary school courses. In the interviews, she thought that the idea of the MBL guide is excellent, but that there are too many unfinished sections. For example, she argued that the electricity and magnetism section contains too many recipe-like instructions. She noted that the photos and text should be connected more explicitly and she was not quite sure what the use was of specific experiments.

The hierarchical index confused the teacher. She suggested that three horizontal steps are enough to make the index understandable. The teacher also found many errors that she marked in a print version of the MBL guide. We corrected these and rewrote the instruction texts so they were more open and gave more opportunities to integrate them in various pedagogical approaches.

The student teachers carried out a pilot test of the MBL guide we designed. All the student teachers felt that it helped them plan teaching sequences.

> "The WWW instructions on teaching methods for practical work and the use of the MBL gave me lots of new ideas for my teaching sequences." "There [in the designed material] were some videos which helped me start studying phenomena."

The student teachers felt that the MBL guide was too concise. They needed more extensive instructions, more specific background theories and more detailed idealisation tips.

> "The Theory unit could be more extensive. I have got used to the style of teachers' guides which sometimes present things very carefully." " The body needs to be heavy enough in the friction investigation." "I need more information on materials, because I don't know which materials are forbidden in school laboratories."

Student teachers had difficulty focusing on the appropriate level *"I need more examples for the upper secondary school level"* Overall, to plan teaching sequences, student teachers needed more sample investigations. Furthermore, we noticed that unfinished experiments tempted the student teachers to make them better than in the guide. Student teachers needed precise information, for example, to find the optimum mass of the object in friction research and the optimum capacitance of a capacitor in AC-circuit investigations. Every phenomenon that the teacher in the training school pointed out as interesting was either fast or weak. Such phenomena are difficult to inquire into without MBL tools.

5. CONCLUSIONS AND IMPLICATIONS

In this study, we followed the guidelines of research-based designing and we used a phenomenographical approach to focus on teachers' needs for an MBL guide. During the designing research process, we specified the properties of a versatile MBL guide. Teachers need MBL guides that offer guidance for different pedagogical approaches. Teachers wished that sample experiments were not recipes, but well-stated chemically, physically and, especially, pedagogically allowing freedom for planning teaching. The guide should also introduce novel – computer-aided – ways of doing practical work.

It seems that teachers need much support for increasing the use of the MBL in science teaching. Technical support is just a starting point, as Newton (2000) argues. A well-designed Web guide is one approach. Traditional in-service training and personal consulting may also play important roles for teachers' pedagogical development. With a well-designed guide, we could enlarge teachers' awareness of the MBL and affect teachers' attitudes toward the MBL.

According to Sassi (2001), teachers complain that there are few or no guides for the MBL, especially for computer-based real-time experiments. Thus, the objective of this research and development process was to find the kind of help teachers need and then design a teacher's MBL guide that meets these needs. The research process shows that a phenomenographical approach forces developers to pay attention to teachers' opinions and ideas concerning the developed product. This case study also shows how a research-based design can be implemented. In our design process, the next phase is to improve the MBL guide further on the basis of field test evaluations. Further research may be interested in studying if a well-designed MBL guide may foster teachers' pedagogical content knowledge; teachers' awareness about pupils' ideas, the hierarchical structure of chemistry and physics and about the technical properties of the MBL, as well as knowledge of the teaching models.

REFERENCES

Adams, P.,E., Krockover, G.,H. & Lehman, J.,D., (1997). Strategies for Implementing Computer Technology in the Science Classroom. In: Rhoton, J. & Bowers, P. (eds), *Issues in Science Education* (pp66–72). Alexandria, VA: National Science Teachers Association..

Arons, A., B., (1997). *Teaching Introductory Physics*. Chichester: Wiley 1997.

Clements, D., H. & Battista, M., T., (2000). Designing Effective Software. In: Kelly, A., E. & Lesh, R, A., (eds), *Handbook of Research Design in Mathematics and Science Education*. London: Erlbaum.

Czerniak, C.,M., Lumpe, A.,T., Haney J.,J. & Beck, J., (1999). Teachers' beliefs about using educational technology in the science classroom. *International Journal of Educational Technology*, 1(2). Available online [http://www.outreach.uiuc.edu/ijet/v1n2/czerniak/index.html].

Driscoll, M., P. & Dick, W., (1999). New Research Paradigms in Instructional Technology: An Inquiry. *Educational Technology Research and Development* 47(2), 7-18.

Driver, R., Guesne, E., Tiberghien, A., (Eds.), (1985). *Childrens Ideas in Science*. Milton Keynes: Open University Press.

ISO 13407. *Human-centered design processes for interactive systems*. International standard ISO 13407:1999(E).

Khan, B., H., (ed.) (1997). *Web-Based Instruction*. Englewood Cliffs: Educational Technology Publications.

Lincoln, S., L. & Guba, E., G., (1985). *Naturalistic Inquiry*. Beverly Hills: Sage.

Linn, M., C., (1996). From Separation to Partnership in Science Education: Students, Laboratories, and the Curriculum. In: Tinker, R., F. (ed.), *Microcomputer-Based Labs: Educational Research and Standards*. Berlin: Springer.

Marton, F., (1981). Phenomenography – Describing Conceptions of the World Around Us. *Instructional Science,* 10, 177-200.

Marton, F., & Pang, M., F., (1999). Two Faces of Variation. Paper presented at 8th European Conference for Learning and Instruction August 24-28, 1999, Göteborg University, Göteborg, Sweden.

Newton, L., R., (2000). Data-logging in practical science: research and reality. *International Journal of Science Education*, 22(12), 1247–1259.

Richey, R., C. & Nelson, W., A. (1996). Developmental Research. In: Jonassen, D., H., (Ed.) *Research for Educational Communications and Technology: A project of the Association for Educational Communications and Technology*. New York: Macmillan.

Sassi, E., (2001). Computer Supported Lab-work in Physics Education: Advantages and Problems. In: Pinto, R. & Surinach, S. (eds). *International Conference Physics Teacher Education Beyond 2000*. Selected Contributions. Paris: Elsevier.

Sassi, E.& Balzano, E., (1996). Learning and Teaching Motion: MBL Approaches. In: Tinker, R., F. (ed.), *Microcomputer-Based Labs: Educational Research and Standards*. Berlin: Springer.

Tinker, R., (ed.), (1996). *Microcomputer-Based Labs: Educational Research and Standards*. Berlin: Springer.

APPLICATION OF A FRAMEWORK APPROPRIATE FOR A MULTILEVEL ASSESSMENT OF EDUCATIONAL MULTIMEDIA SOFTWARE IN SCIENCE (FEVES)

Chrisoula Kounatidou, Georgios Fassoulopoulos, Petros Kariotoglou

School of Education, Aristotle University of Thessaloniki, Greece

ABSTRACT

In this paper we attempt to apply a five –dimensional framework for the analysis and evaluation of educational software in Science, and in particular CD-ROMs. Our study was based on a forgone review of the relevant literature. The framework created includes educational, electronic, social/institutional, epistemological and pupil-focused variables which are not mutually exclusive. An application of this framework to two educational CD-ROMs demonstrated, on the one hand the applicability and flexibility of the framework and on the other the characteristics of the CD-ROMs. The study also revealed some of the teaching implications of incorporating parts of the CD-ROMs into formal and non-formal learning environments.

1. BACKGROUND, AIMS AND FRAMEWORK

The explosion of New Information Technologies at the beginning of the new century has influenced human activity and our everyday lives. In education, the introduction of these new technologies seems to offer an opportunity to restructure schools and curricula and transform the nature of learning (Johnston, 1987). This is an indication of the crucial nature of the quality of educational programs being introduced into schools, as well as of the way this introduction takes place.

Johnston (1987) argues that evaluation of educational software must be based on educational rather than technological criteria. In a similar context, Firdyiwek (1999) points out the danger, in designing courseware systems, of technological convenience in online instruction taking priority over sound pedagogy. For example, Roth et al. (1996) describes how a particular software message to the student-user, designed to help him/her, actually led him/her to take a wrong learning path. She, as well as Squires & McDougall (1994) and Hinostroza & Mellar (2001) also underlines the importance of feedback and familiarization of the user with the program and its components. In addition, much interest is shown in the teacher's role versus the role of the computer (Hinostroza & Mellar (2001), the teaching and learning goals (Squires & McDougall, 1994) and the flexibility of the program (Johnston, 1987; Squires & McDougall, 1994).

By contrast, little interest is shown in either the epistemological or the social dimension. Kelly and Crawford (1996) see computer software in physics, and

D. Psillos et al. (eds.), Science Education Research in the Knowledge-Based Society, 343–351.

especially MBL experiments, as a chance to help students 'talk science' and integrate epistemology into science education. Whereas Berge (1999) underlines that how instruction is designed is based largely on the designer' s interpretation of the world. Similarly, Squires and McDougall (1994) examine whether content is coherent and up to date and whether it encourages sexual, racial or other stereotypes.

Based on this literature, we developed a five-dimensional framework for the analysis and evaluation of educational CD-ROMs in science. In this paper, our aim is to present this framework in brief and to apply it in two Science CD-ROMs. From this implementation the applicability and the flexibility of the framework are expected to be revealed. Additionally, the exploitation of the outcomes (that is the characteristics of the cd – roms) will allow us to form useful teaching implications. We are focusing on "edutainment" (education + entertainment) type CD-ROMs (Mikropoulos, 2000) because we consider that this type of software is more or less 'closed' and is appropriate for individual learning, which is an instance of informal learning.

2. MODE OF INQUIRY

2.1 Our framework (FEVES)

We named our framework "FEVES": F(ramework) of EV(aluation) of E(ducational) S(oftware).

The first, or electronic/technical, dimension looks at either the general technical characteristics or the browsing possibilities of the program: that is, the title, the subject area, the edition, the price, the hardware and software requirements (Squires & McDougall, 1994; Tolhurst, 1992). We also study the way the hypertext is structured and the description of links between data. We are further interested in the qualitative use of graphics, colour, animation and text, and the balance between the user and program control during browsing (Squires & McDougall, 1994; Tolhurst, 1992).

The educational dimension studies the teaching and learning activities, assumptions of the CD-ROM and examines it from the didactical point of view. We evaluate any accompanying tools for the teaching course that may be included in the program, and the role the latter assigns to the teacher (Hinostroza et al, 2000). Other crucial variables are the learning assumptions implied (Firdyiwek, 1999), the degree of flexibility, which depends on the levels of difficulty and the specification of the age and ability range the program was designed for (Johnston, 1987), the students' role and degree of control and the provision of intrinsic versus extrinsic motivations (Squires & McDougall, 1994; Tolhurst, 1992). Many of these variables can be considered as small components of 'interactivity' between a computer and its user. As far as bibliography is concerned, interactivity is translated either as sustained, two – way communication among two or more persons (Liaw & Huang, 2000) or as a means to cause students to consider and reflect on the content and process of learning (Flottemesch, 2000). From the didactical point of view, we examine the goals and objectives stated (or implied) by the program, the accuracy of the content and its simplification or transformation in order to match the pupils' needs

(Mikropoulos, 2000). We also study the teaching methods proposed, the exploitation of students' ideas and, finally, the kind of assessment involved, the format, variety and accuracy of feedback and its correspondence to the content (Kariotoglou & Tselfes, 2000).

The epistemological dimension looks into the epistemological views that are implicit in the content of the program. We examine, for example, the epistemological attitude – positivist, hermeneutic or realistic (pragmatistic) – towards the nature and philosophy of Science (Hacking 1983, Kariotoglou & Tselfes, 2000).

The social content of the program and the characteristics related to the curriculum are the object of the framework's fourth (social / institutional) dimension. We look into the existence of racial, ethnic, sexual and other stereotypes and investigate the extent to which the program is culturally biased (Squires & McDougall, 1994). We examine whether the content is up-to-date, and assess its relationship to the current curriculum and the standard textbooks (Tolhurst, 1999, Mikropoulos 2000).

Finally, the pupil/user aspect (fifth dimension) examines the level of interactivity between the pupil and theCD-ROM, its edutainment characteristics (that is, the pupil's progress through the program, learning difficulties and entertainment value), and compares CD-ROMs to other teaching materials (e.g. textbooks, slides, videotapes). In other words, we examine how a pupil confronts these programs from the technical, emotional and cognitive point of view.

2.2 The CD-ROMs under study

We found it crucially important to apply the FEVES in specific examples, in order to elicit both the educational possibilities of the CD-ROM and the framework's potential for multilevel analysis. For this purpose, we chose 'My Amazing Human Body', and 'Mad About Science – 3. Matter'.

The first of these programs studies the human body from various perspectives. Its main menu consists of game-like activities: 'Me and my day', 'Take me apart' and 'Build me a body', while the fourth one – 'What am I made of?' – encourages the user to explore deep inside the human body. Apart from the main menu there are indexes of 'Fact', 'Did you know?', 'Things to do' files etc., and a 'Secret file', which the user fills in individually. The Biology title is specifically addressed to children of 6 to 10, but appears to be appropriate for older pupils as well, taking into account the texts and the more difficult activities. The second title contains sorting, testing, changing and separating activities plus additional informative texts, and is addressed to 10 to 14-year-old pupils. The environment of all these activities is the kitchen of the hero of the program, where he cooks, tests, sorts and tries to help the user through the above activities. After entering the program, the user has to draw him/herself up and choose three activities. Winning five (5) stars during the activities allows him/her to progress to the next activity.

We chose these CD-ROMs because of the above-mentioned multilevel approaches. Our choice was further oriented by the compatibility of their content with the content of the modern Greek science curriculum and by the age groups targeted, which correspond to Primary (up to 12 years old) and early Secondary School.

2.3 The pupils and the CD-ROMs

Taking all these into account, the selected CD-ROMs were 'presented' in an informal learning environment to three (3) Greek pupils, Thenia (12 years old), Manos (13) and Evaggelia (14). Although all pupils were not native speakers of English, could more or less deal with the language requirements. An educator (one of the authors) was present to intervene in the whole procedure, in case the pupils got 'lost' or found difficulties in translating parts of the programs. Manos and Evaggelia had also attended introductory computer classes at school. Manos had been playing video games for three (3) years, and Thenia had been using simple computer programs and games for some months. Both titles were given to the three pupils to run for as much time as they could, so that they could obtain a global view of them. We consider that since the respective school program include Biology and Chemistry courses all the children may not fully understand, but are familiar or at least do not meet for the first time with the scientific content of both CD – ROMs. We find this clarification necessary since during this step of our research our primary aim is rather to underline and exploitate the characteristics of the cd – roms, which will allow us to form useful teaching implications, than to study the cognitive process of the pupils. At the end of the procedure each child gave the educator a semi-structured interview. The questions proposed by the educator were based on the pupils' response to the programs (fifth dimension), as described in the suggested framework.

3. OUTCOMES

The authors carried out both the analysis of the CD-ROMs and their presentation to the three pupils. In the outcomes, we study the characteristics of the two educational programs according to the suggested framework and we describe the pupils' reactions during and after their use of these CD-ROMs.

3.1 The experts' point of view

With regard to the technical characteristics of the material studied, we can attest to the good quality of the sound, the graphics, the animations and the text, as well as the way these are combined in the program in both CD-ROMs. The environments, however, are rather restricted and neither the teacher nor the pupil can intervene in the 'paths' and change or create his own. In 'My Amazing Human Body' (Biology title), although there is a variety of ways of running the title (file indexes, concept index and main menu of activities), the structure of the hypertext is not sufficiently clear and the user can easily get lost while browsing. On the other hand, in 'Mad About Science' (Chemistry title), the user's choices are significantly oriented and the only choice in running this program is the optional reading of the accompanying scientific texts.

From the educational point of view, neither title offers accompanying tools (worksheets, slides, maps) to support instruction. While the first one is probably designed for an informal learning environment, the second appears to be intended for a formal one, since most of the activities are standardized and correlate to those usually included in school textbooks. There do, however, seem to be many interactive characteristics and activities that support learning such as engaging, questioning, answering, discussing, linking, evaluating, synthesizing. In particular in the biology title there are individualized types of feedback as well as a considerable flexibility, due to the various ways in which it can be used and the multiple levels of difficulty while running it. For example, the 'puzzle' activity is much easier than the 'quiz' or 'everyday life' one. The Chemistry title on the other hand, seems to be addressed to an average pupil level, although there is a great variety in the content and type of activities (testing, sorting and changing). From a didactical scope, in the Biology program the goals are declared as informative and entertaining, whereas in the Chemistry title, which appears to be rather similar, there are no clearly stated objectives. What is more, both programs tend to support, *inter alia*, the traditional mode of learning, giving various types of text that simply inform the pupil/user. The Chemistry title seems to be exploiting the 'science for all' model, taking into account the everyday environment (a kitchen) in the context of which Science is studied and developed (experimenting, sorting, testing etc). However, 'My Amazing Human Body' seems to underline the way Science is connected to Technology and Society, given the use of technological achievements (oven, x-rays, bicycle etc) in an everyday life environment. Thus, there is possible support for the 'pupil as a scientist' model, as well.

In 'My Amazing Human Body' we can assume either the positivistic or the hermeneutic view of the nature of Science: the positivistic one since the program generally encourages the user to explore and find out 'the truth' about the human body (e.g. 'Fact files', 'What am I made of?') and the hermeneutic one on the basis of the general themes it deals with (e.g., 'keeping healthy'), which emerge from the local society and culture (Kounatidou & Kariotoglou, 2001). Approximately the same characteristics in 'Mad About Science' frame a positivistic view of the nature of Science, implicit in the pervasive suggestion that through experiments we will learn about the world around us and the opposite.

Although no stereotypes seem to be promoted, at least not systematically, the activities performed by the heroes of both CD-ROMs (e.g. eating habits, parties, kitchen environment, etc.) reflect the western lifestyle. Finally, the content in both cases seems to be reasonably modern and, though not formally designed for it, relevant to the modern Greek curriculum in the context of which the CD-ROM was studied.

On the whole, we can emphasize the various interactive characteristics and the satisfactory level of flexibility these particular CD-ROMs possess. Moreover, they do not seem to encourage either a single epistemological trend or any specific stereotypes, while the content is reasonably coherent and up-to-date. Comparing the two programs, there appears to be considerable difference either in the age range targeted or the level of flexibility of the various activities, as observed above.

The way the CD – ROMs were analysed in here suggests that in this framework, the expert aims to underline the crucial characteristics which will additionally allow him/ her to predict the way the user will face the program, and not evaluate it according to positive or negative standards.

3.2 The pupils' point of view

Firstly, the pupils encountered no special difficulties in browsing either title. Thenia and Evaggelia found some difficulties during browsing in 'My Amazing Human Body': e.g. in the quiz activity, when they went on to learn more in order to answer a question, they frequently asked: '*how can I get back to the quiz now?*'. Manos, on the other hand, encountered greater difficulty in entering the Chemistry title than in browsing. This experience reflects the experts' comments that significantly oriented browsing in the chemistry title minimizes browsing difficulties: for example, in the Chemistry title, where browsing possibilities are significant, the possibilities of getting lost are minimized, whereas in the Biology one, where the structure is unclear, it is much easier to get lost.

As for the emotional aspect, all the pupils found both programs more or less '*entertaining*' and '*informative*', as the designers suggest in 'My Amazing Human Body'. Thenia said that the texts in this title are '*boring*', perhaps because they lead to a traditional mode of teaching / learning. On the other hand, she stressed that *'it is fun to learn through fun people, like this little hero'*. Evaggelia didn't like the hero 'Seemore skinless' in the Biology title at all and found the Chemistry one better. Manos said that the Chemistry program '*is more 'game-like', more entertaining'*. This difference between the older children and the younger child is possibly due to the fact that Thenia is still young enough to prefer programs with game-type activities, whereas Manos and Evaggelia liked the chemistry program more because it reminded them of school-type activities (counting, sorting) and therefore it helped them to clarify some doubts with regard to content.

The type of knowledge gained during the research seems to be more declarative than procedural. That is, they frequently indicated that they learnt a lot, although sometimes giving poor reasons (e.g. *'I learnt a lot, for example I didn' t know that the largest organ in our body is the liver'* (Thenia) or *'I learnt for example how many bones there are in the spine'* (Manos), etc.). However, they didn't seem to gain access to procedural type knowledge, perhaps because of difficulty with the English language (Thenia and Evaggelia). Thenia found the Chemistry program much more difficult than the Biology one, since in the latter she could enjoy the program dealing only with the activities she found appropriate to her own cognitive level.

Going on to the programs in comparison to other teaching materials used in school (textbooks, slides, maps etc), all the pupils found them much better than the latter, because *'they are much more interesting and entertaining than our textbooks'*.

As far as the familiarity with the computer environment is concerned Manos, with his greater experience in video games, from the very beginning of the program obtained a general view of it and thus he was able to interact with it in a more efficient way. On the other hand, Thenia and Evaggelia were frequently trapped in accidentally wrong browsing movements, a factor which increased their anxiety and prevented them from comprehending the program as a whole. All the children noticed that the programs helped them to learn many things about Biology and Chemistry. Conversely, a certain level of knowledge is taken for granted in order to run each title. Some difficulties in the matter of language also arised, since none of them is a native speaker of English. However, according to the educator who overviewed the whole procedure, this variable didn't in fact prevent them or affect their interacting with the programs.

Taking into account the above, various hints for teaching and learning are proposed. In the next and last part of the paper some conclusions and implications for learning and teaching are discussed. All findings, from the experts' and pupils' point of view, are described in brief in the table below.

Table: Synopsis of the characteristics of the two CD-ROMs

	'My Amazing Human Body'	**'Mad About Science: 3. Matter'**
TECHNICAL	Good quality of sound, graphics, text and animation No possibilities for intervention	
	Not clear enough structure Various ways of browsing	Much clearer structure Significantly oriented browsing
EDUCATIONAL	No accompanying tools Interactive characteristics Various and individualized types of feedback	
	Game- like activities Flexibility – various levels of difficulty	School – type activities Various types of activities with similar level of difficulty Not clearly stated similar ones
	Clearly stated informative and entertaining scopes Possibly supports traditional and discovery mode of learning as well Possibly supports STS model	Possibly supports traditional mode of learning Possibly supports 'science for all' model as well as 'the pupil as a scientist' one
EPISTEMOLOGICAL	Positivistic and hermeneutic	Positivistic
SOCIAL & RELEVANT TO THE CURRICULUM	No systematically supported stereotypes Both programs reflect western lifestyle	
PUPILS' SCOPE	Easy browsing Learning declarative than procedural knowledge Better teaching material than the usual ones Computer experience influences interactivity and browsing	
	Game like activities	School type activities

4. CONCLUSIONS AND IMPLICATIONS

The analysis and evaluation of the two CD-ROMs reveals some teaching implications.

The experts observe that the great level of interactivity in both programs supports learning and corresponds to the pupils' positive attitudes to the CD-ROMs. Similarly, we suggest that programs like 'Mad About Science: 3. Matter' can easily be used in a formal educational environment. This is due to the variety of activities suggested and the fact that both their content (separating, changing, experimenting) and their type (multiple choice, Yes or No) are compatible with typical school activities. On the other side, 'My Amazing Human Body', with its many interactive characteristics as noted above, in combination with the informal activities of the main menu, is more appropriate to be used in informal learning environments, e. g. in a museum or at home. This suggests that programs with high level of interactivity can much more readily be used out of school or at home. But when a program or part of it is used in school, there are other links between teachers' roles and actions in the classroom and the design of educational software that have to be identified (Hinostroza & Mellar, 2001). It is also addressed to a wider age range than the Chemistry title. Not only can it be used in Biology classes, but in other ones as well (interdisciplinary approach): e.g. the puzzle activity can be used in painting courses, whereas the other title can only possibly be used in Physics or Chemistry courses. This implication agrees with the experts' comments and Thenia's attitude: the first one noted that the Chemistry title comprises activities with about the same level of difficulty; in the same context, Thenia prefers to run 'My Amazing Human Body' rather than the other program because it combines both easy and difficult activities.

In both programs the variety of type of activities included suggests an interactive way (as noted in the experts' point of view, §3.2), of evaluating pupils that is very different from the usual ones. Special emphasis can also be given to the formative type of feedback as a respective type of evaluation, since the use of the program is in both cases impersonalised. Similarly, many of their characteristics – e.g. when the user predicts 'what will happen if…' – can enhance the discovery mode of learning, if it is appropriately placed. The traditional mode, according to the experts' comments, can be supported by texts intercalated into the program which would simply inform the user, like the texts the user is led to in the Biology title via the 'Find out more' icon. This is stressed by the fact that in the same case, Thenia and Evaggelia found the texts they met during browsing *'boring'*.

The connection of biology (in the respective title) to everyday life (in the activity 'Me and my day') through technological achievements and everyday material is useful for underlining the relation of Science to Society and Technology (STS model). On the other hand, "science for all" can be enhanced through the use of an everyday environment in the context of which Science is applied, as in the Chemistry title. In a similar way the hermeneutic epistemological paradigm – that is, science as it emerges from local tradition and culture – can be underlined in the instruction in 'My Amazing Human Body', whereas the other program can be used in order to show how we 'learn' the world around us through specific interventions, which at the same time impact upon it (interventional epistemological aspect).

In this paper we tried to apply our framework for the analysis and evaluation of educational CD-ROMs in Science. The framework described above seems to be applicable since it allows us to make assumptions about the characteristics of the program studied. We can also support its integrity because it assesses software from the electronic, educational, epistemological and social aspect, as well as from the pupil's point of view. It also seems to be reasonably flexible, since we can form various hypotheses either about the specific characteristics of a program or the instructional and learning theses of its editors.

REFERENCES

Berge, Z. 1999. *Interaction in Post – Secondary Web – Based Learning*, Educational Technology, January/February, pp. 5 - 11

Firdyiwek, Y. 1999. *Web-Based Courseware Tools: Where Is the Pedagogy?,* Educational Technology, January/February, pp. 29-34.

Flottemesch, K. 2000. *Building Effective Interaction in Distance Education: A Review of the Literature*, Educationl Technology, May/ June, pp. 46 - 51

Hacking, I. 1983. *Representing and Intervening*, Cambridge Univ. Press.

Hinostroza, E., Rehbein, L., Mellar, H. & Preston Chr. 2000. *Developing Educational Software: A Professional Tool Perspective*, Educational and Information Technologies 5:2, pp. 103 – 117

Hinostroza, E. & Mellar, H. 2001. *Pedagogy embedded in educational software design: report of a case study*, Computers & Education, vol. 37, pp. 27 - 40

Johnston, M. 1987. The evaluation of microcomputer programs: an area of debate, *Journal of Computer Assisted Learning*, Vol. 3, pp. 40-50

Kariotoglou, P. & Tselfes, B. 2000. Science Curricula: Epistemological, Didactical & Institutional Approach, *Physics Review* (in Greek), EEF, 31, pp. 19-28.

Kindersley, D. 1997. *My Amazing Human Body*, DK Multimedia (1st CD-ROM used).

Kindersley, D. 1997. *Mad About Science:3. Matter*, DK Multimedia (2nd CD-ROM used)

Kelly, Gr. & Crawford, T. 1996. Students' Interaction with Computer Representations: Analysis of Discourse in Laboratory Groups. *Journal of Research in Science Teaching*, Vol. 33, NO. 7, pp. 693 – 707.

Kounatidou, Chr. & Kariotoglou, P. 2001. The development and application of a framework appropriate for a multilevel assessment of educational multimedia software in Science, in: D. Psillos et al. (eds), *Proceedings of the Third International Conference on Science Education Research in the Knowledge Based Society*, Vol. II, pp. 489 – 491.

Liaw, Sh. & Huang, Hs. 2000. *Enhancing Interactivity in Web – Based Instruction: A Review of the Literature,* Educational Technology, May/ June, pp. 41 - 45

Mikropoulos, T. 2000. *Educational Software,* Kleidarithmos, Athens (in Greek).

Roth, W-M., Woszczyna, C., & Smith, G. 1996. Affordances and Constraints of Computers in Science Education, *Journal of Research in Science Teaching*, Vol. 33, NO. 9, pp. 995 – 1017

Squires, D. & McDougall, A. 1994. *Choosing and Using Educational Software, A Teachers' Guide*, The Falmer Press, London.

Tolhurst, D. 1992. A Checklist for Evaluating Content-Based Hypertext Computer Software, *Educational Technology*, March, pp. 17-21

BRAIN RESEARCH IN SCIENCE EDUCATION RESEARCH

Tassos A. Mikropoulos

Department of Primary Education, University of Ioannina, Ioannina 45110, Greece, amikrop@cc.uoi.gr

ABSTRACT

Recently, neuroscientists, psychologists and pedagogues work on cognition, learning and teaching involving instructional technologies, each of them from their own point of view, towards a biological basis of learning.

This paper proposes the exploitation of neuropsychological methods and non-invasive imaging techniques for the study of mental conditions and especially of cognitive tasks in technologically enriched educational environments for science learning. It proposes the study of brain waves coming from different areas of the cortex and their spectral analysis using Fast Fourier Transforms for the understanding of various cognitive processes taking place during specific tasks of users interacting with computer-based educational environments, and especially virtual reality (VR) systems. The model proposed is supported by recent experimental findings reflecting greater alertness and task engagement in computer-based environments for science learning (Gerlic and Jausovec 1999) and educational virtual environments (Mikropoulos 2000a).

1. BACKGROUND, AIMS AND FRAMEWORK

Cognitive and learning processes are of main research interest to neuroscientists, psychologists and pedagogues. Recent scientific research for the understanding of cognitive and learning processes and the development of a learning theory has resulted in the production of a series of discussions on the modular brain and the biological basis for learning (Leamnson and Betz 1999, Schaverien and Cosgrove 1999, Schaverien and Cosgrove 2000, Rice et al. 1996, Hansen and Monk 2002).

From a neurophysiological point of view, Ray and Cole have reported that electrical brain activity reflects attentional demands, emotional and cognitive processes (Ray and Cole 1985). Lehmann and his associates have stated that mind states can be identified with particular brain functional states measured through brain electric activity, and that characteristics of brain electric field data differ between types of cognitive information processing. In particular, data in the frequency domain have shown that brain alpha activity is a representation of idling, and beta activity is a representation of activation (Lehmann et al. 1993). More specifically, Hillyard and his associates have reported that characteristic electric brain signals have been observed in tasks involving sustained attention, spatial cuing, visual search, selection of features and objects (Hillyard et al. 1996). Riding and Glass reported that different people process the same information to solve the

D. Psillos et al. (eds.), Science Education Research in the Knowledge-Based Society, 353–360.

same mental tasks in different ways, using different strategies and areas of the brain, depending upon their cognitive style (Riding and Glass 1997). All the above references having a neurophysiological approach reporting on cognitive processes, not directly connected with learning processes like those occurring in an educational context.

Results towards these directions have reported from the psychology and pedagogy domain too. Researchers from these fields do not report on brain measurements, but on data coming from the general empirical research methodology and psychological tests. Moreover, they report on learning processes directly connected with the educational context. Rice, Rice and Lovell have proposed that the research on cortical plasticity and enriched environments is promising, and may provide the educational community with a new way of conceptualizing learner development (Rice et al. 1996). McCluskey has reported that right-brain dominant individuals significantly outperformed their left-brain and mixed-brain dominant counterparts, while left-brain dominant individuals outperformed the mixed-brain dominant individuals on designing and developing linear, nonlinear and spatial computer tasks in a hypermedia context. The cerebral hemisphericity of the subjects was determined by checking their eyedness and handedness (McCluskey 1997). Schaverien and Cosgrove have argued that education is profoundly biological, stating that this is a thesis which could make a significant contribution to resolving the philosophical dilemma in which technology and science education research presently find themselves (Schaverien and Cosgrove 1999).

From the above argumentations it seems that neuroscientists, psychologists and pedagogues work on cognition, learning and teaching involving instructional technologies, each of them from their own point of view, having in mind the biological basis of learning.

The goal of this article is to bring together the theoretical models and data coming out from research in human and social sciences with the results from neural sciences research, using brain research, studying brain activity of learners during teaching and learning processes and tasks in sciences using instructional technologies with emphasis on open computer-based educational environments. This idea follows that of Gale and Eysenck's where they emphasised the importance of a multi-disciplinary approach to the study of individual differences, particularly interactions between neurophysiology and cognitive psychology (Gale and Eysenck 1992).

To our opinion the goals of the research needed the exploitation of neurophysiological methodology in science education are the following:

- To associate certain brain activity with specific learning tasks and situations
- To study brain activity in different cognitive styles
- To study brain activity under synthetic versus real educational environments
- To study brain activity under computer-assisted learning environments
- To understand how computer-assisted learning environments influence learning
- To understand how computers affect mind
- To detect differences in information processing between men and women

- To evaluate educational software using brain measurements
- To design computer-based educational and specifically virtual environments leading to specific learning outcomes
- To learn how we learn
- To learn how to learn.

The above topics refer to all disciplines, but they have a particular interest in science education. Issues in science education research involve information and knowledge organization and processing, attentional demands, understanding of abstract concepts and models, problem solving, spatial perception and orientation, visual perception, selection of features and objects, reflection of emotional and cognitive processes, metacognitive abilities, different cognitive styles. These are also among the research interests of neurophysiological research and some significant results have been appeared towards this direction (Ray and Cole 1985, Hillyard et al. 1996, Lehmann et al. 1993, Riding and Glass 1997). Regarding science education research, educational technology provides the tools to support the educational and learning processes, with Information and Communication Technologies covering almost all types of media and learning tools.

Relevant scientific research has shown that computers do affect 'the mind', the way children think, learn, and develop metacognitive abilities (Fletcher-Flinn and Suddendorf 1996). A subject under discussion is the design and evaluation of educational software and the assessment of human-machine interaction for educational purposes. The methodology of educational research, especially when computer-based environments are involved, poses a variety of limitations to the subject under evaluation. For a complete evaluation process with the 'traditional' methodology, there are many factors that have to be taken into account to reach the goals of the research. These are coming from social issues, instructional goals and design, software design and content, technical issues, groups of evaluators, the samples, and evaluation techniques (Mikropoulos 2000b).

Neuroscience research has shown that the imaging of the brain can be used to probe the relation of hemispheric functioning to both emotion and cognition (Ray and Cole 1985, Hillyard et al. 1996, Halgren and Marinkovic 1996) and this might provide a new objective tool for the design of enriched, technology-based educational environments and the assessment of the interaction of students with computer-based educational environments.

This article proposes the use of the models and methodology of neural sciences for the identification and analysis of the emotional and cognitive activities taking place during specific learning activities of users interacting with computer-based educational environments. A further aim of this research is to measure and analyse the cognitive changes that users of educational VR systems experience and to evaluate the consequences of educational software, especially that concerning science education.

2. MODE OF INQUIRY

The human brain operates in a different way when it gets signals coming from synthetic environments such as the computer-generated environments (Mikropoulos 2000a). For the detection and analysis of these signals, there are three main imaging techniques. The first uses Positron Emission Tomography (PET) measuring blood flow on a spatio-temporal scale, recording responses summed over long periods of time. The second technique is functional Magnetic Resonance Imaging (fMRI) that is sensitive to the oxygenation of blood, in a smaller spatio-temporal scale. The third technique is that of electrophysiology, measuring the electric activity of the brain.

We propose the third method for the detection and analysis of emotional and cognitive changes in synthetic educational environments, since it provides event related responses to a single stimulus via evoked potentials with a high spatial and temporal resolution. The electroencephalogram (EEG) is analysed as a time series accompanied by the Fourier transform of the signal, which analyses it into its spectral distribution. Thus, we take information about a given brain state, or a change in state. This is important for science education, since the relevant educational software involves three-dimensional interactive simulations (Mikropoulos 1997), or problem solving situations where neuropsychological results are encouraging (Wei et al. 1998). Moreover, there is much research on computer-based environments for science education and articles on measuring cognitive processes with EEG in synthetic environments for science learning have started to appear (Gerlic and Jausovec 1999). Our choice is also supported by the large amount of research work done using electric brain activity for the study of cognitive tasks (Ray and Cole 1985, Hillyard et al. 1996, Lehmann et al. 1993, Riding and Glass 1997).

This type of brain research seems to be fruitful for science education, since experimental data show considerable differences in electric brain measurements coming from cognitive tasks on different disciplines. Since 1985, results on brain research have shown different brain activity in cognitive tasks concerning performance and cognitive style in stimuli coming from different disciplines (Earle 1985 and references therein). Concerning the important aspect of problem presentation, results showed different EEG power measures related to the form of problem presentation (verbal / figural) with great portions of the brain related to visual and spatial processing during an experiment on the physics of radio and telephone (Jausovec and Jausovec 2000). This study also showed that it is unlikely that individuals solved well- and ill-defined problems by employing similar cognitive strategies being an indication on how we could study problem solving strategies. Skrandies, Reik and Kunze recorded different brain activity during arithmetic and language tasks and found differences between female and male students (1999). Their study indicated that the stimuli coming from the two different task demands (language and arithmetic) formed different mental models. Mölle and associates referred to functional thinking as an important attribute in scientific discourse although less appreciated in science education research and reported that EEG complexity during predicative thinking decreased in comparison to functional thinking (Mölle et al. 2000). They also reported that mental imagery of a moving pendulum resulted in increased prefrontal dimensional complexity of the EEG in comparison to the actual perception of the same object.

3. OUTCOMES

The outcomes of the proposed methodology are based on the characteristics of electroencephalography (EEG). These mainly are its excellent temporal resolution, evoked potentials, spectral analysis, brain mapping, as well as the small cost comparing to the other two techniques (PET, fMRI). Using spectral analysis and studying the different spectral ranges (alpha, beta, theta, gamma rhythms) we get results for different cognitive tasks. With spatial analysis, we may record the areas of the brain where one or more cognitive tasks are taking place.

There are only a few studies reporting results on electric brain activity in virtual worlds. Bayliss and Ballard have reported findings on cognitive activity of subjects who were told to stop their virtual cars at the red traffic light in a VR driving world. The goal of their experiment was to recognize the existence of P300 evoked potential epochs and find the best one of three different signal analysis methods. This work did neither report anything concerning the comparison between real and virtual environments, nor on learning activities (Bayliss and Ballard 1998). Strickland and Chartier have reported on EEG measurements in a virtual reality headset, concluding that differences do exist between real and virtual image processing (Strickland and Chartier 1997). A PET study by Maguire and associates showed parahippocampal involvement in encoding object locations in virtual large-scale space that was not activated during exploration of empty virtual environments (Maguire et al. 1998). The above studies show that there is specific brain activity for cognitive tasks in synthetic environments. These as well as the studies on brain research and learning (Skrandies 1999, Jausovec and Jausovec 2000, Mölle et al. 2000) imply that research on learning activities in educational environments would be very fruitful for science education research. Towards this direction we have started a project on brain research with the following goals:

1. to evaluate information representation during specific learning tasks in both real and synthetic educational environments
2. to detect cognitive load due to different information representation
3. to detect different cognitive styles in order to design different educational environments and activities based on visual and/or verbal representations.

This article presents the results of our first exploratory study emphasising certain attributes of science learning such as attentional demands, spatial perception and orientation, visual perception, selection of features and objects comparing brain electric activity in real and virtual educational environments.

The subjects in our research were 12 university students, undergraduates and postgraduates. They were asked to perform the same cognitive task in both a real and a virtual environment. The virtual environment represented the real laboratory, and the students had to move through a certain path, find and recognize an object.

The virtual environment and task were parts of educational software for science education. During the task in both the real and virtual environments the students' brain signals were recorded and further analysed. Eye-movement measurements showed that all the subjects were more attentive when navigating in the virtual world ($p = 0.003$). The signal diminution at the alpha rhythm (8-13 Hz) and the difference between alpha and beta rhythms for the virtual task indicated that students placed more attention in the virtual environment and were more responsive to the cognitive stimulation ($p = 0.021$). Lower theta activity (4-8 Hz) in the virtual environment demonstrated that all the subjects placed less mental effort in the virtual task signifying that virtual reality provides environments suitable for knowledge transfer ($p = 0.018$). These results and especially the slight eye-movement and alpha signal diminution indicate that virtual educational environments provide opportunities for students to concentrate, perceive and judge more effectively than in real environments. Our results are similar to those of Gerlic and Jausovec, who measured cognitive differences with EEG in a multimedia environment for astronomy teaching involving visualisations (Gerlic and Jausovec 1999). A detailed description of our experiment and analysis of our findings may be found in (Mikropoulos 2000a).

4. CONCLUSIONS AND IMPLICATIONS

The conclusions coming from our first experiments show that EEG techniques can detect and analyse attention, spatial perception and orientation, visual perception, selection of features and objects, cognitive stimulation and mental effort in synthetic science learning environments. This shows the potentials of brain research in science education research. More specifically, our results on attenuated alpha and increased beta frequencies for the virtual environment agree with those of Strickland and Chartier (1997), as well as of Freeman and his associates (2000) reflecting greater alertness and task engagement in the virtual environment. On the contrary, our results on lower theta activity in the virtual environment are in contrast of those of Strickland and Chartier (1997) and Freeman and associates (2000), reflecting a decrease in alertness and task engagement in the real environment. Our results are towards the direction of brain development, structuring of learning and science education, as they have recently reported by Hansen and Monk (2002).

The implications of the proposed methodology for science education are shown by the research questions under consideration. Are there any differences between EEG signals coming from real and synthetic educational environments? What kind of differences are there, if any, in the alpha, beta, theta, and gamma rhythms in real and synthetic environments? Are there differences between EEG signals of male and female subjects? The value of these questions is directly related to our interest in introducing VR for the support of science teaching and learning. It is important to know what cognitive processes are taking place in educational virtual environments since they are mainly based on simulations and visualisations, the key points in science education and enhance users' experiences or provide them with new ones.

These questions are in the context of the evaluation of educational synthetic environments for science learning using the proposed methodology in addition to the standard methodology of social sciences that has been difficult to substantiate the understanding of mental processing differences between real and artificial environments.

REFERENCES

Bayliss, J., D. & Ballard, D., H. (1998). *Single Trial P300 Recognition in a Virtual Environment.* The University of Rochester, Computer Science Department, Technical Report 98.1. Available at www.cs.rochester.edu/trs/robotics_trs.htm.

Earle, J. B. B. (1985). The effects of arithmetic task difficulty and performance level on EEG alpha asymmetry. *Neuropsychologia*, 23(2), 233-242.

Fletcher-Flinn, C., M., Suddendorf, T. (1996). Do Computers Affect 'the mind'? *Journal of Educational Computing Research* 15(2) 97-112.

Freeman, F., G., Mikulka, P., J., Scerbo, M., W., Prinzel, L., J. & Clouarte, K. (2000). Evaluation of a Psychophysiologically Controlled Adaptive Automation System, Using Performance on a Tracking Task. *Applied Psychophysiology and Biofeedback*, 25(2), 103-115.

Friston, K., J. (1997). Imaging cognitive anatomy. *Trends in Cognitive Sciences* 1(1), 21-27.

In:Gale A., & Eysenck, M., W. (1992) *Handbook of Individual Differences: biological perspectives* (Chichester, John Wiley and Sons).

Gerlic, I, & Jausovec, N. (1999). Multimedia: Differences in cognitive processes observed with EEG. *Educational Technology Research and Development*, 47(3), 5-14.

Jausovec, U. & Jausovec, K. (2000). EEG activity during the performance of complex mental problems. *International Journal of Psychophysiology*, 36, 73-88.

Halgren, E., & Marinkovic, K,. (1996). Neurophysiological Networks Integrating Human Emotions. In M. S. Gazzaniga (eds) *The Cognitive Neurosciences* (THE MIT PRESS, Cambridge, MA) 1137-1151.

Hansen, L. & Monk, M. (2002). Brain development, structuring of learning and science education: where are we now? A review of some recent research. *International Journal of Science Education*, 24(4), 343-356.

Hillyard, S., A., Mangun, G., R., Woldorf, M., G., & Luck S. J,. (1996). Neural Systems Mediating Selective Attention. In M. S. Gazzaniga (eds) *The Cognitive Neurosciences* (THE MIT PRESS, Cambridge, MA) 665-681.

Leamnson, R., N., Betz, M., K. (1999). Technology and the Biological Basis of Learning. *Educational Technology & Society,* 2(4), 1-5.

Lehman, D. (1992). Brain Electric Fields and Brain Functional States. In R. Friedrich, A. Wunderlin (eds) *Proceedings in Physics, Evolution of Dynamical Structures in Complex Systems*, 69, (pp.235-248) (Berlin: Springer-Verlag).

Lehmann, D., Henggeler, B., Koukkou, M., Michel, C., M. (1993). Source localization of brain electric field frequency bands during conscious, spontaneous, visual imagery and abstract thought. *Cognitive Brain Research*, 1, 203-210.

MacRae, S. (1994) *Models & Methods for the Behavioural Sciences* (The British Psychological Society).

Maguire, E. A., Frith, C. D., Burgess, N., Donnett, J. G. and O'Keefe, J. (1998) Knowing Where Things Are: Parahippocampal Involvement in Encoding Object Locations in Virtual Large-Scale Space. *Journal of Cognitive Neuroscience*, 10(1), 61-76.

McCluskey, J. J. (1997). An Exploratory Study of the Possible Impact of Cerebral Hemisphericity on the Performance of Select Linear, Non-linear, and Spatial Computer Tasks. *Journal of Educational Computing Research*, 16(3), 269-279.

Mikropoulos, T., A. (1997). Virtual Environments in Science Education. In In M. Bevan (ed.), Proceedings of the International Conference on Virtual Reality in Education & Training, (pp. 43 – 48). Loughborough.

Mikropoulos, T., A. (2000a). Brain Activity on Navigation in Virtual Environments. *Journal of Educational Computing Research*, 24(1), 1-12.

Mikropoulos, T., A. (2000b). Design, Development and Evaluation of Advanced Learning Environments. An Overall Approach. Presented at the HERMES *Advanced systems for teaching and learning over the World Wide Web*, B42-B52, September 1-3 (Samos, Greece).
Mölle, M., Schwank, I., Marshall, L., Klöhn, A. & Born, J. (2000). Dimensional Complexity and Power Spectral Measures of the EEG during Functional versus Predicative Problem Solving. *Brain and Cognition*, 44, 547–563.
Ray, W., J. & Cole, H., W. (1985). EEG Alpha Activity Reflects Attentional Demands, and Beta Activity Reflects Emotional and Cognitive Processes. *Science*, 28, 750-752.
Rice, S., Rice, K., Lovel, L. (1996). Enriched Environments, Cortical Plasticity, and Implications for the Systematic Design of Instruction. *Educational Technology*, March / April, 41-46.
Riding, R. J., Glass, A. (1997). Cognitive Style and Individual Differences in EEG Alpha during Information Processing. *Educational Psychology*, 17(1/2), 219-232.
Schaverien, L. & Cosgrove, M. (1999). A Biological basis for generative learning in technology-and-science Part I: A theory of learning. *International Journal of Science Education*, 21(12), 1223-1235.
Schaverien, L, & Cosgrove, M. (2000). A Biological basis for generative learning in technology-and-science Part II: Implications for technology-and-science education. *International Journal of Science Education*, 22(1), 13-35.
Skrandies, W., Reik, P, Kunze, Ch. (1999). Topography of evoked brain activity during mental arithmetic and language tasks: sex differences. *Neuropsychologia*, 37, 310-329.
Strickland, D., Chartier, D. (1997). EEG measurements in a virtual reality headset. *Presence: Teleoperators & Virtual Environments*, 6(5), 581-588.
Wei, J., Zhao, L., Yan, G., Duan, R., Li, D. (1998). The temporal and spatial features of event-related EEG spectral changes in 4 mental conditions. *Electroencephalography and clinical Neurophysiology*, 106, 416–423.

COMPUTER MODELLING AND SIMULATION IN SCIENCE LESSONS: USING RESEARCH INTO TEACHERS' TRANSFORMATIONS TO INFORM TRAINING

Fani Stylianidou[1], Richard Boohan[2], Jon Ogborn[3]

[1]*Institute of Education, University of London, UK,* [2]*The Open University, UK ,* [3]*Institute of Education, University of Sussex, UK*

ABSTRACT

The curriculum is in a constant state of rapid change. The role of teachers in implementing curriculum innovation is critical, though in the absence of suitable training the intentions of the innovation may be transformed. This paper discusses the research into the nature of teachers' transformations and the development of training materials on computer modelling and simulation, carried out as part of the STTIS (Science Teacher Training in an Information Society) project. For this study eight teacher case studies were carried out. These were concerned to identify possible factors which favour or hinder the take-up of the selected informatic tools in science classes, and to investigate how teachers incorporate these tools in the curriculum. The training materials that have been consequently developed provide activities to enable teachers to learn about the innovations and about the nature of the research into their implementation, and to take account of these ideas in their own implementation of the innovations.

1. BACKGROUND, AIMS AND FRAMEWORK

The central question addressed by the STTIS (Science Teacher Training in an Information Society) research is about the nature of difficulties arising when teachers are asked to adopt an innovative teaching sequence (STTIS, 1998). The starting point of the project is that teachers' interpretation of a proposed teaching sequence may result in transformations of the original didactic intentions. The project's main objective was to investigate these teacher transformations occurring in the implementation of different kinds of curriculum innovations in the context of the science classroom. The outputs from the project include materials to be used in the training of teachers and guidelines intended for policy makers. This paper discusses the research conducted as part of the work of the UK team into the nature of teachers' transformations when using computer models and simulations (Stylianidou and Ogborn, 1999) and presents the training materials that were developed based on this research (Boohan *et al*, 2000).

Nowadays, both in education research and in school practice, increasing attention is paid to modelling activities, and there is wide consensus concerning the educational usefulness of learning how to interpret and build models (see for example Mellar *et al*, 1994). According to Feurzeig and Roberts (1999), acquiring

D. Psillos et al. (eds.), Science Education Research in the Knowledge-Based Society, 361–369.

fluency in model-based inquiry should become an essential goal of pre-university science education. However, as they point out a number of obstacles stand in the way of effective adoption and integration of modelling activities in the school science curriculum. Our choice to study the transformations science teachers make of the use of computer modelling reflects both our acknowledgement of its great educational potential and our resolve to work towards removing some of the obstacles for its incorporation in the secondary curriculum.

2. METHODS AND SAMPLINGS

Eight teacher case studies were carried out for the study. The case studies were concerned to identify possible factors, which favour or hinder the take-up of the selected informatic tools in science classes, and to investigate how teachers incorporate these tools in the curriculum. The main research questions were:

- What are the fits (matches) or gaps (mismatches) between the intended or expected use of the tools and the way the teachers actually use them in practice?
- What do teachers identify as more successful and less successful examples of uses of the tools, and what reasons do they have for identifying them as such?
- Are teachers conscious that they transform the suggested uses of tools in their own practice, and if so what reasons do they give? If not, are there clues about what may determine the teacher's practice?

The teachers in our case studies were selected by their willingness to participate and to attempt to make serious use of computer modelling and/or simulation in their science classrooms. The choices of modelling software as well as the context of its use were left to be made by the teachers.

All teachers were observed making use of the informatic tools. In all instances, record notes were kept for each observation, worksheets and pupils' work were collected and interviews were held with the teachers before and sometimes after the observations took place.

The generation and writing up of the case studies followed a series of steps. Firstly, together with the other partners, we identified the important themes we wanted to address. These were:

- <u>Accepting / changing external goals</u>

The designer of a computer tool may have certain intentions and goals concerning its use. These goals should be seen as related to: the possibility of implementation of the tool; and/or the science learning addressed by the tool. The teachers may choose to commit to these goals, just agree with them, warp them and substitute their goals for them, or plainly reject them. The case studies would attempt to characterise the positions of the teachers in relation to these external goals.

- <u>Rationale vs Practice</u>

The case studies would attempt to identify whether there was any perceived difference between the rationale the teachers offered for their use of the informatic tool and their actual practice, as this was observed and interpreted by the researcher.

- <u>Actual /Realising goals (of teacher)</u>

The case studies would also express an opinion as to whether the teachers succeeded (or not) in accomplishing the goals they had set for the lessons. One would expect to come across some effective practices, some partially effective, and maybe even some ineffective ones. There is a wide range from skilled to awkward teachers.

– Teacher's knowledge

Important information about the teachers' knowledge and relevant experience would be included.

– Effects of physical objective circumstances

The implementation of the computer tools is conditioned by diverse circumstances (technical, physical, subjective, etc.). The case study would make their role, when it was relevant, explicit.Existing practices

Finally, the case study would include mention of the possible effects of the national curriculum, and of other institutionalised practices such as examinations on the use teachers made of the informatic tools.

Based on this pre-determined framework of issues the data for each of the teachers was assembled and a first draft of the case study was drawn up. These drafts were then checked in discussion among the researchers, against the data, and were modified as necessary.

3. AN EXAMPLE OF A TEACHER CASE STUDY

The lesson of one teacher, Albert, is presented in this case study. He is using a spreadsheet to teach about capacitor discharge. It is not intended as a representative case study. However, the case of spreadsheets as a modelling tool is an interesting one to consider, since they can be used to mode a wide variety of situations (both static and dynamic). In addition, they are general-purpose software packages, often taught as part of IT as a subject, and also used in other curriculum areas (Brosnan, 1994).

The two lessons observed had to do with modelling the capacitor discharge curve through a resistor with Excel and were part of a sequence on electric circuits. Before then the class, which had five 17 year-old girls aiming to get between A and C at their Physics A' level the year after, had studied the factors that affect the characteristics of the capacitor charge and discharge curves through data logging.

The teacher, Albert is a biologist by first degree and has taken up teaching A' level physics for the last 2 years. He describes himself in his new role:

> "I'm learning at the same time as the girls are, a lot of the time. I'm more than a few pages ahead in the book, but nevertheless often things come up in lessons that they think of, that I haven't, and so I'm having to think very much on my feet, which I find tremendously stimulating. [...] This is [also] stimulating the girls to be very questioning and I think it's a much better way to learn for all concerned. It's fun, but it's quite a lot of stress."

Albert seems to be a person who is not afraid to experiment, in all sorts of ways. He likes to introduce something new in his teaching (this capacitor modelling activity he used for the first time), even if he risks to fail in doing so. He sees himself as being at the avant-garde in the use of computers for teaching in his school. Not only does he use computers much more than his colleagues - in about a quarter of his lessons - but his department has more recent computers than the rest of the school - *"they had to upgrade to catch up with us"*.

In general Albert sees the use of computers in teaching science *"very closely tied to actual practical work - experimental work"*. His experience using modelling: he had used Excel in relation to teaching about velocity and acceleration once before with another class, not very successfully. The problems he had encountered he attributed to the fact that the pupils were not adept enough at handling Excel. According to him *"it got in the way of physics"*, seeming to suggest that one of his aims using modelling had been to facilitate the delivery of the content. This time he believed he would manage better because the pupils were a bit more practiced with Excel. For Albert pupils' competence with the software is one of the main factors that determine success.

Albert's strongest criterion, however, for choosing Excel to use appeared to have to do with his competence in using it and not with the pupils':

> "I'm very familiar with Excel... so I feel comfortable using this. I wouldn't be diving into this if I weren't adept at solving the sorts of problems that kids come up with."

One of the things Albert was hoping to achieve by using modelling in this lesson was that the pupils began to see mathematically what is going on in the set up.

> "It'll also - I think - introduce them to the mathematics, ... the use of increments to study an exponential curve."

This is a valid 'modelling' aim which however he did not rate as so important because *"they're going to do it in maths a little bit later on"*. He was also hoping that the pupils could see the power of the computer as a tool to investigate what-if questions; that the pupils' physics knowledge, which he believed was good, would be consolidated; and finally, but most importantly for him, that the pupils learnt a new technique/method which he could later use with other topics.

In other words, Albert had short-term 'modelling' aims but also a more basic longer-term aim that he wanted to achieve: to acquaint the pupils with the use of spreadsheets in science, so that he could use them in the future for the teaching of other topics:

> "We' re using the physics in a way as a vehicle to pick up a new technique, a new skill, which I hope will then be used later on."

The first lesson started with Albert explaining to the pupils how one can calculate how much charge has been discharged in a given Dt. To do that he chose to first present them with the final formula and then attempt to deconstruct it to its primary component formulae. The goal he set for them is to make a graph like the one he had given them. Albert took them step by step through what they should write (he had prepared the spreadsheet as a transparency) not only in terms of the formulae, but also in terms of the computational equations they needed to put in each cell. The way he put it to them:

> "I will tell you how to put the equations in. You do not have to invent them this time. Other times I will let you invent them."

A difference between his spreadsheet and the pupils', a mistake in one of the computational formulae, plus the problem of absolute addressing in Excel created confusion. The talk was mainly about software manipulations. The end of the lesson arrived before most girls had time to do the chart Q=f(t).

Albert's account of the lesson in the interview that followed it was:

> "It felt very slow to me. We weren't making the progress that I'd wished to. I thought we would have got the graphs drawn with a few minutes to spare at the end of the lesson, actually, five or ten minutes to spare, so that we could actually start putting in different values and see what happens and then beginning to discuss that. But too many of the girls really hadn't finished doing all the copying and pasting and filling down that needed doing. I felt that they understood fairly clearly the process that we were going through, the calculation of the small charges that flowed in each time interval, I think they got that idea quite well. I think they were beginning to get the idea of how that was put onto the spreadsheet. But I don' t think they made the connection between one row and the next to see how the iteration worked clearly. I think they were just translating my numbers on the screen into their numbers on the sheet without trying to understand where the actual values were coming from; so that needs to be clarified. I'll get them to label up the spreadsheet with that kind of information."

> "From a physics point of view I don't think they learned much they didn't know already, so, the benefits are in the practice of using the computer and in seeing the computer being applied in a physics situation which they haven't seen enough of and the benefits will be gained much later when they're able to do this for themselves on some of the topics like velocity and acceleration. But there's quite a long way to go to get them competent enough to really feel the benefits."

The second lesson unfolded much more smoothly. Albert re-established control of the class by keeping the pupils away from the computers at the start of the lesson and talking them through the first row of the model, clarifying where there was input coming from outside and where there was input generated by the programme itself. He assessed their understanding by getting them to talk to him about the next row.

He then got them to check that their models conformed to his and subsequently asked them to change the values of the resistance and capacitance in order to see what these changes did to the resulting graph. Finally, he asked them to print out their graphs so that they could work out from them for homework the half-life and compare their finding with what they had calculated with the computer.

In the interview after the lesson, Albert seemed much happier about how the lesson had gone. He once more stated what he thought to be the most useful thing that the pupils gained:

> "...familiarity with the machine and with Excel. The physics actually was a vehicle for them to learn about the machine rather than the other way round. Now I'm hoping that they'll become sufficiently familiar with the machine that we can use it as a vehicle to extend their understanding of velocity, acceleration, momentum, kinetic energy and the rest."

4. SUMMARY OF CASE STUDY

– Accepting / changing external goals

Albert clearly warped most external goals concerned with the use of Excel as a modelling tool to teach science. In particular, two main transformations we think we can see in this story. First we see him transforming the message into the medium. He is conscious and explicit about it. This transformation seems to be the product of both his sense of adventure and his sense of safety. He clearly is doing something new he does not have to do, but which he believes in. However, he does not want to take chances. He knows from previous attempts that pupils are not competent enough using the computer, and he believes that their incompetence would get in the way.

So, he chooses a topic, which he is confident that the pupils have understood very well: the charging and discharging of a capacitor. He wants to use the 'physics' as a vehicle to get the pupils familiarised with the process of using a spreadsheet for modelling in science. He also has some aims, which are relevant with the modelling process itself, for example he wants them to understand how an iteration works. On the other hand, any aims which are pertinent to the teaching of the relevant content knowledge, such as to learn to represent mathematically and/or graphically what is going on during a capacitor discharge, seem to be of very little importance to Albert.

– Rationale vs Practice and Teacher's knowledge

The second transformation discloses a discord between his rationale as just described and his practice, and seems to arise from the insecurity he feels about his knowledge of Physics. This insecurity drives him to convert the knowledge to be taught into recipes. As a result the process of modelling becomes a process of following steps and blindly copying a set of formulae on to a spreadsheet.

– Actual /Realising goals (of teacher)

As a consequence of these transformations, we believe that Albert only partially realised his goals and certainly not in the space of one lesson as he had intended to.

– Effects of physical objective circumstances

The fact that the school provides computer laptops for the students makes it easier for Albert to control the class's interaction with them without having to move the pupils to different purpose rooms.

– Existing practices

Finally, the physics examination syllabus is always in Albert's mind, but this does not seem to stop him from trying something 'a little bit off' it, since he believes that in the long term his venture will pay off.

5. DISCUSSION OF TEACHER CASE STUDIES

Variation was expected in how the teachers incorporated the use of computers in their teaching and was indeed detected in the case studies. It mattered a great deal what teachers believed of themselves, who they were, what they could do and how they related to other people, all of which differed from teacher to teacher. Knowing about the school in which the case study took place also was important. As well as differing provision of resources, differences in ethos determined in many cases the expectations both the teacher and the pupils had of each other, and thus of their interaction with the tool.

Concerning the provision of computers and informatic tools, it has not stopped being an issue for schools. Even in schools that on the whole had an acceptable level of provision, the lessons were unquestionably adapted to it, though not heavily constrained by it. Technical difficulties made their presence felt in almost every case. Both the computer and the informatic tool make lots of practical demands that will have to be dealt with. Interestingly, the way and the extent to which this was done differed quite a lot from teacher to teacher.

The most significant difference however, among the case studies was in the way the teachers handled the tension between subject matter and computer experience. Some teachers were clearly interested in using the computer to deliver the science content, some focused more on the experience the pupils would have with the informatic tool, and others were somewhere in between. Overall, we can say that the use of computers in the science classroom still stands out, that is, it is not fully naturalised or blended into the scenery. It has not stopped being an option for the teacher, who consequently has to have reasons to justify it. Still, however, it is not completely innovatory either, and there seems to be an increasing pressure for teachers to incorporate it in their teaching.

6. FACTORS INFLUENCING TRANSFORMATION

The research into teachers' transformations of the use of informatic tools conducted within the STTIS framework (Stylianidou *et al*, 2000), which included the above study, identified a variety of factors, which influence teachers' decisions about what they do in their classrooms, and thus which affect the implementation of an innovation. These can be usefully summarised under the following four headings:

- *Content* – includes the content of the proposed innovation and the content of the existing curriculum, and the perception of the teacher about how the new relates to the old.
- *Beliefs about learning* – include what teachers think that they ought to be doing to support pupils in the classroom, what pupils find easy or difficult and why, and the role and nature of motivation.
- *Values* – include what teachers believe about the nature of their subject, about the purposes of education, about their own role as a teacher, and so on.
- *Contexts, customs and constraints* – includes a wide range from local factors such as classroom layout or the availability of resources, to more global factors such as prescriptions laid down by government, as well as teachers' knowledge of the subject they are teaching, their repertoire of pedagogic strategies, and so on.

7. DEVELOPING THE TRAINING MATERIALS

Having identified the above factors, we proceeded in developing training materials to address them. The approach was to incorporate the teacher case studies into activities that aim to train the teachers to consider the implications of the innovation in their own context. The materials therefore use an approach in which there is an initial and final training session, with a period of some weeks in-between in which teachers can use the ideas in their own practice.

The first session is designed to help teachers to plan and teach a trial lesson using a computer simulation or model. The structure of this session is:

– Exploring current practice

First teachers are asked to review how computer simulations and models are used in their existing science curriculum and to evaluate their existing use of them.

– Learning about the innovation

Then they try out themselves pupil modelling activities of increasing complexity, in which spreadsheets are used for modelling in two contexts – 'Electricity' and 'Forces and Motion'.

– Research into transformations

In the next set of activities they are asked to compare different ways in which teachers may transform the use of computer simulations. These are presented to them in the form of 'stories' made up to illustrate the factors identified in the research findings as affecting teachers' transformations. The teachers are asked to consider these issues in the light of their own teaching experience.

– Planning

Finally, they are asked to plan a trial lesson drawing on the new ideas introduced in the workshop. The second session is designed to help teachers to build on the experience of the trial lesson using computer models in order to plan more extended sequences of lessons within the schemes of work. As part of the session:

– Evaluation

The teachers are asked to report about their experiences in using computer simulation in the classroom and to evaluate these experiences taking account of the factors that affect the transformations of innovations.

– Relating to research case studies

They are then given examples of the teacher case studies to read and relate to the general findings of the research.

– Innovation and transformation

Finally, they are asked to look back on the experiences that they had in the two workshops and the trial lesson, and reflect on what they learned and how their ideas and beliefs have changed (or not). They are also asked to look to the future and consider how to begin to use the ideas in longer term planning.

8. CONCLUSIONS AND IMPLICATIONS

It has been argued that while innovation-focused training is essential, by itself it is too limited, and development needs to take into account the teacher as a person, as well as the purposes and the contexts in which s/he works (Fullan and Hargreaves, 1992). Our approach focuses not just on the innovation but on teachers' beliefs and the contexts in which they work, and recognises that it takes time for teachers to change their practice.

The training materials developed are modest in scope, and can have only limited impact on changing the factors identified in the research. What can be done with these materials, however, is to make teachers aware of some ways in which the curriculum is transformed in implementation, and how these factors affect the transformations that teachers make. By making these factors explicit, teachers can make informed choices in their implementation of new ideas.

There is increasing emphasis on teaching as a research-based profession, but there are difficulties in creating a culture in which communities of researchers and teachers can create a shared body of knowledge (White, 1998). This paper has discussed one possible way in which research findings can be made accessible to teachers involved in curriculum innovation.

REFERENCES

Boohan, R., Stylianidou, F. & Ogborn, J., (2000). *Training teachers for innovation: computer simulation and modelling.* STTIS, UK National Report on WP5 (part 2), EC TSER: CT97 2020 (coord: R. Pintò, Barcelona), Univ. of Sussex, UK.

Brosnan, T., (1994). Using Spreadsheets to Develop Understanding in Science. In: Mellar, H., Bliss, J., Boohan, R., Ogborn, J., Tompsett, C. (eds) *Learning with Artificial Worlds: Computer Based Modelling in the Curriculum (pp. 76-84).* London: The Falmer Press.

Feurzeig, W., & Roberts, N., (1999), (eds). *Modeling and simulation in science and mathematics education.* New York: Springer-Verlag.

Fullan, M., & Hargreaves, A. (1992), (eds). *Teacher development and educational change.* London: The Falmer Press.

Mellar, H., Bliss, J., Boohan, R., Ogborn, J., Tompsett, C., (eds) (1994). *Learning with Artificial Worlds: Computer Based Modelling in the Curriculum.* London: The Falmer Press.

STTIS, (1998). *Outline and justification of research methodology: WP1, WP2 and WP3*, STTIS Report, EC TSER: CT97 2020 (coord: R. Pintó, Barcelona). Synthesis: Ogborn, J. and Stylianidou, F., Univ. of Sussex, UK.

Stylianidou, F. & Ogborn, J., (1999). *Teachers using computer modelling and simulation in the science classroom: the English case*, STTIS UK National Report on WP1.2, EC TSER: CT97 2020 (coord: R. Pintó, Barcelona). Univ. of Sussex, UK.

Stylianidou, F., Ogborn, J., Andresen, O., Balzano, E., Giberti, G., Gutiérrez, R., Kolstø, S. D., Monroy, G., Pérez, O., Pintó, R., Quale, A., Rebmann, G., Sassi, E., (2000). *The nature of use by science teachers of informatic tools.* STTIS Report RW1.2. EC TSER: CT97 2020 (coord: R. Pintò, Barcelona). Univ. of Sussex, UK.

White, R., (1998). Research, theories of learning, principles of teaching and classroom practice: examples and issues. *Studies in Science Education,* 31, 55-70.

PART V

Science Teachers: Knowledge and Practices

EXPLORING SCIENCE TEACHERS' PEDAGOGICAL CONTENT KNOWLEDGE

Onno De Jong

Utrecht University, The Netherlands

ABSTRACT

In this keynote, the development of pedagogical content knowledge (PCK) within (prospective) science teachers is explored and discussed. First of all, the meaning of the term 'pedagogical content knowledge' is examined and some core elements of this concept are described. Secondly, several methods and techniques to explore PCK are reported. Thirdly, important sources for the development of PCK are described. Finally, a naturalistic case study of the development of PCK within a group of prospective science teachers (all M. Sc.) is presented. The group has taken a post-graduate course module, incl. some teaching practice, on a central issue in science education: teaching models. The results indicate that the prospective teachers' knowledge of students' conceptions of models was promoted. Besides, the prospective teachers became aware of their own limited knowledge of appropriate models, and ways to teach this issue. This absence of this knowledge can have caused the observed discrepancy between their teaching intentions (models as constructs) and their teaching practice (models as facts). Implications for (prospective) science teacher courses are discussed.

1. INTRODUCTION

In recent years, there has been a growing interest in the knowledge base of science teachers. In the field of science teacher preparation and inservice training, the interest in the required knowledge base of teachers is particularly promoted by the awareness that the study of an academic discipline may not provide them with the kind of understanding science teachers need to effectively transform their academic knowledge into instructional activities in the classroom (Sanford, 1988). Shulman (1986) has introduced the concept of pedagogical content knowledge (PCK) to acknowledge the importance of the transformation of subject matter knowledge per se (also called: content knowledge; CK) into "subject matter knowledge for teaching" (Shulman, 1986, p. 9).

Research on pedagogical content knowledge (PCK) may contribute to resolving the 'blind spot' that results from a relative lack of research focusing on the content of the lessons taught. In the present lecture, science teachers' PCK and its development is explored and discussed. An important aim is to gain a better understanding of (the development of) PCK. Another aim is to contribute to the design of courses for (prospective) science teachers.

D. Psillos et al. (eds.), Science Education Research in the Knowledge-Based Society, 373–381.

2. CONCEPTUALIZATION OF PCK

In the last decade, a growing number of studies on PCK have been published (c.f. Van Driel et al., 2001; Gess-Newsome and Lederman, 1999; De Jong et al., 1998). Various scholars, elaborating on Shulman's work, have proposed different conceptualizations of PCK, in terms of the features they include or integrate. For example, Magnusson et al. (1999) presented a model of the relationship among the following constituent domains of PCK: subject matter knowledge (e.g. substantive knowledge, syntactic knowledge), pedagogical knowledge (e.g. knowledge of educational aims, knowledge of classroom management), and context knowledge (e.g. knowledge of specific student and school characteristics). However, other scholars, for example Barnett and Hodson, (2001), consider PCK as a constituent of pedagogical context knowledge, together with other components: academic knowledge, classroom knowledge, and professional knowledge. Gess-Newsome (1999) categorized views about PCK as either integrative or transformative. The former view holds that knowledge of teaching is merely the interaction between other forms of teacher knowledge (PCK as a 'mixture'). The latter view stands in contrast by declaring that other forms of teacher knowledge are transformed into a new form of knowledge (PCK as a 'compound'). Both views can be considered as the two ends of the PCK spectrum.

Although these conceptualizations of PCK may be rather confusing, yet it seems that the following elements are central in any conceptualization of PCK with respect of a specified content area:

- Knowledge of representations of subject matter for teaching,
- Knowledge of relevant instructional strategies,
- Knowledge of specific student conceptions,
- Knowledge of student learning difficulties.

Obviously, these elements are intertwined and should be used in a flexible manner. There appears to be agreement on the nature of PCK. First, since PCK refers to particular topics, it is to be discerned from knowledge of pedagogy, of educational purposes, and of learner characteristics in a general sense. Secondly, because PCK concerns the teaching of particular topics, it may turn out to differ considerably from CK. Finally, many scholars suggest that PCK is developed through an integrative process rooted in classroom practice, and that PCK guides the teachers' actions when dealing with subject matter in the classroom.

3. METHODS AND TECHNIQUES TO STUDY PCK

An overview of the methods and techniques, that can be used to study PCK in any context, is given by Baxter and Lederman (1999). They introduce the following classification for the research tools that are used:

- Convergent and inferential techniques, such as Likert-type self-report scales, multiple-choice items and short answer formats,
- Concept mapping, card sorts and pictorial representations,

- Multi-method evaluations with triangulation, including combinations of research tools such as interviews, teacher's log and video-prompted recall.

Convergent and inferential techniques could be used for identifying knowledge of student conceptions and learning difficulties. Concept mapping, cards sorts and pictorial representations may be suitable for investigating content knowledge for teaching, while multi-method approaches may be appropriate for exploring knowledge of relevant instructional strategies.

Investigating PCK is not very easy because PCK is, to a large degree, unarticulated and tacit in nature. In their lessons, teachers have to make many decisions in a split of a second. These actions are guided by 'split-second' thoughts, but these thought are rather unconscious. For that reason, many teachers experience difficulties in explaining what they are doing or what they know. Some research tools are very appropriate to take the 'tacit' character of PCK into account, such as the stimulated recall interview, the lesson preparation method, and the story-line method.

In a *stimulated recall interview* (Calderhead, 1981), teachers explicate their cognitions while they are looking at a videotape of a lesson that has been given shortly before. This interview technique can be considered as a substitute for the 'thinking-aloud' technique, which is not very appropriate because of the interrupting effect in the classroom situation. The videotape is used as a tool for teachers to remember what he or she thought during the lesson and to stimulate the teacher to experience their lesson again.

Meijer (1999) has indicated that the technique of stimulated recall interview can function as a fruitful tool in the context of teacher preparation. Prospective teachers can use the technique to explore the PCK of their mentor at their teaching practice school. She found that prospective teachers very much appreciate the opportunity to gain access to the thinking of expert classroom teachers in this way. For example, from course workshops, prospective teachers learn the rule to attract attention by asking questions to a great variety of students because this questioning approach is more effective than asking questions to a single student or a small group of students. When the prospective teachers observe their mentors in the classroom, they see that these experienced teachers spread out their questions in all directions. It looks if these teachers apply this questioning approach as a rule. However, by stimulated recall interviewing, the prospective teachers found that their mentors have specific reasons to ask a particular question to a particular student. The decision to select a particular student is influenced by their knowledge of the competences and personal characteristics of the student under consideration. In conclusion, the technique of stimulating recall interview is a fruitful tool for making explicit the thinking underlying expert teaching practice.

The initial PCK of teachers can be explored in the context of *preparing lessons*. As research has shown, the content to be taught is the most important issue in the process of short-term planning of new instructional units (Sanchez and Valcarcel, 1999). Teacher educators can investigate the initial PCK of their students by making use of the lesson preparation method. This method is described by Van der Valk and Broekman (1999) and adapted by De Jong (2000).

The adapted version consists of two main parts. The first part includes an individual lesson preparation task: prospective teachers are invited individually to prepare 'introductory' lessons about a specified topic. The lessons (of a given duration) have to be aimed at a class of certain age and level. The prospective teachers are not allowed to consult any textbook and they are told to assume that a number of specified curriculum topics had been taught beforehand. The second part includes a group meeting interview & discussion: prospective teachers are invited to explain their written lesson plans and to lead a plenary discussion of them. In the context of this discussion, prospective teachers have to answer a number of specific open questions, for example, about their knowledge of the curriculum topic. In the group meeting setting, the teacher trainer as well as the prospective teachers receive almost instantaneous information and this can be discussed immediately in order to develop promising teaching strategies. In conclusion, the adapted lesson preparation method provides an opportunity to teacher educators to act as researchers in improving their teaching practice. Moreover, this method allows prospective teachers to become aware of their own initial PCK.

The *story-line method* (Gergen, 1988) provides information about teachers' subjective evaluation of their experiences regarding particular aspects of teaching. By drawing a story-line, teachers describe their evaluation, on a scale from positive to negative, as a function of time. They are also asked to clarify the main ups and downs in their story-line. Beijaard et al. (1999) have reported about this method, and they concluded that the story-line method can be used efficiently.

4. SOURCES FOR DEVELOPMENT OF PCK

Within the domain of science teaching, several studies have been performed on the development of science teachers' PCK, in the context of both preservice and inservice teacher education. From these studies, the main sources for PCK development can be described as follows:

Knowledge of subject matter. The CK that prospective teachers have acquired during disciplinary education usually contains alternative conceptions and deficiences (Smith, 1999). For instance, Gess-Newsome and Lederman (1993) noted that the subject matter structures of prospective biology teachers, who had nearly completed the requirements for a B.S. in Science Education, was often vague and fragmented at the start of their teacher education program. During this program, the prospective teachers developed more coherent and integrated subject matter structures.

Teaching experience with respect to specific topics. According to Lederman et al. (1994), the development of PCK among prospective science teachers is promoted by the constant use of CK in teaching situations. Initially, prospective teachers separate CK from general pedagogical knowledge. As a result of teaching experiences however, these types of knowledge are being integrated.

Knowledge of students' conceptions and learning difficulties. By getting acquainted with the specific conceptions and ways students reason, prospective teachers may start to restructure their CK into a form that enables productive communication with their students (Lederman et al., 1994). In addition to field-based experiences, prospective teachers may benefit from studying students' preconceptions with respect to a specific topic during teacher education courses, and comparing and discussing these preconceptions in relation to their own conceptions (Geddis, 1993). Such activities may stimulate prospective teachers to generate transformations of CK and topic specific teaching strategies.

Participating in specific workshops. Clermont et al. (1993) have studied the effects of a short, intensive workshop on specific teaching strategies (i.e., chemical demonstrations). They found that the PCK of prospective science teachers participating in this workshop developed towards that of expert teachers. On the other hand, Adams and Krockover (1997) found that workshops can have a negative effect because they can stimulate prospective teachers to copy conventional instructional strategies, stressing procedures rather than student understanding.

Comparing these factors with the sources for PCK development described by Grossman (1990), we may suggest that (a) disciplinary education, naturally, constitutes the basis for CK, (b) observation of classes may promote the knowledge of students' conceptions, (c) classroom teaching experiences may stimulate the integration of CK and PCK, thus contributing to the development of PCK, and (d) specific courses or workshops during teacher education have the potential to affect PCK. It will be clear that the given description of sources for development of PCK reflects an analytical view on PCK, based on the four core elements that are mentioned before.

5. A STUDY OF DEVELOPING PROSPECTIVE SCIENCE TEACHERS' PCK

5.1 Research context and research question

A common research project was launched at Utrecht University (O. De Jong) and Leiden University (J. Van Driel), The Netherlands. The research aims to explore the development of PCK of a central topic in science education, viz. the teaching and learning of scientific models.

Many scholars have conducted studies on the teaching and learning of the content of specific models, for instance, studies on corpuscular models, in which learning difficulties with respect to understanding molecules and atoms are investigated (e.g., De Vos and Verdonk, 1987; Harrison and Treagust, 1996). Grosslight et al., (1991) have reported on students' conceptions of models and their use in science. They describe the following levels of understanding:

Level 1: In this level of understanding, models are thought of as either toys or simple copies of reality.

Level 2: The idea is included that models are designed for particular purposes, especially to help communication, however, the main focus is still on the models and the reality modelled, not the ideas portrayed per se.

Level 3: This level of understanding is characterized by the belief that models are constructed in the service of developing and testing ideas and explanations about phenomena.

From their research, Grosslight et al., (1991) have found that middle and high school students' conceptions of models in science can be considered as level 1 and/or level 2 understanding, while experts reason at level 3.

Up till now, little is known about teaching activities concerning models, which are actually applied by science teachers, and for what reasons. It seems that there have been no investigations of prospective teachers' knowledge base in this domain. For that reason, the following research question was addressed: What changes in prospective teachers' PCK of models in chemical education can be identified?

5.2 Design and procedure

The present study was situated in the context of a post-graduate teacher education program (one-year), qualifying for the teaching of chemistry at pre-university level (especially for grades 10-12 of secondary education. It was decided to develop a new course module as a part of the teacher education program, which is aimed at focusing specific attention to the teaching of scientific models. This module was taught about halfway the teacher education program over a period of about ten weeks. A group of 12 prospective science teachers (all M. Sc.) was involved in this course. In order to monitor the development of their PCK, a multi-method approach was chosen (cf. Baxter and Lederman, 1999). Data were collected at specific moments:

- In the beginning of the module, all prospective teachers responded to a written questionnaire, aimed at exploring their initial PCK of models.
- During some workshops, the participants discussed articles on models and the teaching of these issues, and expressed teaching intentions.
- After teaching a chosen curriculum topic focusing on models at their teaching practice school, they reflected on their teaching experiences by writing individual reports.
- During a final workshop, these reports were discussed.

All workshop sessions were recorded on audiotape and subsequently transcribed. These data, together with the other collected data, were independently read and analyzed in an iterative way by the two researchers. The results were compared and discussions were focused on consensus.

5.3 Results of the study

Two main categories of changes that occurred in the prospective teachers' PCK could be identified. First, changes in the knowledge of students' difficulties associated with the learning of models. Second, changes in the knowledge of teaching activities aimed at promoting students' understanding of models.

Changes in knowledge of difficulties associated with the learning of models. The pre-teaching responses could be classified into three categories of student learning difficulties, viz. abstract nature of atomic models, the relationship between models and phenomena, and the process of modelling. The post-teaching responses could be classified into the same categories, but an increased number of prospective teachers indicated learning difficulties. Moreover, there responses were more explicit, more detailed and more precise.

Obviously, the prospective teachers' responses were related to the topic they had focused on during teaching practice. For instance, a prospective teacher described in great detail the conceptions of some of his students about the relationship between molecular structure and the boiling point of the corresponding substance. Also, he suggested explanations for these conceptions, such as: "Because the students had used 3D models, they were able to compare the properties of the molecules".

Another prospective teacher gave a detailed description of a discussion between two of his students about a molecular explanation for the vaporization of water. One student reasoned that during vaporization, water molecules would fall apart in separate hydrogen and oxygen atoms, whereas the other student argued that the water molecules remained intact.

Changes in knowledge of teaching activities aimed at promoting students' understanding of models. The pre-teaching intentions could be classified into four categories, viz. discussing models as man-made inventions, addressing the relationship between models and phenomena, addressing multiple models, and inviting students to build models. The post-teaching reflections could be classified into the same categories, but were more rich and detailed. An increased number of prospective teachers focused on one category only, viz. addressing the relationship between models and phenomena. A decreased number of prospective teachers focused on the three other categories.

Discrepancies between the pre-teaching responses and the post-teaching responses were found among the majority of prospective teachers. These prospective teachers intended to pay attention to models as constructs, invented by scientists, but in their teaching practice, they addressed models as objects or facts that are given.

Some prospective teachers also formulated new teaching intentions. Nearly all of these intentions referred to the usefulness of models and modelling for teaching and understanding. For example, a prospective teacher concluded that in the future he would try to visualize as many concepts as possible using models and audiovisual aids.

6. CONCLUSION AND DISCUSSION

Despite the fact that all prospective teachers had a M. Sc. degree, their knowledge of models and modelling was not very pronounced. Some inconsistencies in their knowledge could be indicated. For example, some prospective teachers wrote that a model represents a target that cannot be observed or measured directly. However, nearly all of them categorized a car toy as a model too. The *prospective* teachers were not aware of these inconsistencies. This may be caused by the partly 'tacit' or implicit character of their knowledge.

By teaching topics that include models, the prospective teachers had promoted their awareness of student learning difficulties. However, despite intentions to pay attention to models as constructs, most of them focused in their teaching on models as objects or facts that are given. This kind of discrepancy is also found at experienced teachers (Koulaidis and Ogborn, 1989). It may be caused by absence of knowledge of appropriate strategies for teaching models as man-made inventions.

The results of the present study can be used as a starting point for the revision of the module. In this module, much more attention should be paid to *prospective* teachers' own initial PCK of models and modelling, for example in the context of the preparation of lessons including the nature and role of models and their use. It will be also important to pay attention to the presentation of models to the students and stressed the need for the teacher to present a model carefully, for example by developing it through posing scientific problems, and discuss relevant and irrelevant attributes of the target under consideration. (cf. Harrison and Treagust , 2000).

Regarding the professional development of science teachers, the following implications can be addressed:

- It is important that teachers try to become aware much more about their implicit initial PCK of specific science issues and topics, for example in the context of the preparation of lessons.
- Own classroom teaching experiences have a strong impact on the development of PCK. For that reason, it is important to ask teachers to reflect on their PCK after teaching.
- The mentor (or colleagues) at school can also function as an important source for PCK development. Prospective teachers could be stimulated to inquire their mentor's PCK of specific science issues and topics. Experiences teachers should be stimulated to explore the PCK of their colleagues at school.

ACKNOWLEDGEMENT

I wish to thank my colleague Jan Van Driel, Leiden University, The Netherlands, for the fruitful cooperation and stimulating discussions.

REFERENCES

Adams, P.E. & Krockover, G.H. (1997). Beginning science teacher cognition and its origins in the preservice secondary science teacher program. *Journal of Research in Science Teaching,* 34, 633-653.

Barnett, J. & Hodson, D. (2001). Pedagogical context knowledge: toward a fuller understanding of what good science teachers know. *Science Education*, 85, 426-453.

Baxter J. A. & Lederman N. G. (1999). Assessment and measurement of pedagogical content knowledge. In: J. Gess-Newsome & N. G. Lederman (Eds.), *Examining pedagogical content knowledge* (pp. 147-161). Dordrecht: Kluwer Academic Publishers.

Beijaard, D., Van Driel, J. & Verloop, N. (1999). Evaluation of story-line methodology in research on teachers'practical knowledge. *Studies in Educational Evaluation*, 25, 47-62.

Calderhead, J. (1981). Stimulated recall: a method for research on teaching. *British Journal of Educational Psychology,* 51, 221-217.

Clermont, C.P., Krajcik, J.S., & Borko, H. (1993). The influence of an intensive in-service workshop on pedagogical content knowledge growth among novice chemical demonstrators. *Journal of Research in Science Teaching*, 30, 21-43.

De Jong, O., Korthagen, F. & Wubbels, Th. (1998). Research on science education in Europe: teacher thinking and conceptual change. In: B. J. Frazer & K. G. Tobin (Eds.), *International Handbook of Science Education* (pp. 745-758). Dordrecht: Kluwer Academic Publishers.

De Jong, O. (2000). The teacher trainer as researcher: exploring the initial pedagogical content concerns of prospective science teachers. *European Journal of Teacher Education*, 23, 127-137.

De Vos, W., & Verdonk, A.H. (1987). A new road to reactions, part 4: The substance and its molecules. *Journal of Chemical Education*, 64, 692-694.

Geddis, A.N. (1993). Transforming subject-matter knowledge: The role of pedagogical content knowledge in learning to reflect on teaching. *International Journal of Science Education*, 15, 673-683.

Gergen, M. M. (1988). Narrative structures in social explanations. In: C. Antaki (Ed.). *Analysing Social Explanations* (pp. 94-112). London: Sage.

Gess-Newsome, J. (1999). Pedagogical content knowledge: an introduction and orientation. In: Gess-Newsome, J. & Lederman, N. G. (Eds.) (1999), *Examining Pedagogical Content Knowledge* (pp. 3-17). Dordrecht: Kluwer Academic Publishers.

Gess-Newsome, J., & Lederman, N.G. (1993). Preservice Biology teachers' knowledge structures as a function of professional teacher education: A year-long assessment. *Science Education*, 77, 25-45.

Gess-Newsome, J. & Lederman, N. G. (Eds.) (1999), *Examining Pedagogical Content Knowledge.* Dordrecht: Kluwer Academic Publishers.

Grosslight, L., Unger, C. & Jay, E. (1991). Understanding models and their use in science: conceptions of middle and high school students and experts. *Journal of Research in Science Teaching*, 28, 799-822.

Grossman, P. L. (1990*). The making of a Teacher: Teacher Knowledge and Teacher Education.* New York/London: Teachers College Press.

Harrison, A.J., & Treagust, D.F. (1996). Secondary students' models of atoms and molecules: Implications for teaching chemistry. *Science Education*, 80, 509-534.

Harrison, A.J., & Treagust, D.F. (2000). A typology of school science models. *International Journal of Science Education*, 22, 1011-1027.

Koulaidis V. & Ogborn, J. (1989). Philosophy of science: an empirical study of teachers' views. *International Journal of Science Education*, 11, 173-184.

Lederman, N.G., Gess-Newsome, J., & Latz, M.S. (1994). The nature and development of preservice science teachers' conceptions of subject matter and pedagogy. *Journal of Research in Science Teaching*, 31, 129-146.

Magnusson, S., Krajcik, J., & Borko, H. (1999). Nature, sources and development of pedagogical content knowledge. In: J. Gess-Newsome & N. G. Lederman (Eds.*), Examining pedagogical content knowledge* (pp. 95-132). Dordrecht: Kluwer Academic Publishers.

Meijer, P. C. (1999*). Teachers' practical knowledge: teaching reading comprehension in secondary education.* Ph. D. Thesis. Leiden University, The Netherlands.

Sanchez, G. & Valcarcel, M. V. (1999). Science teachers' views and practices in planning for teaching. *Journal of Research in Science Teaching*, 36, 493-514.

Sanford, J. P. (1988). Learning on the job: conditions for professional development of beginning science teachers. *Science Education*, 72, 615-624.

Shulman, L.S. (1986). Those who understand: knowledge growth in teaching. *Educational Researcher*, 15, 4-14.

Smith, D. C. (1999). Changing our teaching: the role of pedagogical content knowledge in elementary science. In Gess-Newsome, J., & Lederman, N.G. (Eds.), *Examining pedagogical content knowledge* (pp. 163-197). Dordrecht, The Netherlands: Kluwer Academic Publishers.

Van Der Valk, A. E. & Broekman, H. G. B. (1999). The lesson preparation method: a way of investigating pre-service teachers' pedagogical content knowledge. *European Journal of Teacher Education*, 22, 11-22.

Van Driel, J., Beijaard, D. & Verloop, N. (2001). Professional development and reform in science education: the role of teachers' practical knowledge. *Journal of Research in Science Teaching,* 38, 137-159.

RELATING RESEARCH IN DIDACTICS AND ACTUAL TEACHING PRACTICE: IMPACT AND VIRTUES OF CRITICAL DETAILS

Laurence Viennot

LDSP University Paris 7

ABSTRACT

This address is focused on the actual impact of research-based suggestions on teacher practice. It envisages the transformations that teachers are likely to effectuate on teaching strategies that they are led to implement. It is argued that some decisive elements for a resonant interaction between teachers and researchers are situated at the micro-level of "critical details". The designing of a sequence, the way it is evaluated and the corresponding training are discussed from this viewpoint.

1. INTRODUCTION

"Why is the impact of research on practice apparently so light?" (Millar 2000). It might be that what is often called the products of research, typically research-based teaching sequences, do not keep their promises, i.e. do not foster the learning processes they are intended to. But, most probably, the gap between research and practice can also be attributed to a non-resonant implementation of such "products" by the teachers. Communication means in any case that the message is reinterpreted, i.e. transformed, by the receiver. Teachers indeed have been found to share some determining transforming trends.

It has been argued that a good means to fill the gap between research and practice could be to emphasize more strongly the theoretical bases of the "products": "The way research influences practice in education is not through discovery of a detailed and specific mode of teaching but through substantiation of principles which pervade thinking about teaching and learning" (Gunstone & White 2000).We will discuss this statement in the light of what has been observed in teacher practice when these are supposed to implement an innovative sequence. But even upstream in the story of such a "product" – a research-based teaching sequence - some forks may occur. As Lijnse (2001) already said, speaking of general learning theories: "…on application such theories only result at best on heuristic rules." An example of diverging suggestions made by researchers sharing the same constructivist "principles" - i.e. educational constructivism, see Ogborn (1997), Gil-Perez (2001) - concerns what is often called "the materialised ray". As will be seen in the next section, letting think that a rectilinear series of lighted points represents a ray of light is a very questionable strategy, all the more so if it is used at the outset

D. Psillos et al. (eds.), Science Education Research in the Knowledge-Based Society, 383–393.

of a sequence on light and vision. The use of such a strategy is advised without any particular *caveat* by some authors (Driver *et al.* 1994) whereas it is presented as needing much caution by others (Kaminski 1989, Hirn & Viennot 2000, Colin *et al.* 2002), a fact which indicates clearly that other factors than general learning theories are at stake. Among these factors, I would like to stress the consideration of small aspects of practice, in relation with general principles, as well as with the content and the corresponding learners' common approaches. Whenever such details are likely to have important learning effects, they deserve the label of "critical details".

First, a few examples of teacher transforming trends will be analysed in relation with some critical aspects of practice. Then, the impact of such details on the links between research and practice will be discussed, this at three levels: design of teaching sequences, evaluation of such products, teacher training.

2. TEACHER TRANSFORMING TRENDS: SOME CRITICAL DETAILS

In order to anticipate how teachers in normal institutional contexts are likely to react to an innovative research-based teaching sequence, it is of crucial importance to analyse the main transforming trends that may intervene in such a communication process. There has been a great many studies concerning curriculum implementation and teacher training (see for instance Gil-Perez & Pessoa de Carvalho 1998) which show in particular the problems linked to a deficient knowledge of the content, or to inappropriate views of Science and of teaching/learning processes. Some recent studies (Hirn & Viennot 2000, Hirn 1995,1998, Viennot *et al.* 1999) provide more specific information about the ways teachers interpret the messages received more or less directly from research; they have shed light on very precise aspects of teacher strategies concerning the teaching of particular science contents.

In particular, the transforming action of teachers often takes one or several of the following aspects, among others.

2.1 Fragmentation

The links between different concepts, or between different aspects of an activity proposed to pupils, are commonly neglected (Hirn & Viennot 2000, Hirn 1995, 1998, Viennot *et al.* 1999, Pinto *et al.* 2000). This suggests that the global rationale is lost sight of, as if teaching meant only piling up unconnected episodes.

For instance in a syllabus for grade 8 launched in France in 1993, it was recommended to introduce rectilinear propagation via impacts of light on screens and the corresponding shadows, then to punch some holes in various areas of such screens so as to analyse, observe and discuss what part of the source is visible from behind each hole. The conceptors' idea was to treat these two kinds of phenomena as explicitly distinct, starting from the one which is the most easy to analyse (Viennot & Chauvet 1997) then using it as a building block to introduce the non trivial rule: some light from the seen object necessarily enters the observers' eye.

In her study about teachers' transforming trends, Hirn (1995,1998; Hirn & Viennot 2000) analysed some assessment tasks used by the consulted teachers. One of these tasks is about an extended source, an extended obstacle and a screen (fig. 1): the pupils are asked to describe the lighting of the different areas of the screen and to determine their limits; then, to predict what part of the source will be seen from behind three holes punched in areas of various lighting. In each case, they have to draw rectilinear lines – limits of zones - to answer correctly. What is missing for this task to be in line with the innovators' intentions, is a question about what *links* the two phenomena analysed: lighting and vision. From behind a fully lighted zone, where some light is received from the whole source, this source can be entirely seen, whereas from a shadowed area, it is only partly visible. Neglecting an opportunity to discuss this type of relationship is not an innocuous attitude, according to us. But it is very frequent.

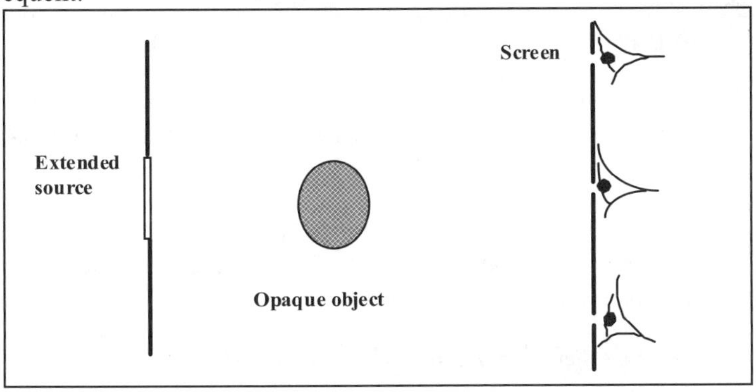

Figure 1. An experimental set-up for the questions: what is the lighting of the different areas of the screen (before the holes are punched); what will you see through each hole ; how are these phenomena related ?

2.2 *Seeing is understanding*

A common and often implicit view is that seeing is enough to understand. One can expect that this well-known trend will act as a determining factor in teacher practice. Among the possible outcomes of this empiricist position, in terms of teaching strategies, not only does this view exclude learning cycles of the type "prediction-experiment-debate", but it can lead to outright inconsistency and also reinforce previous naïve ideas. For example, in order to introduce the rectilinear propagation of light, a frequent strategy is to show a ray of light "materialised" by a sheet of paper – in fact, a rectilinear series of lighted points. Figure 2 shows the kind of situation that demystifies the demonstrative value of such a device. A lamp behind a slit may well generate a rectilinear track that just stops at the foot of a mirror, supposedly in charge of "reflecting" what is presented as a "ray": if this lamp is situated higher than usual, a rectilinear path of light that passes at the top of the slit may reach the horizontal support of the mirror just at its foot (fig. 2a), or slightly further (fig. 2b), which corresponds to a seemingly incongruous stop for the luminous track observed. But, beyond this tricky situation, the main point is that, in such a setup, the source is not in line with the so-called "ray", which constitutes a patent example of inconsistency. In addition, this strategy is likely to reinforce the idea that light is visible *per se*, from the side (fig. 2).

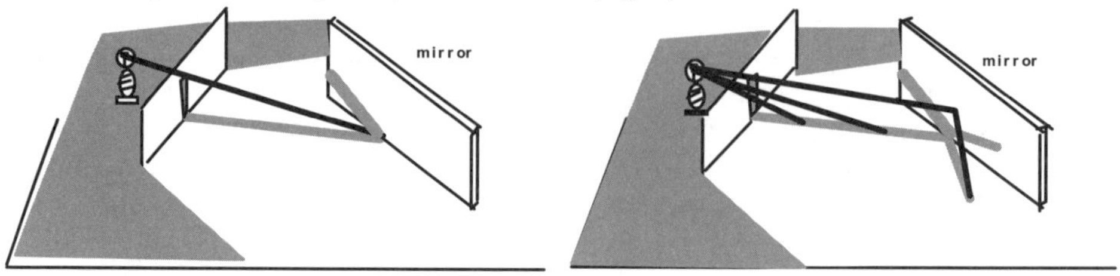

Figure 2. What is seen on the horizontal support is only a series of lighted points: two variants of what appears as "a defective mirror". In black: rays of light; in grey: lighted points and their images (Kaminski 1989)

It is difficult to find a better example of a detail of practice that, if not cautiously accompanied by sound explanations, may lead to the idea that interpreting hastily what is seen prevails on mere conceptual coherence.

2.3 Exclusive linkage between a material device and a concept

According to the same logic, a given set-up is understood as emblematic of a given phenomenon, and reciprocally, a given item of the content matter is firmly attached to a demonstrative set-up (Hirn & Viennot 2000, Hirn 1995, 1998, Viennot *et al.* 1999). A classical demonstrative set-up being itself often linked to particular ways of teaching, this results in two one-to-one relationships: one set-up/one phenomenon, and one set-up/one strategy.

For instance, during a class session observed by Hirn (ibid.), a teacher strongly suggested that a filter could only transmit light, not diffuse it. In such cases, a pupil might seriously wonder how such a filter may be visible without diffusing some light, all the more so when the geometrical condition for vision – briefly put: light in the eye – is a major conceptual goal of the sequence.

Similarly, any work about imaging is thought to require an optical bench, teaching colour synthesis is held to mean using a Newton disk (Chauvet 1994, 1996). When inconsistency, and/or the reinforcement of naïve ideas, is the price to pay, it is important to be able to analyse these risks in detail.

2.4 Using images in teaching: non problematic?

Among the aspects of teaching practice that deserve much attention, according to us (Pinto *et al.* 2000, Kress & van Leuwen 1996, Colin 1997, 1999), the role of images is widely underestimated by teachers. For example, this time at university level, figure 3 outlines a type of diagram very commonly used to present Young's experiment. In fact several problems may be raised. First, only a path of light is drawn after each hole, which might suggest that no other path of light could be considered, whereas as soon as another arrival point on the screen is considered, one must draw other lines "emerging" from each hole. Second, the same symbol – an arrow - is put on all the drawn paths of light, before and after the holes. However, the light starting from the primary source can be represented by rays of light, themselves appropriate to indicate the routes followed by *energy*. By contrast, the lines drawn beyond the holes only represent some paths useful to calculate the *phases* of the two superposed waves.

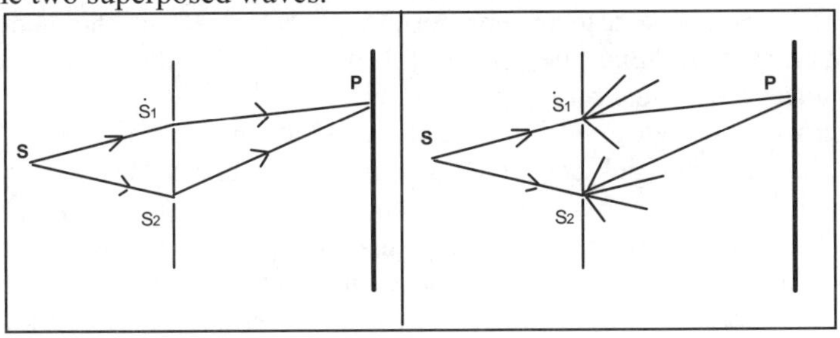

Figure 3. Young's experiment: a) a common diagram; b) a suggestion to avoid students' comments such as "light is deviated"

This important distinction is not mirrored by any difference in the symbols used. These two "details" concur to reinforce a story-like reading of the image, as if a unique hero – a ray – started from the source to arrive somewhere on the screen. No wonder if some comments given by students confronted to this image are clearly in terms of geometrical optics: "light is deviated"(Colin *et al.* 2002, Pinto *et al.* 2000, Colin 1997, 1999).

Such results, and others not reported here (Viennot *et al.* 1999, Pinto *et al.* 2000) do not merely confirm that implementing an innovative sequence is not a straightforward process. They throw a very crude light on the fact that the critical details of practice that might efficiently convey the didactic intentions of a sequence are the most difficult for teachers to take up. Yet, if neglected, these "details" may ruin the message (Hirn & Viennot 2000, Hirn 1995, Viennot *et al.* 1999).

3. INTERACTION BETWEEN RESEARCH AND TEACHING PRACTICE: THE ROLE OF CRITICAL DETAILS

The potential importance of such apparently small aspects of teaching practice, and the need to reinforce the links between research and practice lead to re-examining three main aspects of these two fields.

3.1 Fine-grained design of teaching sequences

Some authors (Millar 1989, Lijnse 1994) have long been calling attention on the importance of designing teaching sequences in detail. These claims contributed to take some distance from views on the teaching-learning processes focused on two aspects only: pupils common ideas and situations thought of as appropriate to engender a conceptual change. What might be considered, in line with these founding claims, concerns the knowledge we have, or do not have, about teachers and what might be called, adopting a physicists' metaphor, their "passing band". Several studies about teacher transforming trends lead to the conclusion that anticipating on these might be very useful, if resonant implementation of research-based sequences is a goal seriously aimed at.

When some teacher trends are already well known, for instance the tendency to establish an exclusive link between a material device and a particular approach to a concept, this should be, we suggest, taken into account in advance, in the very design phase of a sequence. Following this example, the use of an optical bench might be explicitly advised against if the main target of a sequence was the principle of image formation, and not positions and size of images. But also, taking this point positively now, it seems especially appropriate to try and find a new device that might convey the idea of a research-based, innovative teaching strategy. For instance, the punched screens described above (fig. 1), the use of which was suggested for teaching optics at grade eight in France, turned out to be an efficient vehicle for at least some of the didactic intentions of the sequence (Hirn & Viennot 2000)

But teacher probable reactions may be unknown in advance, especially when a finer level of analysis is needed. Then, a specific investigation should be conducted, ideally, in the first stages of the process of sequence design, in parallel with at least two other - intricate - preparatory studies, i.e. rethinking the content and investigating students' common ideas and ways of reasoning. Thus, at university level, some facilitating factors have been found by Colin (1999a, 1999b, see also Viennot 2002), in this case to help teachers pay greater attention to the difference in status between the different lines drawn in geometrical and/or wave optics (Colin & Viennot 2001). Among these factors, to give an idea of how specific such a point may be, it seems preferable to discuss the status of paths of light with the example of a small number of discrete diffracting sources than with an extended source or with a grating. The latter situations indeed often elicit from teachers comments like " there is a plane wave in the direction of each maximum", or "red is more deviated than blue".

Most probably, an extended search for such facilitators should be carried out when any teaching sequence is launched with a given group of teachers. Our knowledge in this regard is quite embryonic.

3.2 *Precisely targeted and quantitative evaluation*

Another factor that is probably decisive for the optimised adoption of a research-based product is providing teachers with the means of its detailed, meaningful *and* quantified evaluation. On the one hand, quantified evidence of conceptual attainment in learners is probably necessary to convince teachers and policy-makers of the value of what is being proposed. As Millar *et al.* (2000: 7) wrote: "What possible future can such research have if it cannot be shown to lead to measurable improvement in students' learning ?". On the other hand, for the evaluation to be meaningful, detailed aspects of understanding have to be considered. All too often, if this is not done, one might think that a teaching goal has been reached whereas the pupils' conceptual achievement is in fact very partial. For instance, Saltiel and Kaminski (1996) searching to evaluate the innovative teaching in optics already commented on, collected some comments about the "visibility" of laser beams that seem, at first sight, to show a proper understanding, for instance: " ..., to see the laser beam, you should throw some sand or dust in the air", " the light emitted by the laser reflects on these particles (*with a drawing showing this light entering an eye*) " or " when there is some small particles as chalk, these turn into receptors that can diffuse light. ". But, as recalled by these authors, several links are necessary to analyse the visibility of what is often called a " materialised ray ":

a) There are some diffusing objects (first comment quoted above);
b) The diffusing objects send back some light in a large zone of space (third comment);
c) Some diffused light reaches the observer's eye (second comment).

Once sorted out and counted according to these criteria, the responses displayed a less positive balance. In fact, the only benefit of the pupils who participated in the new sequence, as compared with other ones that had been taught the preceding one, was an increased awareness of the need for dust, or equivalent material agents, for "rays" or "beams" to be "visible". But there was no hint that the conceptual target of this teaching, i.e. the chain of transformations on light involved in vision, was understood any further. The quasi unanimous lack of explanations mentioning light and the observer's eye was observed in both samples (table 1).

Table 1. Percentages of more or less complete comments about the visibility of a "laser beam" at grade 8, collected before and after the introduction in 1993 of a new syllabus in France; two different questions, " Laser ", and " Rock concert " have been used (Saltiel & Kaminski 1996).

	New	syllabus	Old	syllabus
Comments mentioning…	Laser N=272	Rock N=258	Laser N=409	Rock N=421
dust, …	26%	27%	13%	15%
dust+ diffused light	2%	7%	2%	4%
dust+diffused light+eye	2%	1%	1%	1%
other	4%			

Now, if a detailed appreciation of learners' conceptual achievement is needed, one might object that "meaningful-quantitative" evaluation becomes an impossible task and may result in an excessively complicated process. Multiplicity, indeed, has to be analysed in order to find some "pattern". An example is the attempt by Rainson (1995, Viennot & Rainson 1999) to establish some meaningful profiles of conceptual attainment, by grouping some indicators of students' understanding into "families" (see also Chauvet 1994, 1996). Concerning the question of superposition of electric fields and corresponding difficulties, her investigation showed that some patterns were very stable in the profiles obtained, corresponding to "classical knowledge" (drawing correctly the electric fields in a given static situation, for several points) or "causal analysis" (reckoning that in every case the sources of the field were all the charges of the universe) or "synthesis" (responding correctly on both aspects, for each considered point). Several critical aspects of practice had been injected in successive versions of this sequence and their respective outcomes in terms of learning were analysed with this tool. The results are, in brief: no effect at all when awareness and motivation of the teacher (the researcher herself) were the main ingredients, student destabilisation and defective synthesis in an intermediate version, noticeable effects in terms of causal analysis and synthesis without negative side effects on classical knowledge in the final version (fig. 4).

This type of quantitative and multidimensional evaluation provides us, we think, with a sound basis to discuss a sequence with teachers and facilitate its resonant implementation or adaptation. In this case, it gives us some reason to think that several coherently oriented details of practice may, even in a standard institutional context and at practically no additional cost, significantly affect the way learners apprehend a conceptually demanding physical situation.

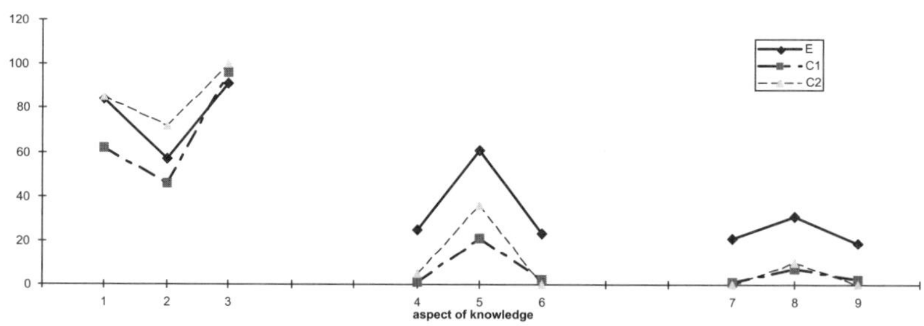

Figure 4. Percentages of answers for questions concerning a system at electrostatic equilibrium

Classical knowledge, 1, 2, 3: right answer to "represent the electric fields at ..(several points)";

Causal analysis,4, 5, 6: right answer to "what are the sources of these fields (same points)";

Synthesis: 7, 8, 9: right answer to both questions for, respectively, same points.

E: Experimental group; C1, C2: control groups of , resp., normal and higher *a priori* ability.

3.3 Detail-oriented training

Teacher training is classically advocated to remedy the improper implementation of sequences. Although it is far from being the only thing to consider, sound training is obviously needed.

As underlined before, teachers, if mainly presented with general views on Science Education and on Science, might not clearly understand the corresponding stakes. The worst thing would be a purely verbal adaptation to academic training. In this regard, a considerable research effort remains to be made, in order to go beyond the high rate of observed failures and the scarcity of research-based and validated training materials presently available.

An important condition, in this perspective, seems to keep a sharp eye on those small aspects of practice commented upon since the beginning of this address, and which seem difficult for teachers to consider. Two questions arise. First, are all details critical? Obviously not, but it is less obvious to decide which ones are relevant in this respect. Investigations and crosschecking constitute the non-miracle response. Another factor to be considered is the communication process. Trainee teachers are in a position, which is analogous to that of learners at school: they cannot be seen as passive receivers. Keeping to a parallel with the case of younger learners, a problem posing approach seems a priori appropriate to orientate the design of training episodes and materials. The problem in question, I suggest, is precisely how to substantiate great principles in detailed actions. More specifically, it may take the form of a question: given this global rationale for such and such sequence, is this particular aspect of practice compatible with this rationale? Or, more abruptly: are we coherent if we think this and do that? A sensible appreciation of coherence between global views and small actions is not straightforward, and it requires education.

Some investigations show the benefits of training in which some critical details, coherently oriented by a common rationale, have been stressed. For instance, a better awareness of the impact of an image has been observed in teachers after a training session during which *other* critical details had been carefully worked upon (Colin *et al.* 2002, Viennot *et al.* 1999).

4. CONCLUDING REMARKS

The results discussed above support the following claims. As regards research-based teaching sequences, a fine-grained design is a necessary condition for an appropriate implementation. Anticipating teachers' reactions is called for at this level. Although a favourable factor, a detailed and quantitative evaluation of sequence effectiveness is not sufficient to ensure an optimal interaction between the research outcomes and the teachers concerned: teacher training is decisive in this process. In line with the preceding results and claims, such training should be conducted with a double concern: an anticipation of teachers' reactions, again, and the specification of a set of critical details closely linked to the global rationale of the proposed sequence. A considerable amount of appropriate documents should be designed (Sassi *et al.* 2000, Colin *et al.* 2001). Ideally, feedback loops might inspire a series of converging improvements, in the spirit of "developmental research" (Lijnse 2001: 312). Researchers in didactics will be all the more convincing when they seek to reduce the gap between research and practice, I suggest, if they provide substantial results and materials based on the preceding concerns. In this effort, it may be useful to be aware not only of teachers' trends, themselves partly determined by their institutional environment, but also of researchers' own tendencies: no one has the privilege of infallible awareness and lucid flexibility. In particular, researchers' views on the content might be, despite their efforts, strongly affected either by a lack of familiarity or, seemingly at the opposite, by a lack of distance with respect to commonly accepted approaches, both factors that may lead to disregard critical details.

Taking seriously the programme outlined above is demanding, but we cannot settle for partial measure in this field.

REFERENCES

Chauvet, F., (1994). *Une séquence d'enseignement sur la couleur*. Unpublished thesis, LDPES, University Paris 7.

Chauvet, F., (1996). Teaching colour: designing and evaluation of a sequence, *European Journal of Teacher Education*, 19(2), 119-134.

Colin, P. & Viennot, L., (2001). Using two models in optics: Students' difficulties and suggestions for teaching, *Physics Education Research, American Journal of Physics Sup*. 69 (7) S36-S44.

Colin, P., (1997). Passage de l'optique géométrique à l'optique ondulatoire. Unpublished "Mémoire de tutorat", LDPES, University Paris 7.

Colin, P., (1999a). Deux modèles dans une situation de physique: le cas de l'optique, Unpublished thesis, LDSP, University Paris 7.

Colin, P., (1999a). Two Models for a Physical Situation: The Case of Optics. Students' Difficulties, Teachers' Viewpoints and Guidelines for a 'Didactical Structure'. In: *Proc. of ESERA*, M. Komorek et al. (Eds.), Kiel, Vol 2, 143-145; and 2001: *Research in Science Education, Past, Present and Future*, H. Behrendt, H. Dahncke, R. Duit, W. Gräber, M. Komoreck, A. Kross, P. Reiska (Eds), Dordrecht: Kluwer. 241-246.

Colin, P., Chauvet, F. & Viennot, L. (2002). Reading Images in Optics: Students' Difficulties, and Teachers' Views. *International Journal of Science Education*. Special issue, 24 (3), 313-332.

Colin, P., Chauvet, F., Rebmann, R. & Viennot, L., (2001). Designing research-based strategies and tools for teacher training – Reading images in optics. In: Psillos, D. et al (eds), *Science Education Research in the Knowledge Based Society* (pp. 1498-151). Proceedings of the Third International and materials on web site: www.ccr.jussieu.fr/ldsp/sttis_p5/index_uk.htm

Driver, R., Asoko, H., Leach, J. Mortimer, E. & Scott, P. (1994). Constructing Scientific Knowledge in the Classroom, *Educational Researcher*, 23 (7), 5-12.

Gil-Perez, D. & Pessoa de Carvalho, A., M., (1998). Physics Teacher Training: Analysis and Proposals. In: Tiberghien, A., Jossem, E.–L. & Barojas, J. (Eds), *Connecting Research in Physics Education with Teacher Education. International Union of Pure and Applied Physics*, ICPE Publications: Ohio. http://www.physics.ohio-state.edu/~jossem/ICPE/BOOKS.html.

Gil-Perez, D., (2001). Constructivism in Science Education: the need for a clear line of demarcation. In: Psillos, D. et al. (Eds), *Science Education in the Knowledge Based Society* (pp. 10-15). Proceedings of the third International Conference (ESERA).

Gunstone, R. & White, R. (2000). Goals, Methods and Achievements of Research in Science Education, in [1], 293-307. (p 302).

Hirn, C.(1995). Comment les enseignants en sciences physiques lisent-ils les intentions didactiques des nouveaux programmes de quatrième?, *Didaskalia* 6, 39-54.

Hirn, C.,(1998). *Transformations d'intentions didactiques par les enseignants: le cas de l'optique élémentaire en classe de quatrième*, Unpublished thesis, LDPES, University Paris 7.

Hirn, C. & Viennot, L., (2000). Transformation of Didactic Intentions by Teachers: The Case of Geometrical Optics in Grade 8 in France. *International Journal of Science Education*, 22, 4, 357-384.

Kaminski, W. (1989). Conceptions des enfants et des autres sur la lumière. *Bulletin de l'Union des Physiciens*, 716, 973-991.

Kress, G. & Van Leeuwen, T., (1996). *Reading Images: the Grammar of Visual design*, London: Routledge & Kegan Paul.

Lijnse, P., (2001). Didactics of Science: The Forgotten Dimension in Science Education Research?. In: [1] 309-326. (p 310).

Lijnse, P.,L., (1994). La recherche-développement: une voie vers une "structure didactique" de la physique empiriquement fondée, *Didaskalia*, 3, 93-108.

Millar, R., (1989). Constructive criticisms, *International Journal of Science Education*, Special issue, 11, 587-596.

Millar, R., Leach, J. & Osborne, J. (Eds), (2000). *Improving Science Education*. Open University Press, Buckingham. See Introduction by the Editors.

Ogborn, J., (1997). Constructivist metaphors of learning science,.*Science & Education*, 6 (1), 121-133.

Pinto, R., Ametler, J., Chauvet., F., Colin, P., Giberti, G., Monroy,, G., Ogborn, J., Ormerod, F., Sassi, E., Stylianidou, F., Testa, I., Viennot , L., (2000). *Investigation on the Difficulties in Teaching and Learning Graphic Representations and on Their Use in Science Classrooms*, STTIS Transversal Report (WP2).

Rainson, S., (1995). *Superposition des champs électriques et causalité* , Unpublished thesis, LDPES, University Paris 7.

Saltiel, E. & Kaminski, W., (1996). Un exemple d'évaluation des nouveaux programmes: problèmes liés à l'évaluation elle-même et à la formation des maîtres. *Bulletin de l'Union des Physiciens,* 786, 1271-1287.

Sassi, E., Monroy G., Testa, I., Ametller, J., Andresen, O., Balzano, E., Boohan., R., Chauvet, F., Colin, P., Couso, D., Giberti, G., Gutierrez., R., Hirn-Chaine, C., Ogborn, J., Pintó, R., Quale, A., Rebmann, G., Stylianidou, F., Viennot, L., (2000). *Teacher training materials favouring the take-up of innovations*. RW5, STTIS Transversal report.

Viennot, L. & Chauvet, F., (1997). Two dimensions to characterise research-based teaching strategies, *International Journal of Science Education* , 19 (10), 1159-1168.

Viennot, L. & Rainson, S. (1999). Design and Evaluation of a Research-based Teaching Sequence: The Superposition of Electrics Fields. *International Journal of Science Education* , 21 (1), 1-16.

Viennot, L., (2000). Anticipating Teachers' Reactions to Innovative Sequences: Examples in Optics. In: Pinto, R. & Surinach, S. (Eds). *Physics Teacher Education Beyond 2000* (pp. 65-72), Paris, Elsevier,.

Viennot, L., Chauvet, F., Giberti, G., Gomez, R., Hirn, C., Monroy, G., Ogborn, J., Pinto, R., Sassi, E., Stylianidou, F., Testa, I., (1999). *Investigation on Teacher Transformations when Implementing Teaching Strategies*, STTIS Transversal Report (WP3).

STTIS ("Science Teacher Training in an Information Society") is a research project founded by EU, DG XII, within the framework of the TSER programme (coord. R. Pinto, UAB, Barcelona), http://www.blues.uab.es/~idmc42

TRANSFORMING THE STANDARD INSTRUMENT FOR ASSESSING SCIENCE TEACHER'S SELF-EFFICACY BELIEFS (STEBI) FOR USE IN DENMARK

Annemarie Møller Andersen, Søren Dragsted, Robert H. Evans, Helene Sørensen

The Danish University of Education

ABSTRACT

The context of this paper is a Danish/American study that addresses the influence of school values and traditions on new teachers by looking at their science teaching self-efficacies both before they enter into a new school environment and at intervals after they have begun to teach. The purpose of the work reported here was to adapt the Science Teaching Efficacy Beliefs Instrument (*STEBI*) (Riggs & Enochs, 1990) for use in Denmark. Because the 'new' instrument had to be crafted to assess self-efficacy among science teachers in another culture, mere translation was not possible. Initial versions of the instrument were written, back translated and rewritten, based on differing cultural values. The reliability of both scales of the new instrument (*STEBI-DK*) were favourable (self-efficacy α=.93; outcome expectations α= .86), when compared to previous versions of *STEBI*. Cultural, attitudinal or preparatory differences between Danish and American pre-service elementary teachers were revealed in significantly different scores on our trial version of our *STEBI-DK* instrument. The longer term educational utility of *STEBI-DK* will be to sample and track the self-efficacies of pre-service elementary teachers of science during their first years of teaching.

1. SIGNIFICANCE

This work is important, not merely for our current Danish/American study, but as a potential model for the adaptation of science education research instruments from one culture to another. The internationalisation of science education research inevitably involves the use of well-accepted instruments in many countries and languages.

The Science Teaching Efficacy Beliefs Instrument, *STEBI* (Riggs and Enochs, 1990 and Enochs and Riggs, 1990), is a good example of the type of instrument that can be internationally useful. *STEBI* has found wide use in English-speaking North America (for examples, see: Cannon and Scharmann (1996) and Ramsey-Gassert, Shroyer, and Staver (1996)). However, it may be applicable across many other learning communities.

Research Objectives. The objective for this project is to find out whether science teaching self-efficacies amongst pre-service Danish and American elementary teachers interact significantly with 'science cultures' and collective efficacies in their schools to affect science instruction. The specific purpose of the work reported here

D. Psillos et al. (eds.), Science Education Research in the Knowledge-Based Society, 395–403.

was to adapt the *STEBI* instrument for use in Denmark and concurrently to develop a model for such international instrument adaptation. The transformed instrument will be the primary quantitative tool for the Danish study and its subsequent American replication.

Theoretical Underpinnings The appropriateness of adapting the *STEBI* instruments lies in the universality of the construct. Of self-efficacy, Bandura (1997, p.3) said: "Perceived self-efficacy refers to beliefs in one's capabilities to organize and execute the courses of action required to produce given attainments." Bandura pioneered the view that beliefs are intimately associated with behaviour. He showed that based on life experiences, people have specific expectancies about their action-outcome contingencies. They act not only because they believe their actions will result in specific outcomes but also because they believe in their own ability to perform those actions. This is why *STEBI* includes both self-efficacy and outcome expectation scales.

1.1 Relevant Research

Self-efficacy and Teaching. The relationship of self-efficacy to teaching was pioneered by a number of researchers (e.g. Ashton & Webb, 1986; Gibson & Dembo, 1984; Enochs & Riggs, 1990). This research has shown that teaching behaviours such as persistence at tasks, risk taking and the use of innovations are all related to self-efficacy (Ashton & Webb, 1986). For example, in science teaching, teachers with high self-efficacies are more likely to use inquiry and student centred teaching methods while those with low efficacies are more likely to be teacher directed. Research that is more recent has found that teacher self-efficacy beliefs strongly influence the nature of a teacher's role, planning and consequently curriculum and student learning (Tobin et al. 1994).

STEBI Instruments. Riggs & Enochs (1990) developed a valid and reliable instrument to assess the science teaching self-efficacy of in-service elementary teachers (*STEBI-A*) and subsequently a pre-service version, *STEBI-B* (Enochs & Riggs, 1990). The original *STEBI* instruments are consistent with Bandura's (1997) assertion that such beliefs are situation specific by focusing on expected behaviour with regard to science teaching. The model for the *STEBI* items was the more general instrument designed by Gibson and Dembo (1984).

The *STEBI* instruments used Likert scales and positively and negatively worded items to assess elementary teachers. Following are typical items from the original *STEBI-A* and *STEBI-B*, with the latter's changes to the future tense, in parentheses. The first example is from the Personal Science Teaching Efficacy Scale and the second from the Science Teaching Outcome Expectancy Scale.

- I am (will) continually finding better ways to teach science.
- When a student does better than usual in science, it is often because the teacher exerted a little extra effort.

International Adaptations. Rich, et al (1996) adapted a self-efficacy instrument for use in Israel by following a strict protocol including back-translations and juried reviews. Their adapted instrument maintained the factor structure and reliability of the instrument (Rich, et. al., 1996).

The Third International Mathematics and Science Study (TIMSS) and the Programme for International Student Assessment (PISA) studies of the past decade, encountered large adaptation challenges as they sought to fit original instruments to many languages and cultures (Martin and Gregory, 2000 and OECD, 2000). Consequently, their protocols were useful to us in designing our own approach.

The primary translators in TIMSS had expertise in the content of the tests and were familiar with the target student populations. These translators had to have excellent knowledge of both English (the original language) and the target language (Martin and Gregory, 2000). To assure reliability and validity, the *TIMSS* 1999 had two independent translations made for each target language, and then had a review team justify the versions. Finally, an independent international translation company verified the final resulting versions.

2. METHODOLOGY

Instrument Development. Since our goal was to adapt *STEBI-B* for use in Denmark (and later in America), we began by translating it into Danish.[1] In dealing with an abstract construct, we knew that every nuance of our modified instrument was important. We also knew from the experiences of *TIMSS* and *PISA* that mere translation was not likely to result in an instrument with both reliability and validity (Martin and Gregory, 2000 and OECD, 2000). We used a protocol similar to *TIMSS.* Our team consisted of three Danish science teaching educators, one of whom served on the Danish translation team for *PISA*, and one American science teacher educator. The Danes had a high level of competence in English and all had extensive experience in science teacher education and were familiar with test development. Because modifications in the wording of most items was necessary, the use of Cronbach's coefficient alpha was essential to monitor and maintain the internal consistency of the instruments. Using our professional knowledge was our most important guarantee of validity. Consequently, we sought to base our validity assessments on both strong theoretical as well as empirical foundations (Stevens, 1996).

With these objectives and tools in mind, our team made several versions of the instrument and discussed each change in both English and Danish. As our evolving Danish version of *STEBI* was not a direct translation of the original English version, we needed a new and parallel English form. To establish the validity of our modified *STEBI* instruments, they were submitted to one of *STEBI*'s original authors (Enochs, 2000) for comment and subsequent change. Then, before testing, the Danish version was given to an out-of-field American university researcher who speaks good Danish and who is familiar with Danish culture, for 'back-translation'. Based on this input our evolving instrument was revised before pilot testing. After the trials in Denmark and the United States, the parallel instruments were adjusted, based on written feedback. To be consistent with the other versions of *STEBI*, we chose the

[1] A full sample of the different versions of the instrument may be obtained from the authors.

name *STEBI-DK* to associate it with its ancestors and to indicate that it was developed to be consistent with Danish culture.

Subjects. Since the goals are to use *STEBI-DK* in both Denmark and the US, two separate populations were used for development, one from each country. Both groups were recruited from convenient pre-service teachers.

The Danish sample of 23 were mostly from a general science course in a teacher training college in Copenhagen where they were in their last year of preparation for teaching. A few were enrolled in a course to learn to teach elementary school science at The Danish University of Education. The ages of the Danish sample were typical for pre-service teachers; mostly in their twenties with a few older and predominately they were females. The US sample consisted of 43 education majors from the state of North Carolina, who were nearing completion of their preparations to teach. Most were just over twenty years of age, as is typical for US education students, with a few older than that and most were also females. All of those solicited responded to the questionnaire with both answers and comments on any ambiguities or other problems they encountered while completing it.

3. ANALYSIS

Reliability. *STEBI-DK* was simultaneously tested in Denmark and the United States on a combined population of 66 (Denmark 23, USA 43). A *t-test* was performed and an item analysis was done separately on both sets of results. The samples were then combined, since the goal was to develop a modified version of *STEBI-B* that would function well in both countries, and to take advantage of the larger combined sample. Cronbach's alpha and item-total correlations were used to make decisions about instrument reliability.

Validity. Since the original validity of the *STEBI-A* instrument and it's descendents (*STEBI-B*, *STEBI-CHEM* and *MTEBI*) had been well established by Enochs and Riggs, 1990 and Riggs and Enochs, 1990, we submitted our original version to Larry Enochs (2000) for comment. Next, we attempted to use confirmatory factor analysis to see if our revisions significantly altered the two content validity constructs already established. We examined the factor structure of *STEBI-DK* with the *AMOS* statistical package (Arbuckle, 1999).

4. RESULTS

4.1 Instrument Development

Changes to item 4 illustrate the cultural evolution of the test items.

The original *STEBI-B* included a reference to 'grades':

When the science grades of students improve, it is often due to their teacher having found a more effective teaching approach.

The concept of 'grades' was inappropriate in Denmark since students are not graded at the elementary level. Hence, 'grades' was changed to 'interest' in our first Danish version:

Når elevernes interesse i natur/teknik øges, er det ofte fordi læreren har fundet frem til en bedre tilrettelæggelse af undervisningen.

'Grades' was also changed to 'interest' in the first modified English version:

When students' interest in science improves, it is often due to their teacher having found better ways to teach science.

However, the original change from 'grades' to 'interest' (in both Danish and English) was not consistent with 'grades' in English (Enochs, 2000), so the English version was changed to 'learning', which made it less 'grade' oriented.

When students' learning in science improves, it is often due to their teacher having found a better way to teach science.

The remaining discrepancy between 'interest' in Danish and 'learning' in English was resolved by changing the Danish to 'udbytte', which translates as 'profit from'.

Når elevernes udbytte af natur/teknik øges, er det ofte fordi læreren har fundet frem til en bedre tilrettelæggelse af undervisningen.

One English item (number 7) retained the Danish concept of 'interest' to help keep both instruments parallel.

The comments and feedback from the American subjects also resulted in minor changes in wording. The Danish students found two test items (numbers 7 and 8) so similar to each other that they showed resistance to answering both. Since both items functioned well in other *STEBI* instruments, and Larry Enochs said that there wasn't any significance to their sequence, we decided to split up the items for the final version. The Danish students also objected to a long sequence of 'I' statements so again we split them up for our final version.

4.2 Reliability

Item analyses for the Danish, American and combined pilot studies were conducted. Three items had combined Item-total correlations below .30, which is generally considered the least a useful item should have (Robinson, Shaver and Wrightsman, 1991). Consequently, we dropped two of the items from the final version of the instrument. These same items had poor Item-total correlations, on the recently completed *MTEBI* (Enochs, Smith and Huinker, 2000). One item was not dropped because it performed exceptionally well in the Danish sample (unlike the other two) and it survived all the other permutations of *STEBI.* The alpha values achieved among other members of the *STEBI* family were commensurate with the final values for *STEBI-DK* (see Table 1).

Table1: Reliability Coefficients for the STEBI family of Instruments

Origin	Instrument	Self-efficacy Reliability Scale (α)	Outcome Expectations Reliability Scale (α)
Riggs and Enochs, 1990	*STEBI*-A	.92	.77
Enochs and Riggs, 1990	*STEBI*-B	.90	.76
Rubeck and Enochs, 1991	*STEBI*-CHEM	.88	.80
Enochs, Smith, and Huinker, 2000	MTEBI	.88	.77
This Paper	*STEBI*-DK	.93	.86

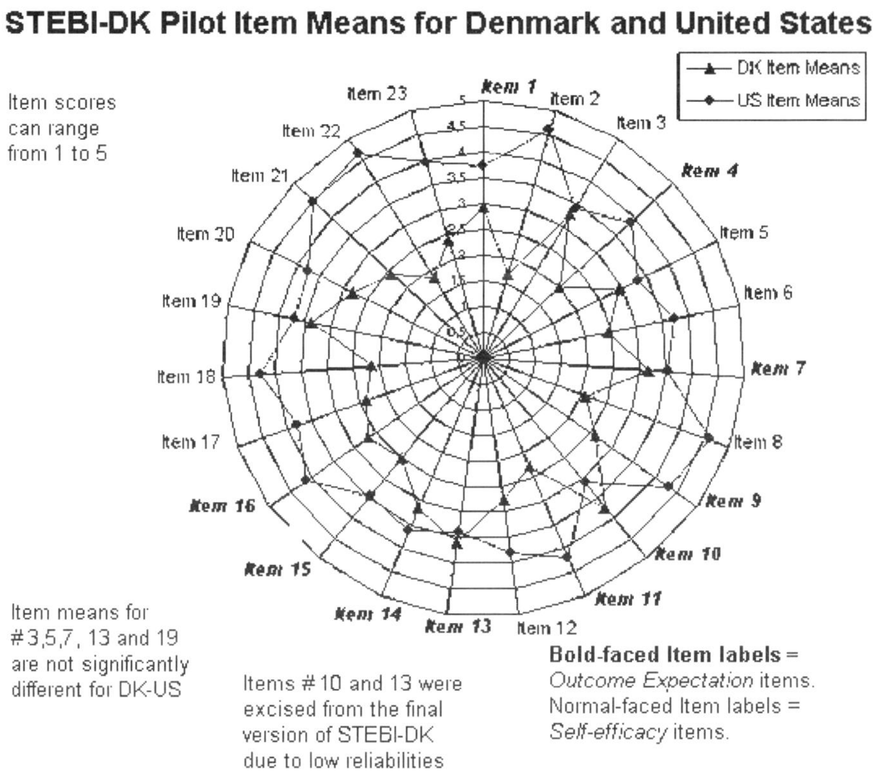

Figure 1: *STEBI-DK* Pilot Item Means for Denmark and the United States

The *t-test* showed that the Danish and American samples were significantly different (t= -14.774, df 64, sig. =.00) and that with all of the items except for numbers 3,5,7,13 and 19, the Americans showed a significantly higher (> self-efficacy and outcome expectations) score. Figure 1 plots the mean item scores of the

Danish sample against that of the American sample, revealing the significantly different scores.

4.3 Validity

By using professionals close to the field of elementary teacher education and by making as few changes as possible to the original instrument we reduced potential problems with validity. The feedback we got from Larry Enochs, added to our confidence (2000). However, due to our small samples we were unable to run a successful confirmatory factor analysis to see if the constructs of self-efficacy and outcome expectations remained separate and distinct in *STEBI-DK* (Stevens, 1996). Consequently, for the present, we must rely on the history of the *STEBI*, our development protocols and our own professional experiences with our modified versions in regards to validity.

5. DISCUSSION

The results of this work suggest that *STEBI-DK* within each country is a reliable instrument whose English and Danish forms based on back and forth translations and reliability performances, are very similar. Since our pilot sample sizes were much smaller than necessary to use confirmatory factor analysis to check on their validity, we can only presume it, based on *STEBI-DK*'s descent from and close resemblance to the *STEBI* family of instruments. The somewhat lower reliability for the Science Teaching Outcome Expectancy scale as compared to the Personal Science Teaching Efficacy Belief scale (.86 vs..93) is consistent with similarly lower Cronbach's alpha values for all other *STEBI* instruments (see Table 1). Larger samples in both the United States and Denmark will be needed to assess validity with confirmatory factor analysis.

Importantly, this study has revealed significant differences in responses of similar populations in Denmark and the United States to both scales of the *STEBI-DK*. We have hypothesised that this overall difference in level may be attributable to a Danish self-concept known as *Janteloven*. *Janteloven* is a characteristic of Danish culture that is usually expressed as awareness that others expect you not to express the belief that you are better than the group (Sandemose, 1999).

While this cultural tradition may be responsible for the overall lower Danish scores, an alternative explanation would be that a lack of teaching preparation may explain the lower scores on *STEBI-DK*. Until this year (after our sample was drawn), Danish students did not have specific preparation to teach elementary science. This would be consistent with Bandura's belief that self-efficacy is specific to each situation (Bandura, 1997).

A third hypothesis for these significant differences is that new Danish teachers do not have the same kind of professional expectation for change that their American counterparts have. This difference is not seen as a problem for tracking changes in self-efficacy and outcome expectations within either the Danish or the

American cultures, since the reliabilities of both versions of the instruments are high ($\alpha = .86$ and .87 respectively).

The implications are that while the instrument can be used to assess changes in self-efficacy and outcome expectations among pre-service elementary teachers within the two countries, any comparisons between the students of Denmark and America, using *STEBI-DK* would be inappropriate. By inference, it may also be problematic to use other *STEBI* instruments to make comparisons between these two countries, and perhaps others as well. More work is needed to determine the potential for this kind of bias in other cultures. We are particularly interested in gathering more data, including interviewing subjects, to more fully assess the reasons for differences between Danish and American students on *STEBI-DK* and to test our three hypotheses about the causes of these differences.

The longer term educational utility of *STEBI-DK* and it's English version will be to sample and track the self-efficacies of pre-service elementary teachers of science during their first years of teaching. We thereby hope to achieve two objectives. Firstly, we want to establish the degree to which strong self-efficacies are associated with excellent science teaching. Concomitantly, we are looking at the degree to which self-efficacies are ameliorated by school cultures.

Secondly, since strong self-efficacies are related to the ability to apply the 'best' teaching practices learned during teacher preparation (Ashton & Webb, 1986) as well as to defining a teacher's role in school (Tobin et al, 1994), progressive measures of self-efficacy can provide an indicator of how new teachers are managing in their new school environments. This can lead to support and assistance from teacher educators, hopefully resulting in greater early success for new teachers as well as better teaching for their pupils. We are continuing this research by applying *STEBI-DK* to new groups of Danish elementary science teachers.

Danmarks Pædagogiske Universitet); Emdrupvej 101; 2400 København NV
E-Mail Contact: annemari@dpu.dk

REFERENCES

Arbuckle, J.L, (1999). *Amos: version 4.* Chicago, IL: SmallWaters Corporation.

Ashton, P., & Webb, R. (1986). *Making a difference: Teacher's sense of efficacy and student achievement.* New York: Longman.

Bandura, A. (1997). *Self-efficacy: the exercise of control.* New York: W.H. Freeman and Company.

Cannon, J.R. & Scharmann, L.,C., (1996). Influence of a cooperative early field experience on pre-service elementary teachers' science self-efficacy. *Science Education, 80,* 419-436.

Enochs, L.G. (2000). Personal communication with the authors.

Enochs, L.G. & Riggs, I.M. (1990). Further development of an elementary science teaching efficacy belief instrument: A pre-service elementary scale. *School Science and Mathematics*, 90, 695-706.

Enochs, L.G., Smith, P.L. & Huinker, D. (2000). Establishing factorial validity for the mathematics teaching efficacy beliefs instrument. *School Science and Mathematics*, 100(4), 194-202.

Gibson, S. & Dembo, M.H. (1984). Teacher efficacy: A construct validation. *Journal of Educational Psychology,* 76(4), 569-582.

Martin, M.O. & Gregory, K.D., (Eds.), (2000). *TIMSS 1999 Technical Report.* Chestnut Hill, MA: Boston College.

OECD, (2000). *Measuring student knowledge and skills: The PISA 2000 assessment of reading, mathematical and scientific literacy.* Paris: OECD Publications.

Ramsey-Gassert, L., Shroyer, M.G., & Staver, J.R. (1996). A qualitative study of factors influencing science teaching self-efficacy of elementary level teachers. *Science Education, 80,* 283-315.

Rich, Y., Smadar, L, & Fischer, S. (1996). Extending the concept and assessment of teacher efficacy. *Educational and Psychological Measurement,* 56(6): 1015-1025.

Riggs, I.M. & Enoch, L.G., (1990). Toward the development of an elementary teacher's science teaching efficacy belief instrument. *Science Education,* 74(6). 625-637.

Robinson, J.P., Shaver, P.R., & Wrightsman, L.S., (Eds.), (1991). *Measures of personality and social psychological attitudes.* (Vol. 1). New York: Academic Press.

Rubeck, M. & Enochs, L., (1991). *A path analytic model of variables that influence science and chemistry teaching self-efficacy and outcome expectancy in middle school science teachers.* Paper presented at the annual meeting of the National Association for Research in Science Teaching, Lake Geneva, WI.

Sandemose, A., (1999). *En flygtning krydser sit eget spor (A fugitive crosses his tracks).* Århus: Schønberg.

Stevens, J. (1996). *Applied multivariate statistics for the social sciences* (3rd ed.). Mahwah, NJ: Lawrence Erlbaum Publishers.

Tobin, K. Tippins, D.J., & Gallard, A.J., (1994). Research on instructional strategies for teaching science. In D. Gabel (Ed.), *Handbook of Research on Science Teaching and Learning,* (pp. 45-93). New York, NY: Macmillan Publishing Company.

TEACHERS' CONFIDENCE IN PRIMARY SCIENCE AND TEACHER-STUDENT INTERACTIONS

[1]Darrell L. Fisher, [2]Bruce G. Waldrip

[1] Curtin University, Perth, Australia, [2] La Trobe University, Australia

ABSTRACT

This study describes the validation and application of an instrument to assess teacher-student interactions in primary science classrooms. The Questionnaire on Teacher Interaction (QTI) was modified to suit primary classrooms and used for the first time with a large sample of 2,371 Australian primary science students and their 121 teachers. The reliability of QTI scales ranged from 0.62 to 0.83. Teachers, who understood their student needs, were helpful, friendly, demonstrated strong leadership and allowed some student responsibility and freedom, were more likely to have students with more favourable attitudes towards science. The teachers' responded to items about their perceived self-competence and their use of Student Outcome Statements (SOS). Where teachers had greater confidence in their ability to explain concepts, their students perceived them as being more helping/friendly and understanding, and less dissatisfied and admonishing. Teachers who were more confident with using SOS were perceived as being more helping/friendly and understanding.

1. INTRODUCTION

An Australian government funded inquiry, the Discipline Review of Teacher Education in Mathematics and Science (1989), commonly known as the Speedy report, stated that science teaching in primary schools was more noticeable for its absence than its presence. The report noted that primary science was more honoured in its breach than by its presence. While a number of primary teachers implement a highly effective science program in their schools, the majority appear to avoid or minimise the amount or type of science teaching that they undertake. A report of the National Board of Employment, Education and Training, Five to fifteen - Reviewing the compulsory years of schooling (1993) noted that science teaching is not receiving the importance that it warrants because primary teachers lack competence in teaching science. Many primary teachers view the preparation and teaching of science as being too difficult and too demanding (Goodrum, Cousins & Kinnear, 1992). A recent study, Matching Science and Technology to Future Needs: 2010 (Australian Science, Technology and Engineering Council, 1996) has identified the need to increase science literacy within the Australian community. It is during the primary school experience that many students begin to form ideas and attitudes towards science. Hence it is important to examine the effect that student-teacher interactions and teachers' self-perceived competence has on students' attitudes. A

D. Psillos et al. (eds.), Science Education Research in the Knowledge-Based Society, 405–413.

recent study, Foundations for Australia's Future - Science and Technology in Primary Schools (Stocker, 1997, p1), stated that *much has been achieved in primary science and technology education over the past ten years but much more needs to be done*. Australia's performance in the Third International Mathematics and Science Study (Lokan, Ford & Greenwood, 1996) indicated that, since the previous study, primary students' science ability performance had improved. Students develop ideas and attitudes towards science at the primary school level. This study focused on the important primary school level.

2. STUDENT-TEACHER INTERACTIONS

Recent reviews (e.g. Fraser, 1998; Fraser & Walberg, 1991) show that science education researchers have led the world in the field of classroom environment over the last two decades, and that this field has contributed much to understanding and improving science education. For example, a key to improving student achievement and attitudes is to create learning environments that emphasise those characteristics, which have been found to be linked empirically with student outcomes. However, classroom environment research has been somewhat limited at the primary level compared with the secondary level.

One particular development in classroom environment research occurred in The Netherlands where the focus was on the interactions that occurred between teachers and students. Wubbels, Creton and Hooymayers (1985) developed a model to map interpersonal teacher behaviour extrapolated from the work of Leary (1957). The model maps interpersonal behaviour with the aid of an influence dimension (Dominance, D - Submission, S) and a proximity dimension (Cooperation, C - Opposition, O). These dimensions are represented in a coordinate system divided into eight equal sectors. Every instance of interactional teacher behaviour can be placed within this system of axes (Figure 1).

This model has been used in The Netherlands in the development of an instrument, the *Questionnaire on Teacher Interaction* (QTI), to gather students' perceptions of their interactions with their teacher (Wubbels & Levy, 1993). The QTI contains eight scales based on the eight parts of the model. A description of each of these scales together with a sample item is shown in Table 1.

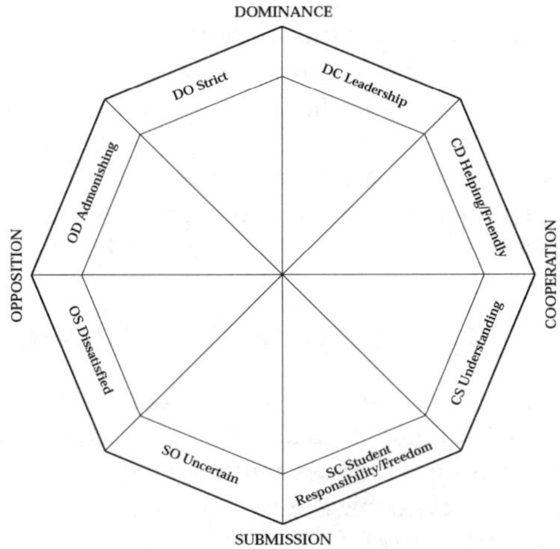

Figure 1. The model for interpersonal teacher behaviour.

Table 1. Description of Scales and Sample Items for each Scale of the QTI

Scale Name	Description of Scale (The extent to which the teacher...)	Sample Item
Leadership	...leads, organises, gives orders, determines procedure and structures the classroom situation.	This teacher knows what is going to happen next in this class.
Helping/Friendly	...shows interest, behaves in a friendly or considerate manner and inspires confidence and trust.	This teacher helps us with our work.
Understanding	...listens with interest, empathises, shows confidence and understanding and is open with students.	This teacher trusts us.
Student Responsibility/ Freedom	...gives opportunity for independent work, gives freedom and responsibility to students.	This teacher allows us to take responsibility for what we do.
Uncertain	...behaves in an uncertain manner and keeps a low profile.	This teacher allows us to tell him/her what to do.
Dissatisfied	...expresses dissatisfaction, looks unhappy, criticises and waits for silence.	This teacher thinks that we cheat.
Admonishing	...gets angry, express irritation and anger, forbids and punishes.	This teacher gets angry quickly.
Strict	...checks, maintains silence and strictly enforces the rules.	This teacher is strict.

An Australian secondary school version of the QTI has been used in studies involving upper secondary science classes across Australia (Fisher, Fraser & Wubbels, 1993; Fisher, Fraser, Wubbels & Brekelmans, 1993; Fisher, Henderson & Fraser, 1995; Fisher, & Waldrip, 1999). These studies strongly supported the validity and potential usefulness of the QTI within the Australian context, and suggested the desirability of conducting further and more comprehensive research involving the QTI. In one study in Singapore (Goh & Fraser, 1996), the QTI was modified and made suitable for use in primary schools. This paper reports the first large-scale adaptation of the QTI to the primary level in Australia and adds to our understanding of primary school classroom learning environments.

The QTI also exists in a Teacher version which assesses the teacher's own perceptions of their interpersonal behaviour in a particular class (Fisher, Rickards & Fraser, 1996). This Teachers' version also was used in this study.

3. STUDENT OUTCOME STATEMENTS

Most Australian states are currently implementing curricula in which Student Outcome Statements are prominently featured. Student Outcome Statements reflect what the state system considers to be essential knowledge, skills and processes for students to achieve in each subject area (Education Department of Western Australia, 1994). The adoption of Student Outcome Statements in Australia has implications especially for primary teachers, most of whom were trained initially as generalists, to teach towards specified standards across a range of subjects in ways that are open to public scrutiny and verification. The Education Department of Western Australia's (1996) two-year trial of Student Outcome Statements provided evidence that utilisation of Student Outcome Statements can improve student learning. Therefore, it was timely to investigate the extent to which primary teachers believed they were using Student Outcome Statements (SOS).

4. STUDENT ATTITUDES

Successful implementation of teaching strategies to teach primary science is likely to result in the establishment and maintenance of positive student attitudes towards science and consequently, achievement. A number of researchers have shown that students' attitudes towards science are related to achievement (Klopfer, 1992; Schibeci, 1984). Accordingly, this study identified associations between student-teacher interactions and students' attitudes towards science. To assess students' attitudes towards science a ten-item scale selected based on the *Test of Science -Related Attitudes* (Fraser, 1981) was used.

5. METHODOLOGY

This study builds on previous one learning environment studies, but is distinct in that it incorporates classroom environment theory and research to examine teachers' implementation of SOS and student attitudes towards science in primary schools.

Specifically, the research tasks were: to modify the QTI and determine its reliability and validity for use with students and/or teachers in primary science classrooms; investigate associations of student-teacher interactions with teachers' use of

Students Outcome Statements; and investigate associations of student-teacher interactions with students' attitude to science.

The study involved a sample of 2,371 science students in 121 classes in 46 Australian primary schools. Each student in the sample responded to the QTI and the attitude scale. The QTI contained 48 items which have been construct and content validated by teachers, students and fellow researchers (Waldrip, & Fisher, 2000). Each scale contained five items which were responded to on a five-point scale with the extreme alternatives of Never - Always. Students were asked to indicate how often their teacher demonstrated that behaviour in their classrooms during science lessons.

The teachers in the sample responded to a teacher version of the QTI. The wording in the teachers' version was altered slightly. For example the item "This teacher helps us with our work" in the student version became "I help students with their work" in the teacher version. The teachers also responded to a set of items indicating the extent to which they felt they used the SOS in their teaching, planning, student learning and assessment procedures. In this study, the Cronbach alpha reliability of this SOS scale was 0.89.

6. RESULTS

Instrument Reliability

The first research question explored involved the reliability and validity of the QTI instrument to the primary classroom. The internal consistency/reliability (Cronbach alpha reliability coefficient) and scale item mean of each of the QTI scales are shown in Table 2. The table shows that for the sample of students, the alpha coefficients ranged from 0.59 to 0.83 suggesting that each QTI scale has acceptable reliability, especially for scales containing a relatively small number of items. For the teacher version all scales had a reliability of greater than 0.6 with the exception of the Student Responsibility/ Freedom scale. This scale would need to be revised before it was used again in the teacher version. The scale item means (range 0 to 1) showed that the scales of Leadership, Helping/Friendly, Understanding, Student Responsibility/ Freedom and Strict behaviour were the most strongly perceived scales while their were lower perceptions of the less desirable scales of Uncertain, Dissatisfied and Admonishing. A desirable characteristic of the QTI is that it is capable of differentiating between perceptions of students in different classes. This characteristic was explored using a one-way ANOVA with class membership as the main effect. using the individual as the unit of analysis. The results in Table 2 indicated that each scale differentiated significantly ($p<0.01$) between classes. The eta^2 statistic represents the amount of variance in student-teacher interaction scores accounted fro by class membership and in this study it ranged from 0.22 to 0.37.

Table 2. Item Mean, Cronbach Alpha Reliability and Ability to Distinguish between Classes for Each Scale of the QTI

Scale	No of Items	Unit of Analysis	Alpha Reliability	Scale Item Mean	ANOVA Results Eta^2	
Leadership	6	Individual	.66	.72	.25	*
		Teacher	.66	.74		
Helping/ Friendly	6	Individual	.83	.77	.35	*
		Teacher	.70	.74		
Understanding	6	Individual	.78	.76	.28	*
		Teacher	.67	.81		
Student Responsibility/ Freedom	6	Individual	.59	.55	.34	*
		Teacher	.32	.54		
Uncertain	6	Individual	.65	.22	.22	*
		Teacher	.67	.16		
Dissatisfied	6	Individual	.71	.23	.32	*
		Teacher	.67	.20		
Admonishing	6	Individual	.73	.30	.37	*
		Teacher	.73	.22		
Strict	6	Individual	.72	.41	.29	*
		Teacher	.60	.53		

n=2,371 students and 121 teachers *$p<0.05$

7. ASSOCIATIONS

Teacher-student Interpersonal Behaviour

The second research question explored associations between the QTI scales and teachers' perceptions of their use of SOS using simple correlations and multiple regression. Table 3 shows the results of these simple correlations and the more conservative multiple regression analysis. between the QTI, and teachers' use of SOS. Overall, as depicted in Table 3, five scales of the QTI, were found to be associated with teachers' use of SOS.

Table 3. Associations between QTI Scales and Teachers' Use of SOS - Simple Correlations (r) and Significant Multiple Correlation (ß)

QTI Scales	Use of SOS r	ß
Leadership	.20*	
Helping/ Friendly	.34**	
Understanding	.40**	.05*
Student Responsibility/ Freedom	.01	
Uncertain	.04	
Dissatisfied	-.24**	
Admonishing	-.31**	
Strict	-.01	
Multiple R		.47**
Correlation, R^2		.22

*$p<0.05$ **$p<0.01$ n=2,371 students and 121 teachers

Leadership, Helping/Friendly, and Understanding behaviours were positively correlated with the teachers use of SOS, while the Dissatisfied and Admonishing scales were negatively correlated. It is noteworthy that where the teachers perceived they used the SOS more, their students perceived them as more understanding, helping and friendly and leaders in the classrooms. Also, teachers who used the SOS more apparently expressed less dissatisfaction with their students and admonished them less. The multiple regression showed that 22% of the variance in teachers' use of SOS could be attributed to their interaction with their students. When associations between the QTI scales were controlled for, it was the Understanding scale that retained its significance.

Attitudes

Past environment research has often investigated associations between student outcomes and the nature of the classroom environment (Fraser, 1994). In order to permit examination of the predictive validity (i.e., the ability to predict student outcomes) of the QTI, students completed a simple Likert-type questionnaire, which assessed students' attitudes towards science (Fraser, 1981). In this study, the Cronbach alpha reliability of this attitude scale was 0.90.

Simple correlation analyses were used in examining the degree of association between each of the QTI scales and attitude to science. Overall, as depicted in Table 4, most of the scales of the QTI were found to be associated with students' attitudes. Furthermore, it can be seen that all of the significant correlations were positive for Leadership, Helping/Friendly, Understanding and Student Responsibility/Freedom Behaviours and were negative with Uncertain, Dissatisfied, Admonishing and Strict behaviours. That is, these latter behaviours were associated with lower scores on attitude towards science. Teachers who displayed understanding of their students, a helpful and friendly manner, demonstrated strong leadership, and allowed a certain measure of student responsibility and freedom, were more likely to have students with more favourable attitudes towards science.

Table 4. Associations between QTI Scales and Attitudes - Simple Correlations (r) and Multiple Correlation (b)

QTI Scales	Attitudes	
	r	ß
Leadership	.35*	.23*
Helping/ Friendly	.31*	.11*
Understanding	.30*	.01
Student Responsibility/ Freedom	.30*	.16*
Uncertain	-.05*	.03
Dissatisfied	-.20*	-.03
Admonishing	-.15*	-.07#
Strict	-.09*	.04
Multiple Correlation, R		.41

*p<0.05 **p<0.01 n=2,371

These associations were further investigated using multiple regression. The magnitude and statistical significance of the regression coefficient provides a measure of the association between the outcomes and input variable when scores on the other input variables are held constant. Beta weights and significance levels are reported in Table 4 for each QTI scale and there is a high degree of congruence with the results of the simple correlations. Table 4 shows that the number of significant regression weights for the multiple correlation analysis was eight for attitudes. An examination of the signs of the significant beta weights in Table 4, reveals that the regression weight for attitudes was positive for Leadership, Helping/Friendly, and Student Responsibility/ Freedom Behaviours and were negative with Admonishing.

8. DISCUSSION

This paper has described the validation of a questionnaire (QTI), which assesses eight scales of teacher and student interactions of upper primary school students. This exploratory study also examined associations between primary students' attitudes to their science classes and their perceptions of teacher-student interactions. Additionally, it explored the associations between students' perceptions of student-teacher interactions and their teachers' perceptions of their use of SOS.

The reliability for each scale of the QTI was obtained and for the Student Version the Cronbach alphas ranged between acceptable values of 0.59 and 0.83 for this adaptation. For the Teacher Version, all scales except Student Responsibility and Freedom had satisfactory reliability.

Associations between students' perceptions of student-teacher interactions and their attitudes were found. Correlation analysis suggested that more positive student attitudes are associated with teachers who display understanding of their students, a helpful and friendly manner, demonstrate strong leadership and allow a certain measure of student responsibility and freedom. Given the evidence of associations of positive attitudes with cognitive achievement, it would seem worthwhile to encourage teachers to be classroom leaders who are understanding and helpful, and friendly towards their students. This has implications for the training of teachers. It is also apparent that when teachers perceive they used the SOS more, their students perceive them as more understanding, helping and friendly, and leaders in the classrooms. Also, teachers who use the SOS more apparently expressed less dissatisfaction with their students and admonished them less. This would suggest that there is value in teachers becoming familiar with SOS. This study does not allow us to state anything about causation and this has implications for further research.

REFERENCES

Australian Science, Technology and Engineering Council, (1996, May). *Matching science and technology to future needs: 2010. Summary Report.* Canberra: Australian Government Publishing Service.

Education Department of Western Australia. (1996). *Profiles of student achievement. Science in Western Australian Government Schools.* Perth: Education Department of Western Australia.

Fisher, D., Fraser, B., & Wubbels, Th., (1993). Interpersonal teacher behavior and school environment. In Th. Wubbels & J. Levy (Eds.), *Do you know what you look like: Interpersonal relationships in education.* London: Falmer Press.

Fisher, D., Fraser, B., Wubbels, Th. & Brekelmans, M., (1993). Associations between school learning environment and teacher interpersonal behaviour in the classroom. In: Fisher, D., (Ed.), *The Study of Learning Environments,*(pp. 32-41) *Volume 7*. Perth, WA: Science and Mathematics Education Centre, Curtin University of Technology.

Fisher, D., Henderson, D., & Fraser, B., (1995). Interpersonal behaviour in senior high school biology classes. *Research in Science Education, 25*(2), 125-133.

Fisher, D., Rickards, T., & Fraser, B.,J., (1996). Assessing teacher-student interpersonal relationships in science classes, *Australian Science Teachers Journal, 42*, 28-33.

Fisher, D.,L. & Waldrip, B.,G. (1999). Cultural factors of science classroom learning environments, teacher-student interactions and student outcomes, *Journal of Science Education and Technology, 17(1),* 83-96.

Fraser, B., J., (1981). *Test of science-related attitudes (TOSRA).* Melbourne: Australian Council for Educational Research.

Fraser, B., J., (1994). Research on classroom and school climate. In: Gabel, D., (Ed.), *Handbook of research on science teaching and learning,* (pp. 493-541). New York: Macmillan.

Fraser, B.,J., (1998). Science learning environments: Assessment, effects and determinants. In Fraser, B.,J. & Tobin, K.,G., (Eds.) *The international handbook of science education*. Dordrecht, The Netherlands: Kluwer.

Fraser, B., & Walberg, H., (Eds)(1991). *Educational environments: Evaluation, antecedents and consequences*. Oxford: Pergamon Press.

Goh, S.,C., & Fraser, B.,J., (1996). Validation of an elementary school version of the Questionnaire on Teacher Interactions. *Psychological Reports, 79*, 512-522.

Goodrum, D., Cousins, J. & Kinnear, A., (1992). The reluctant Primary School Teacher, *Research in science education, 22*, 163-169.

Klopfer, L.,E., (Ed.) (1992). A summary of research in science education - 1990. *Science Education, 76*(3), 239-338.

Leary, T., (1957). An interpersonal diagnosis of personality. New York: Ronald Press Company.

Lokan, J., Ford, P., Greenwood, L., (1996*). Maths and science on the line:* Australian junior secondary studentsí performance in the Third International Mathematics and Science Study. Melbourne: Australian Council for Educational Research.

National Board of Employment, Education and Training (1993). *Science, Technology Education: Foundation for the Future*. Canberra: Australian Government Publishing Service.

Schibeci, R.,A., (1984). Attitudes towards science: An update. *Studies in Science Education, 11*, 25-59.

Stocker, J., (1997). Foundations for Australia's Future. *Science and technology in primary schools.* Canberra: Australian Government Publishing Service.

Waldrip, B., & Fisher, D. (2000). Teacher-Student Interactions in Primary Science: Validity and Application of a Questionnaire. In D. L. Fisher & Jong-Hsiang Yang, (Eds.), *Proceedings of the Second International Conference on Science, Mathematics and Technology Education* (pp. 467-476). Perth: Curtin University of Technology.

Wubbels, Th., Creton, H. & Hooymayers, H., (1985). *Discipline problems of beginning teachers*. Paper presented at annual meeting of American Educational Research Association, Chicago, IL. (ERIC Document 260040)

Wubbels, Th. & Levy, J., (Eds.), (1993). Do you know what you look like: Interpersonal relationships in education. London: Falmer Press.

TEACHERS' VIEWS AND ATTITUDES TOWARDS THE COMMUNICATION CODE AND THE RHETORIC USED IN PRESS SCIENCE ARTICLES

Krystallia Halkia

Department of Primary Education, University of Athens

ABSTRACT

This article is focused on the ways press science articles affect schoolteachers in their everyday science classroom practice. The sample of the research consists of 82 primary school general teachers (PSGT) and 70 secondary school science teachers (SSST). The study showed that teachers (PSGT and SSST) consider press science articles as prime sources for their self-education in science matters. They update their knowledge in contemporary scientific issues or gain completely new knowledge in some, unknown for them, regions of science. Teachers trust the information found in newspaper science articles and they believe that they constitute very attractive, reliable and effective educational material for their everyday science lessons.

1. BACKGROUND, AIMS AND FRAMEWORK

The present article investigates the possible interactivity of formal and informal learning in science education. In particular, it is focused on how much and on which way schoolteachers are affected by press science articles in their everyday science classroom practice. While many researchers are involved in studying the learning of science through formal education, only few of them have pointed out the role played by informal sources in shaping the public's understanding of science. Nowadays, this role seems to be quite crucial, since learning in science can result from a synthesis of many different sources all around us. The informal sources designed to promote the understanding of science, appear in a great variety of contexts and through an increased number of media, like television, newspapers, magazines, books, museums, etc. (Kariotoglou et al. 1999, Wellington 1991, Collins and Bodmer 1986). Thus informal science is easily accessible to students, to their teachers and to the public in general.

Science knowledge coming from informal sources usually deals with contemporary science issues, or with science issues which affect people's everyday lives. Maybe that is one of the main reasons why it seems more meaningful to people than the knowledge coming from formal education. This kind of knowledge decisively contributes to shaping the public's attitude towards science (Schibeci 1989), while at the same time it also largely contributes to creating "citizen science", a "science that relates in reflexive ways to the concerns, interests and activities of citizens as they go about their everyday business" (Jenkins 1999).

D. Psillos et al. (eds.), Science Education Research in the Knowledge-Based Society, 415–423.

Among the great variety of informal sources, it is now widely recognized that "press science" promotes scientific literacy to society. This is done by "those professionals that provide educational and interpretative opportunities for the general public to better familiarize itself with science, in addition to those who report science as "news" and write about science in general" (Laugksch 2000).

The two approaches of science education represented by "press science" and by "school science" do not follow similar routes in the process of science learning and they do not share the same aims. Press science on one hand and formal science on the other, may both serve as "intermediaries" between scientists and the public (Wellington 1991). However, although each of these two media contributes in its own way to the public understanding of science, they are characterized by completely different methodological tools and contrasting modes of interaction with their receivers. Press science functions in a context of free choice where the receiver is personally attracted by an article and chooses to read it, while formal school science is addressed to the student in a context of compulsory requirements (Wellington 1991). Thus, press science presents an extensive variety of attractive, but incidental scientific issues which are barely confined by rules of systematic approach and inter-article consistency, while formal school science is usually shaped around a carefully structured, strict curriculum.

As recently as the 60's, the decisive role of professionals who could reveal to the public the secret beauty of science, creating an overall positive attitude towards science, has been traced. Thus great science fiction authors (like Isaac Asimov), as well as renowned "popularisers" of science (such as Richard Dawkins, Paul Davies and Stephen Hawkins) are largely responsible for building the public's awareness of science, as contrasted to the rather limited contribution provided in this direction by the formal science education taught in schools (Fensham 1999). The problem faced by formal science education is that "citizen science" often is not compatible with "school science".

Recently, scholars and science educators from all over the world, have urgently warned that literacy in science is remarkably low (Physics on Stage 1999). In Greece, as well as in other countries, research addressed to the student population, reveals that students characterize physics as one of the most difficult and less attractive subjects of secondary school programs of study (Halkia and Karanicas 1999). But at the same time an increased interest by the public in science matters is being manifested. This is reflected in the high frequency with which science subjects are covered in the pages of newspapers and magazines (Koulaidis 2002). Indeed, lately, newspapers have developed special insets dedicated to science, which are largely addressed to students and teachers. These science articles and insets seem to elicit a large response from their audience. This might be an indication that the press has developed a successful science communication code with the public (Lewenstein 1992). This code seems to be acceptable to most people's cognitive schemata, since people are neither accustomed to, nor easily accept, the strict formalistic code of science researchers.

In addition to that, the science topics covered by the press articles seem successfully to address people's special interests. This informal science education has proved to be flexible in addressing different groups of people with special interests. In that way, each citizen has the opportunity to construct a body of science knowledge adapted to his own needs, which will enable him to cope with his every day problems (Jenkins 1999).

It therefore seems reasonable to suppose that today students gain a great deal of their overall science knowledge through informal sources of learning. Sometimes, such non-formal science knowledge comes to students directly from informal sources, while some other times it comes to them indirectly through their teachers or through their parents, their friends and their relatives, who are influenced by these informal sources.

In the framework of formal education, teachers decisively determine the process of science learning. Greek science curriculum demands the quick processing of a large amount of content, leaving very little freedom to teachers to use other sources apart from the provided science textbook in their everyday practice. However, teachers' own scientific awareness can often be influenced by non-formal sources of scientific knowledge. Recently, more and more pictures, information, data etc from such non-formal sources, make their way into Greek classrooms. The present research is designed to investigate the reaction of primary school general teachers (PSGT) and secondary school science teachers (SSST) to the communication code and the rhetoric that the press uses in science articles: it attempts to trace which communication techniques can be of potential use in science education, as well as teachers' attitudes towards the science articles published in the press.

2. METHODS AND SAMPLES

The research target population is that of Greek primary and secondary school teachers. The sample comes from 82 randomly selected primary and 60 randomly selected secondary schools of the region of Central and South Greece and covers regions of different socioeconomic status. One general teacher out of each primary school as well as one science and technology teacher out of each secondary school was randomly selected. Thus, the sample consists of 82 primary school teachers and 70 secondary school teachers.

The research instrument is a questionnaire consisting of four (4) parts:

- The first part (A) collects demographic data;
- The second part (B), consisting of 9 items, traces the teachers' habits and responses in reading press science articles (whether they read newspapers and magazines, the kind of articles they choose to read, the kind of elements that attract their attention and interest in articles etc.);
- The third part (C), consisting of 7 items, explores teachers' attitudes towards the science and technology articles which are published in the press and their everyday practice towards the use of these articles in their science lessons and
- The fourth part (D) analyses teachers' responses to an experimental treatment: Four (4) articles, published on the same day (Sunday) in Greek newspapers with a large circulation, have been presented to teachers. Teachers are asked to:

1. Choose and read an article among those four (4) newspaper science articles, which refers to different aspects of science and technology and utilizes different communication techniques;
2. Explain the reasons for their choice;
3. Choose the article's section (a paragraph or a picture or both) which impressed them most and
4. Choose the article's section (a paragraph or a picture or both), which could be of use in their science lessons.

   (The three last questions of part (D) are open-ended questions).

The four articles we are referring to are briefly the following:

a) "How an Earth catastrophe can be avoided, in view of a possible collision with an asteroid or a comet" (Astronomy): A scientist, who often writes popular articles for the press, writes the specific article. He adopts a communication code, characterized by common stereotypes of an outside danger threatening the earth. The article contains a lot of information simulating a possible disaster, accompanied by evocative pictures.
b) "The rapid change of climate and how it affects Greece" (Environment): Similarly, a scientist writes this article. The article adopts a popularised code, which however is strictly limited to the presentation of facts and data. It contains a lot of scientific information and is accompanied by many abstract graphs.
c) "Robotics: In a few years time robots will replace unskilled workers" (Technology): This time, a journalist signs the specific article. He adopts a journalistic presentation, containing attractive trivial information and social references. It is accompanied by a lot of descriptive photographs.
d) "Strings, the music of the universe" (Physics Theories): An eminent scientist is the writer of this article. He attempts a conscious popularisation of a difficult subject, using a lot of analogies, metaphors and scientific models. Evocative pictures accompany the article.

The completion of the questionnaire lasted one hour. We introduced the questionnaire to the research subjects by telling them that its main aim was to answer a question raised within the scientific community. The question was whether press science articles affect science education or not. Relevant argumentation expressed among members of the scientific community was strongly divided. Thus, to get a clear answer, we were addressing the people qualified to know that is the teachers themselves.

The teachers' responses were collected mainly using Likert scale items and items ranking in order of importance. For the fourth part responses, a content analysis was conducted. All data were analyzed using the SPSS program.

3. RESULTS

The main results of the research are the following:

(Part B of the questionnaire) Do teachers read science articles?

- The great majority of teachers read newspapers regularly (every day 36,8% and every Sunday 48,7%).
- They read magazines too (85,5%), but they choose magazines, which cover a wide variety of subjects and topics and are accompanied by many impressive photographs. Thus, in the list of ten magazines, which include science subjects exclusively or partially, 46,7% of teachers prefer to read (1st) the "National Geographic" (Greek translation), while only 5,3% of them read (9th) the "Scientific American" (Greek translation).
- For primary school teachers, science is one of the least favoured subjects they choose to read (6th among 9), while, for secondary school science teachers, science is the most favoured one (1st among 9).
- Most teachers are mainly attracted by a science article's title and its accompanying pictures.

(Part C of the questionnaire) Do press science articles affect school science?

- Teachers report that they utilize press science articles in their everyday teaching (95% PSGT and 98% SSST).
- They also report that press science articles:
  a) Constitute a reliable source of information (83% of PSGT and 64% of SSST).
  b) Present complex science concepts in an effective way (82% of PSGT and 54% of SSST).
  c) Contain more contemporary knowledge than the corresponding science textbooks (87% of PSGT and 83% of SSST).
  d) Contain more effective and attractive information for their students than the corresponding science textbooks (89% of PSGT and 74% of SSST).
- Teachers utilize press science articles in their science lessons in one or more of the following ways:
  a) As students' motivation in the introduction of a subject (85% PSGT and 76% SSST);
  b) As supportive educational material, by using the article's pictures (80% PSGT and 60% SSST);
  c) As a source of contemporary scientific knowledge for further self education (61% PSGT and 70% SSST).

(Part D of the questionnaire) Which press science article and why?

The content analysis of teachers' responses to the fourth part of the questionnaire (the one concerning with their choice and reading of the specific newspaper science articles) revealed the following:

I) Which science article do teachers choose to read?

The sample teachers chose to read the following science articles (table 1).

Table 1

Teachers	Astronomy "How an Earth catastrophe…"	Environment "The rapid change of climate…"	Technology "Robotics: In a few years time…"	Physics theory "Strings, the music of the universe"
Primary	12%	56%	15%	17%
Secondary	20%	31%	9%	40%

One first remark about the distributions in table 1 shows that teachers prefer to read articles written primarily by scientists. The primary school general teachers (PSGT) prefer to read articles that refer to phenomena affecting people's lives (environment), while the majority of the secondary school science teachers (SSST) prefer to read articles, which either use a scientific language and refer to phenomena affecting people's lives (environment) or present a complex theory in a simple manner (physics theory).

Content analysis on the open-ended questions revealed no significant differences between the PSGT and the SSST. Therefore for reasons of conciseness these answers (tables 2, 3 and 4) are being reported as representing one single group of teachers (adding PSGT and SSST together). Below, only answers that elicited frequencies higher than 10% are being reported.

II) Why do teachers choose each particular science article?

The common reasons for their choice, as they emerge from their answers, are presented in the following table 2.

Table 2.

a/a	***Reasons for choosing an article***	%
1	Subject (Interesting – Up to date)	37.5
2	Fear-concern about coming **danger** (collision of earth with a meteorite/comet, rapid change of climate, robots replace human beings)	35.5
3	For their personal reading and **self education**	17.8
4	**Writer** (known scientist)	17.1
5	Article's fascinating title- subtitle (Key phrase)	13.2
6	Subject touching **students' interests**	11.2

Reading the above data (table 2), some interesting points came out. First, teachers are attracted by the thrilling element of a coming danger, which threatens people's lives. Second, they choose to read an article for their self-education, a response that emerged again in the results of part B of the questionnaire. Third, they prefer to read articles written by scientists and not by journalists. In addition, some teachers choose to read articles touching their students' interests.

III) What part of the articles, do they personally like most?

The content analysis of each suggested paragraph or sentence, which the teachers personally liked, reveals some common features, which are presented in the following table 3.

Table 3.

a/a	***The part of the article that I personally like most***	%
1	Data confirming high probability of coming **danger** for human life, or society's structure, due to: a) The earth's collision with a comet or a meteorite, b) The degradation of natural resources, c) Human beings' replacement by robots	29.6
2	Information testifying that **human activities** affect the climate	19.7
3	Didactic tools: use of **analogies, metaphors, simulations** for space-time, strings, greenhouse effect, earth's possible disasters	19.1
4	Scientific **models** (weather prediction, greenhouse effect, atomic structure)	18.4
5	Article **illustration**: a) Evocative pictures b) Descriptive pictures c) Mental pictures: conceptual maps and/or graphs	13.8
6	Scientists' views and activities: a) Views for the inner structure of matter b) Prediction and prevention of coming dangers c) Construction of robots which replace human beings	10.5
7	Epistemological issues: Historic evolution The limits of scientific thought	10.1

The data (table 3) reveal that primarily the element of danger motivates the interest of teachers and makes the article's content meaningful to them. They also appreciate the use of "didactic tools" (analogies, metaphors and simulations) as well as the use of models in the particular science articles, which enable them to understand the underlying science concepts presented in the text. They also liked the pictures (mostly evocative) accompanying the specific article.

(IV) What part of the particular articles, do they believe that could be used in their science lessons?

The content analysis of each suggested paragraph or sentence that the teachers consider as having the potential to be used in science classrooms, reveals some common features, which are presented in the following table 4.

Table 4

a/a	***The part of the article that I could utilize in my science lessons***	%
1	Data confirming high possibility of coming **danger** for human life, or society's structure, due to: a) the earth's collision with a comet or a meteorite, b) the degradation of natural resources, c) human beings' replacement by robots	34.9
2	Article **illustration**: a) evocative pictures b) descriptive pictures c) mental pictures: conceptual maps and/or graphs	24.3

3	Didactic tools: use of **analogies, metaphors, simulations** for space-time, strings, greenhouse effect, earth's possible disasters	21
4	Scientific **models** (weather prediction, greenhouse effect, atomic structure)	16.4
5	Information testifying that human activities affect the climate	15.1
6	Epistemological issues: Historic evolution The limits of scientific thought	10

These main common features are the narrative elements in a science story, the illustrations (mostly evocative), the use of some didactic tools (analogies, metaphors and simulations) and the use of models, which are crucial for helping students to understand the underlying science concepts (table 4).

4. CONCLUSIONS AND IMPLICATIONS

Press science articles have been recognized to be a prime medium for the continued science education of non-scientists (Phillips et al. 1999, Korpan et al. 1997). Unexpectedly, this present study showed that teachers (PSGT and SSST) consider press science articles as prime sources for their self-education in science matters. They update their knowledge in contemporary science matters or gain completely new knowledge in some, unknown for them, regions of science. Maybe this is the reason why they appear expressing great confidence in articles written by scientists and presented with an attractive code. In that respect, university scholars too have to seek their own role and consider their responsibility in popularising science since, through their intervention, science knowledge comes back to the science classrooms (Dunwoody 1992, Thomas et al 1987).

Thus through teachers, press science crosses the borders of formal education and now reaches the students formally in two ways: Firstly, the use of press science articles in science classrooms helps students to understand the public's discourse about science, as well as to develop several communication skills (Hand 1999). Secondly, such articles can be used as educational material. This is supported by the findings of this present research, which reveal that teachers trust the information found in newspaper science articles and they believe that they constitute very attractive, reliable and effective educational material for their everyday science lessons. Similarly, as other researchers have also pointed out, the rhetoric used in science articles seems to be very appealing to teachers (Koulaidis 2002). Thus, teachers seem to choose sections from the articles that use an emotional - «poetic» language with a lot of metaphors and analogies to introduce difficult scientific concepts, e.g. they highly appreciate the narrative elements of a popularised article. The appeal of this narrative element seems to reflect a deep psychological need. From earliest times, man felt the need to listen to stories, which give shape to his fears. Teachers seem to discriminate between the boring content of science textbooks used in formal education on the one hand and the charming science "stories" presented in the press on the other. These findings may suggest the directions for further research about the structure of science textbooks in formal education. A "hypertext" form of science textbooks, where occasional press science articles, of current concern, could be the trigger for motivating students' and teachers' interest for science matters. This could lead them to further searching and conscious studying of the underlying subject matter, according to their personal abilities and interests. Thus, we strongly share Wellington's (1991) argument that, if press science is used carefully and critically, it can help and be supportive to formal education.

The press seems to be a very powerful intervening component in formal science education. This must not be underestimated when the designers of science education policy plan the coming reforms. Some of the communication strategies used in these articles could be of potential use in science education: in designing science textbooks and science educational software or in planning teachers training science

ACKNOWLEDGMENTS

We would like to thank the postgraduate students Ms S. Theodoridou and Miss K. Malamitsa for their valuable contribution regarding the systematic distribution of the questionnaire to the participating teachers and the University of Athens (Research Committee) for its financial support during this research.

REFERENCES

Collins, P. & Bodmer, W. (1986), The public understanding of science. *Studies in Science Education*, 13, 96-104.

Dunwoody, S. (1992). The challenge for scholars of popularised science communication: explaining ourselves. *Public Understanding of Science*, 1, 11-14.

Fensham, P. (1999). School science and public understanding of science. *International Journal of Science Education*, 21(7), 755-763.

Halkia, K. & Karanicas, I. (1999). Attitudes and views of Secondary Education students towards the subject of Physics. *Proceedings of the 8th Greek Conference of Greek Physicists' Union* (pp. 215-219). 28-31 January 1999.

Hand, B. (1999). A writing in science framework designed to enhance science literacy. *International Journal of Science Education*, 21(10), 1021-1035.

Jenkins E., W. (1999). School science, citizenship and the public understanding of science. *International Journal of Science Education*, 21(7), 703-710.

Kariotoglou, P. & Papasotiriou, C. (1999). *The educational aspects of informal science education programs*. Paper presented in the Conference: Science as Culture, Como, Italy.

Korpan, C.,A., Bisanz, G.,L. & Bisanz J. (1997). Assessing literacy in science: Evaluation of scientific news briefs. *Science Education*, 81, 515-532.

Koulaidis, B. (2002). Towards a redefinition of the Didactic of Science: Pedagogic and Communication. *Proceedings of the 2nd Greek Congress: The Didactic of Science and the application of New Technologies to Education* (pp. 3-17). 3-5 May 2000, Cyprus.

Laugksch, R. (2000). Scientific Literacy: A conceptual overview. *Science Education*, 84(1), 71-93.

Lewenstein, B. (1992). *When science meets the public*. Washington: American Association for the Advancement of Science.

Phillips, L.,M. & Norris, S.,P. (1999). Interpreting popular reports of science: what happens when the reader's world meets the world on paper?, *International Journal of Science Education*,. 21(3), 317-327.

Physics on stage (a joint program by CERN, ESA and ESO), Project. descr., 1999.

Schibeci, R.,A. (1989). Home, school and peer group influences on student attitudes and achievement in science, *Science Education*. 73(1), 13-24.

Thomas, G. & Durant, J. (1987). Why should we promote the public understanding of science? *Scientific Literacy Papers*, 1, 1-14.

Wellington, J. (1991). Newspaper science, school science: friends or enemies? *International Journal of Science Education*, 13(4), 363-372.

Science Teachers' Perceptions Of The Current Situation Of Planetary Emergency

[1]Vilches, A., [2]Gil-Pérez, D., [3]Edwards, M., [4] Praia, J.

[1,2,3] *Universitat de València, Spain,* [4]*Universidade de Porto, Portugal*

ABSTRACT

During the United Nations Conference on Environment and Development held in Rio in 1992 *educators of every subject* were asked to contribute to public awareness and understanding of the problems and challenges related to our planet's future, in order to make possible citizens' participation in well grounded decision-making. The purpose of our contribution is to analyse what science teachers' perceptions are, because without correct teacher perception of the planetary crisis and its possible solutions, we can't expect their effective implication in this dimension of education.

1. BACKGROUND, AIMS AND FRAMEWORK

Until the second half of the 20th Century, our planet seemed very large, practically without limits, and the effects of human activities remained locally compartmentalised (Fien 1995). But these compartments have begun to dissolve during the last decades, and many problems (greenhouse effect, ozone depletion, acid rain…) have acquired a global dimension (Bybee 1991; Orr 1995). The "state of the world" has become, for this reason, an object of growing concern. This is the reason why educators are being asked to contribute to public awareness and understanding of the problems and challenges related to our planet's future, in order to make possible citizens' participation in well grounded decision-making (United Nations 1992).

Our hypothesis is that, in spite of such dramatic appeals, attention paid by science teachers to the state of the world is still very poor and constitutes a serious *missing dimension* in science teachers education and, more generally, in science education research and innovation (Gayford 1993; Gil-Pérez 2000). Our purpose is, very precisely, to analyse what science teachers' perceptions are, because without correct teacher perception of the planetary crisis and its possible solutions, we can't expect their effective implication in this dimension of education. This analysis demands, firstly, to clarify what we understand by a correct view of the state of the world.

If we want to understand correctly the current situation of planetary emergency and how to act upon it, it seems necessary to go beyond the consideration of some concrete or local environmental problems. As Tilbury (1995) signals, "environmental and development problems are not solely caused by physical and biological factors" and "an understanding of the parts played by aesthetic, social,

D. Psillos et al. (eds.), Science Education Research in the Knowledge-Based Society, 425–432.

economic, political, historical and cultural elements is required". We need to construct a global picture of the state of the world and study thoroughly the possible causes and remedies. With this aim:

- We have made an effort to take into account the contributions of different studies elaborated with an explicit global point of view, and very particularly the Worldwatch annual dossiers on the state of the world (Brown et al. 1984-2001);
- We have also analysed papers published on the subject in science education and environmental education journals;
- We have undertaken a Delphi study (Gil-Pérez 2000) implicating several dozens of science teachers of all levels and from different countries.

These studies have permitted us to construct a network (**Table 1**) which summarises the ensemble of *related* problems, causes and challenges that characterise, in our opinion, the state of the world (Gil-Pérez 2000).

2. METHODS AND SAMPLES

In order to find out to what extent teachers' conceptions on the state of the world respond to a correct perception of the problems to be faced and the measures to be adopted, we have conceived several experimental designs as, for instance:

- To pose an open question to teachers in training and in service, about "*problems and challenges that humanity has to face*", in order to see if they make reference to the different aspects that are relevant in this respect;
- To analyse the content of science textbooks (looking for any reference to the state of the world and its future);
- To analyse the content of papers published in science education journals (looking for researches related to the state of the world as, for instance, analyses of students' or teachers' conceptions).

The open question has already been proposed to wide samples of science teachers, in service and in training from Spain, Portugal and Latin America (Argentine, Brazil, Chile, Cuba, Mexico, Panama…) with a total of 327 teachers in service and 521 in training. On the other hand, we have analysed more than 300 Spanish science textbooks. Finally, we have analysed 9932 papers published between 1992 and 1998 in 31 science education journals.

The information collected with these different designs has been analysed using the network elaborated to summarise the ensemble of *related* problems, causes and challenges that characterise the state of the world (table 1).

Table 1. Problems and challenges that characterise the present state and near future of the world

0) The main aim should be to lay the bases of a sustainable development .

This directs our attention towards an ensemble of interconnected aims and actions:

1) To put an end to socio-economic growth, guided by particular interests in the short term, that seriously damages the environment and is particularly dangerous for living beings

This economic growth produces, among other things, these particular problems:

1.1. A growing, disordered and speculative urbanisation
1.2. Environmental pollution and its consequences (greenhouse effect, acid rain, global climatic change…)
1.3. Depletion of natural resources (fossil energy resources, fertile soil, drinking water…)
1.4. Ecosystems degradation and destruction of biological diversity (cause of illness, ravenous hunger…)
1.5. Destruction, in particular, of cultural diversity

2) To put an end to the following causes (and, at the same time consequences) of an unsustainable socio-economic growth:

2.1. Over-consumption of "developed" societies and dominant groups .
2.2. Demographic explosion in a limited planet
2.3. Social inequalities between human groups
2.4. Conflicts and violence associated with these inequalities (military conflicts, Mafia's activities, speculation of transnational enterprises that escape any democratic control…)

3) To adopt positive measures in the following fields:

3.1. Political measures at the planetary level capable of promoting and controlling the necessary protection of social and physical environment, before the current degradation processes become irreversible

3.2. Educational measures to overcome the usual tendency to behave attending to particular interests in the short term, making possible changes in personal values and life style choices to promote solidarity

3.3. Technological measures to better satisfy human needs without damaging the environment, capable of favouring a sustainable development. This includes, for instance, the search for new energy sources, the improvement of efficiency in food production, the prevention of illness and catastrophes or the diminution and recycling of waste)

4) To associate the precedent measures with the need to universalise and enlarge human rights

4.1. Democratic civil rights (opinion, association…) for everybody as a condition sine qua non for citizens' decision-making about current and future environmental and social problems.

4.2. Economic, social and cultural rights (to a satisfactory job, to health, to education and culture…)

4.3. Right, in particular, to investigate any kind of subject (life's origin, genetic manipulation…) without ideological limitations, but with a social control that takes into consideration the social and environmental consequences and prevents the hasty application of non sufficiently tested technologies.

4.4. Solidarity rights (right to a healthy environment, right to peace and right to a sustainable development)

3. RESULTS

The results obtained with the different samples of teachers are highly concordant and show a quite reductionist perception of the science education community about the state of the world, which is also reflected in science textbooks and in papers published in science education journals.

3.1 References to sustainability

When thinking about problems and challenges concerning the future of humanity, the basic aim signalled by experts is to lay as near as possible the bases of a *sustainable development,* that is to say, of a development that "meets the needs of the present without compromising the ability of future generations to meet their own needs" (World Commission on Environment and Development 1987). "*Sustainability is the central unifying idea society most needs at this point of human history*" (Bybee 1991).

From the evidence of our surveys, references to global considerations such as the idea of sustainable development are very rare: only 7.3% of the 327 science teachers in service and 4.1% of the 521 teachers in training make any references to sustainability.

But, how are we to move towards a sustainable society (or sustainable development)? Experts (United Nations 1992) refer to the necessity of putting an end to a series of *interconnected* facts, each one having a particular importance and deserving particular attention, but completely linked to the rest: none of them can be understood or treated without taking into account the whole ensemble. It is not enough, for instance, to criticise -as is usually done- environmental pollution and its consequences (greenhouse effect, acid rain...) or depletion of natural resources: there are other problems to be taken into account.

3.2 Attention to the problems that affect our survival and their causes

Quite high percentages of science teachers –both in service and in training- signal environmental pollution, depletion of natural resources and destruction of biological diversity among the main problems humanity has to face. On the contrary, there are few references to related problems such as growing and disordered urbanisation or the destruction of *cultural* diversity (less than 20 % make any reference to these aspects). This very frequent omission is a clear example of the reductionism that characterises science teachers' views -and even environmental educators' views- about our planet's problems (Fien 1995).

The same reductionism appears in relation to the *possible causes of the Earth's degradation*: About 40 % of science teachers denounce economic growth, guided by particular interests in the short term, as the foundation of the current degradation processes. This is a correct appreciation, according to the analyses of, for instance, the World Commission on Environment and Development or the Worldwatch Institute (Brown et al. 1984-2000). But these problems are intimately related to other phenomena that should be considered as well. Firstly, we draw attention to the over-consumption of so called "developed countries" and of dominant groups in each society (United Nations Development Program 1998). A consumption that keeps growing as if the Earth's capacities were infinite is indefensible (Folch 1998).

Secondly, it is necessary to halt the demographic explosion on a planet, which has limited resources (Ehlrich and Ehlrich 1990). In a planet where, since the second half of the 20th century, more human beings have been born than in the whole of humanity's history and where the *present* population would need the resources of three Earths to generalise the standard of living of the developed countries (United Nations 1992).

In short, over-consumption and demographic explosion determine an economic growth which is extremely corrosive of the physical *and cultural* environment. Astonishingly, in our survey science teachers do not seem to be aware of the importance of these two determining factors; only about 20% make any reference to demographic growth, and over-consumption is mentioned by less than 10%(!).

On the other hand, over-consumption in developed countries and demographic explosion in others provoke grave inequalities. Billions of fellow humans are scarcely able to survive in non-developed countries and there is the exclusion of wide segments of the "first world"... while a fifth of the human population offers its high-consumption model (United Nations Development Program 1997; Folch 1998). Extreme poverty in undeveloped countries, that is a consequence of the demographic explosion and of the imposition of particular interests and values (through military conflicts or through the activities and speculations of trans-national enterprises which seek to avoid any democratic control) lead inexorably to an unsustainable exploitation of natural resources, in a desperate attempt to pay back interest, satisfy external debts and gain some benefit. These inequalities and derived conflicts receive certain attention in the responses of our survey of science teachers (about 50% make reference to them). But, as we have already indicated, no relationship is established between these problems and over-consumption or demographic explosion. There seems to be a serious lack of understanding of the relevance of globalisation to the problems outlined.

3.3 What positive measures can be adopted?

A holistic approach to the state of the world demands more than diagnosing the problems; it is also necessary that teachers study the possible solutions to the described planetary crisis, in order to help their pupils to explore alternative approaches and to participate in actions aiming to favour particular alternatives. We can structure the different proposals made by researchers and institutions into the following three groups:

- *Technological* measures to better satisfy human needs without damaging the environment;
- *Educational* measures to make possible the necessary changes in personal values and life-style choices;
- *Political* measures at a planetary level to avoid the imposition of particular interests and values harmful for other people or future generations.

About 30% of science teachers in service and 50% of science teachers in training refer to the need for these technological measures. Incidentally, this is one of the few aspects where we find strong differences between teachers in training and in service.

Nevertheless, although technology has an important role to play, it is necessary to question the widespread and erroneous idea that the solution to the serious problems which humanity has to face today depends solely on a better knowledge and on more advanced technologies: options and dilemmas are essentially matters of ethics (Tilbury 1995). This conclusion directs us in part to the educational measures we have to consider.

About 60% of science teachers in service and 50% of science teachers in training signal the need for educational measures, in accordance with the recommendations of sustainability experts (United Nations 1992). The educational measures proposed to contribute to a sustainable society put the accent on global analyses and solidarity (Delors et al. 1996). We need an education that contributes to a correct perception of the state of the world and prepares citizens for decision-making, generating responsible attitudes and behaviours (Bybee 1991; Fien 1995; Tilbury 1995) oriented to the attainment of a culturally plural and physically sustainable development.

On the other hand, individual contributions can and must go beyond the private domain and extend to professional, social and political activities. It is necessary, in particular, that individual and collective actions avoid local or partial approaches and contemplate many-sided environmental questions (pollution, resources depletion…), and other related problems such as social inequalities and conflicts, from a planetary perspective. The ecologists slogan "to think globally and to act locally" has its limitations; we now know that it is also necessary to *act globally* as well, adopting *political measures at a planetary level*, capable of avoiding the imposition of particular interests and values harmful for other peoples or for future generations.

In short, a new world order is required, based on co-operation and solidarity, with institutions capable of avoiding the imposition of particular vested interests harmful to other people or to future generations (Folch 1998). Unfortunately, only about 20% of science teachers in training or in service contemplate the need for political measures to guarantee the defence of the environment and life on Earth.

The ensemble of measures just discussed, appears nowadays to be associated with the need to universalise human rights. The next section is dedicated to clarifying this relationship.

3.4 Sustainable development and human rights

It may seem strange to establish such a direct relationship between human rights and sustainable development. In fact, only 6% of teachers in service and in training consider that overcoming the current degradation processes and inequalities is a question of human rights. For this reason, we shall try to clarify what is understood nowadays by *human rights*, a concept that has been growing and now contemplates three "generations" of rights (Vercher 1998).

We can refer, firstly, to *democratic civil rights (opinion, association...) for everybody, without social, ethnic or gender limitations*. They constitute a condition *sine qua non* for citizens' decision-making about current and future environmental and social problems (Folch 1998). They are known nowadays as "first generation human rights", because they have been the first rights to be demanded and obtained (not without conflicts and frequent violations) in a growing number of countries.

We refer, secondly, to *economic, social and cultural rights* or "second generation human rights" (Vercher 1998) as, among others, the right to a satisfying job; to an adequate dwelling; to appropriate nourishment; to health; to family planning and free enjoyment of sexuality (the only limitation being the freedom of others; to an education of quality, *throughout* one's life, without social, ethnic or gender limitations; to investigate any kind of subject (life's origin, genetic manipulation...) without ideological limitations but with a suitable degree of social control to prevent the hasty application of insufficiently tested technologies.

Finally, we refer to third-generation human rights, known as *solidarity rights* "because they tend to preserve the integrity of the whole population" (Vercher 1998). They incorporate the right to life in a suitable environment, the right to peace and the right to a sustainable development for all people and future generations.

The *ensemble* of these rights appears to be a requisite (and, at the same time, an objective) of a sustainable society; *they are all interconnected.* We cannot conceive, for instance, the interruption of the demographic explosion without the recognition of the right to family planning and free enjoyment of sexuality... and these are connected also to the right to education. In short, then, achieving sustainable development is synonymous with universalising human rights in its widest sense. Unfortunately, we notice, most science teachers in our survey did not make any reference to human rights: only 6% of teachers (both in service or in training) consider that overcoming the current degradation processes and inequalities is a question of human rights; references to democratic civil rights are made by less than 2% of science teachers and less than 7% refer to rights of solidarity; the only rights that are mentioned by a significant percentage of science teachers –about 10% - are socio-economic ones.

4. CONCLUSIONS. A GLOBAL ANALYSIS OF THE RESULTS AND PERSPECTIVES

Any attempt to face up to the problems that affect human survival should contemplate *the ensemble* of challenges that we have pointed out (table 1). We have already shown that high percentages of science teachers' perceptions (both in service and in training) about the state of our planet ignore each one of the aspects studied. In fact the mean of the aspects identified by teachers in service and in training is, respectively, 4.7 and 4.9 (from a total of 19). The similarly poor perceptions of both groups emphasises our finding that the concern for the state of the planet is a missing dimension in science education.

These results explain why we have referred to the state of the world as a missing dimension in science teachers' education. A dimension to be urgently incorporated. We have already begun to implement workshops to facilitate the study of these problems by science teachers (Gil-Pérez et al. 2000). The first results obtained are quite encouraging.

REFERENCES

Brown, L., R. et al. (1984-2001). *The State of the World.* New York: W.W. Norton.

Bybee, R.,W., (1991). Planet Earth in Crisis: How Should Science Educators Respond? *The American Biology Teacher*, 53(3), 146-153.

Delors, J. et al. (1996). *L'Éducation: un Trésor Est Caché Dedans.* UNESCO. Commission Internationale sur l'Education pour le vingt et unième siècle. Paris : Jacob O.

Ehlrich, P.,R. & Ehlrich A.,H., (1990). *The Population Explosion.* London: Touchstone.

Fien, J., (1995). Teaching for a Sustainable World: the Environmental and Development Education Project for Teacher Education. *Environmental Education Research*, 1(1), 21-33.

Folch, R., (1998). *Ambiente, Emoción y Ética.* Barcelona: Ariel.

Gayford, C., (1993). Where Are We Now with Environment and Education? (Editorial) *International Journal of Science Education*, 15 (5), 471-472.

Gil-Pérez, D., (2000). The state of the world. A missing dimension in physics teacher education. In: Pinto, R. & Surinach, S. (eds), *Physics Teacher Education Beyond 2000.* Paris: The Data Science Library.

Gil-Pérez, D., Vilches, A., Astaburuaga, A. & Edwards, M., (2000). La Atención a la Situación del Mundo en la Educación de los Futuros Ciudadanos y Ciudadanas. *Investigación en la Escuela*, 40, 39-56.

Orr, D.,W., (1995). Educating for the Environment. Higher Education's Challenge of the Next Century. *Change*, May/June, 43-46.

Tilbury, D., (1995). Environmental Education for Sustainability: Defining the New Focus of Environmental Education in the 1990s. *Environmental Education Research*, 1(2), 195-212.

United Nations, (1992). *UN Conference on Environmental and Development.* Agenda 21 Rio Declaration, Forest Principles. Paris: Unesco.

United Nations Development Program, (1997). *Human Development Report. Human Development to Eradicate Poverty.* New York: Oxford University Press.

United Nations Development Program (1998) *Human Development Report. Consumption for Human Development.* New York: Oxford University Press.

Vercher, A., (1998). Derechos Humanos y Medio Ambiente. *Claves de razón práctica*, 84, 14-21.

World Commission on Environment and Development, (1987). *Our Common Future.* Oxford: Oxford University Press.

PART VI

International Research and Development Projects

A EUROPEAN RESEARCH PROJECT FOR NEW CHALLENGES IN SCIENCE TEACHER TRAINING

Roser Pintó

Departament de Didáctica de la Matemática i les Ciències Experimentals, Universitat Autònoma de Barcelona, Spain

ABSTRACT

This paper intends to give an overall view of a European project devoted to analyse teachers' classroom practices when they are faced with some specific curricular innovations and to detect differences between the proposed approaches and the implemented ones. The goal was to describe in detail teachers' transformations in order both to be able to better address teacher training courses or materials introducing curriculum innovations and, in addition, to make the intentions of curriculum designers more suitable and reasonable to teachers. A research project, where five Universities with some common and some particular research purposes are involved, needs to be carefully planned if the target is to achieve comparable results. The paper aims at presenting, on the whole, the procedure followed during this three years research work and, at the same time, some very general results.

1. THE STTIS PROJECT

The STTIS Project (Science Teacher Training in an Information Society) is an R+D project funded by the European Commission's Directorate-General for Research through the TSER program, in the 4th framework period. Five European universities from five countries were involved. Research teams worked together in the area of science teacher education, for more than three years. The leaders of the research teams:

Laurence Viennot from Paris VII University (FR)
Elena Sassi from the Univ. Federico II of Naples (IT)
Andreas Quale from the Univ. of Oslo (NO)
Jon Ogborn, from Sussex University (UK)
Roser Pintó from the Univ. Autònoma de Barcelona (SP), being the co-ordinator.

In the forthcoming pages, contributions from all the research groups are present. Since, it is not possible to summarise all the research results, now our attempt is to offer a global picture jointly with the process to get them.

2. RATIONALE OF THE STTIS PROJECT

The Information society has brought different innovations within the school that have converged with other changes impelled by research in education: changes of

D. Psillos et al. (eds.), Science Education Research in the Knowledge-Based Society, 435–446.

curricula, changes in the priority of objectives, changes of the perception of education and learning, changes in teaching methods, etc. (BLACK, P. et al. 1996).

Specifically, research into teaching and learning processes, and into processing information within the current technological and changing society have decisively lead to new guidelines to be followed by teachers when selecting and structuring teaching content or in selecting textbooks for their students; when approaching teaching strategies for their classes (new technological resources, learning activities based on the use of media or IT tools, experimental work using computers, etc.) or when following given didactical instructions (Bell, De Jong et al., Van den Akker, 1998). Failing to understand the effects in proposing educational innovations could result, and is already resulting, in a very inefficient way.

In the STTIS project, the focus was on the teachers' adaptation to the challenges that the reforms and curricular innovation presuppose. Teachers obtain knowledge of curricular innovations through different ways such as in-service courses organised for educational authorities, university departments or professional groups. In this transmission of information by different sources, the content of such information is transformed. The information receiver is not a passive agent. Rather, whether consciously or not, he/she selects, categorises, and interprets the incoming information. «Communications are not simply 'received' but are re-made, re-constituted and transformed by the receiver. Communication must be regarded as action, as minds acting on other minds, which then act in response. Those acts of response are necessarily transformative, making new meanings relate to previous ones» (STTIS, 1998).

The STTIS research teams involved in the research presented in the forthcoming pages have adopted this *transformative view of the nature of communication*. Given this view, teachers should be considered as information transformers, given that they transform current new information.

3. GENERAL AIMS

The STTIS research teams were interested in determining how teachers transform the innovations encountered and in seeking certain trends by which these transformations can be attained. By knowing how teachers carry out the process of adaptation to an innovation, it is assumed that certain drawbacks concerning adequate implementation can be avoided. In other words, the conditions for science teachers to successfully implement certain curricular innovations in their classes, the obstacles that must be removed and the appropriate materials for reducing the effect of unfavourable factors have all been at the core of the STTIS research project.

Throughout the STTIS project, the implementations of specific curricular innovations as well the distance between expectation and real actions in classrooms have been carefully analysed. From the large variety of curricular innovations, only the implementation of three types has been selected: symbolic representations, computer tools and teaching sequences for certain scientific content.

4. SHARING A RESEARCH METHODOLOGY

A research project involving five European teams with their particular idiosyncrasies needs to accurately define the methodology that it will be followed if general results have to be obtained from the national ones. In the STTIS project:

- The research methodology adopted is *qualitative* in nature, because we aimed at a qualitative diagnosis and description of the problems.
- The research is fundamentally *applied in character*; that is, it is directed to informing and so to potentially improving practical decision-making and action.
- The problems selected to characterise are of *general importance and significance*; that is, they are not trivial in nature and are not highly localised or specific, (relevant to only a few contexts).
- The reliability of the results was achieved by:
  - Parallel research in different European countries in order to distinguish difficulties, which are more (or less) strongly related to local cultures and circumstances.
  - Focusing on teachers who are willing to innovate, thus focussing on effects that are likely to appear when teachers are seriously trying to implement an innovation.
- Such research places a heavy *demand on crosschecking - on 'triangulating'* - data and the interpretation of these data. We aimed at relying on *multiple data sources* (interview, class observation) and also to control, *variety* in the specific content and contexts of teaching that were under consideration.
- There is also the matter of cross checking or triangulating the terms and concepts used in analysing data and forming hypotheses or interpretations. The methodology provided for cross-comparison not only of 'results' but also of methods used and conceptualisations formed in order to 'produce' those results. Data from one country needed to be re-inspected through using the interpretative schemes used by another country, for example.
- Since not every required kind of innovation exists in every country, the *sampling* needed to be (and was) adapted to the specific circumstances with a *controlled variation.*
- *National cultures, structures and practices* help to determine the nature of innovation that could be practicable or was thought desirable. Thus, innovation should not be 'held constant' whilst national culture and context varied. *Research strategies could vary* only for principled reasons, but different methodologies adopted should address the *same* overall questions.
- Since the terms in which results are reported is a fundamental and essential part of writing a trans-national report, special care was taken for the *homogenisation of the language used.*

5. STEPS INVOLVED IN THE RESEARCH

The research plan designed to accomplish the above aims has three phases or stages:

Phase 1. Parallel empirical research studies on the implementation of innovation in five countries

Phase 2: Crossing national results on a particular innovation, in each country

Phase 3: Crossing results on parallel innovations across countries

Schema 1 graphically represents the overall of the project:

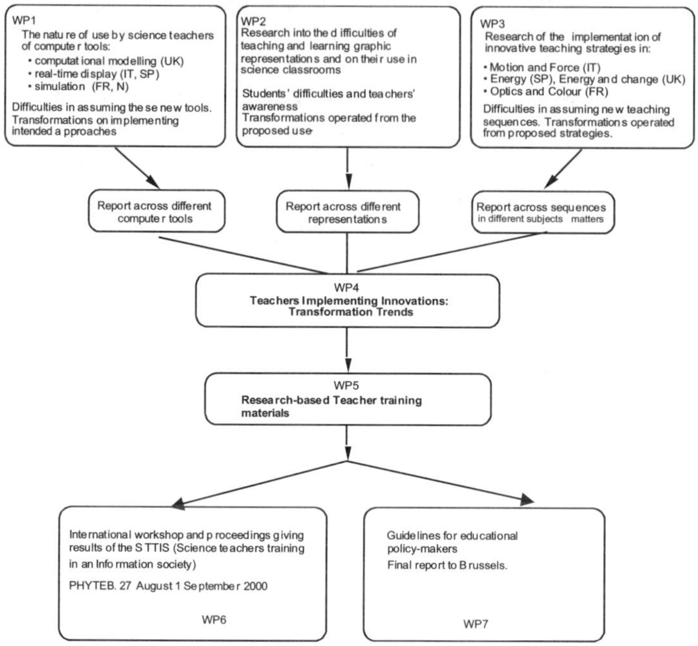

Figure 1

RESEARCH TASKS: THE IMPLEMENTATION OF THREE CURRICULAR INNOVATIONS

In the Phase 1, the five research teams involved in the research, working in parallel, have carried out on-site research, aiming at the identification of the difficulties involved in implementing three types of specific curricular innovations and of the transformations involved in implementing intended approaches. The paragraphs below provide a brief account of these tasks.

6. RESEARCH INTO THE NATURE OF USE BY SCIENCE TEACHERS OF INFORMATIC TOOLS. WORK PACKAGE 1

The work package[1] was devoted to new technologies for science classes. It constitutes the so-called Work Package 1, and was co-ordinated by Jon Ogborn, from University of Sussex.

The purpose was to investigate *the difficulties to assume the use of informatic tools* and to determine transformations undertaken on implementing intended approaches to use them. The set of research studies intended to provide an answer to the following research questions:

> What are the fits (matches) or gaps (mismatches) between the intended or expected use of the tools and the way that teachers actually use them in practice?
>
> What do teachers identify as more l and less successful examples of uses of these tools, and what reasons do they have for identifying them as such?
>
> Are teachers conscious of transforming suggested uses of tools in their own practice, and if so what reasons do they give? If not, are there clues about what may determine the teacher's practice?

Initially, the state of the art on the use and value of informatic tools in the five partner countries (France, Italy, Norway, Spain and the UK) was examined. (Subwork package WP1.1). Here we present some results from the research focused on the actual use of selected informatic tools in the science classroom. (WP1.2). The main data collected were analysed as a set of case studies of teachers using selected computer tools in science classrooms and, were concerned with identifying factors that favour or hinder their up-take.

The case studies themselves are each a report of a complex of human action in context, each part to be understood in relation to its whole. As a whole, they show that, despite certain obvious national differences, the issues which teachers face in constructing viable classroom events using IT ('lessons') have much in common. This appears despite the schools range, teachers' circumstances, variation in the supply and availability of informatic tools, differences in teaching practices or differences in national school structures.

In any of the countries, informatic tools have become an 'obvious' or 'natural' and 'taken for granted' tool for science teachers to use. Informatic tools are not fully 'naturalised'. In each set of national case studies, there are teachers who are new, or almost new, to the tools they are using. On many occasions, teachers reflect on "How I will do it differently another time".

The newness or novelty of the use of informatic tools is widely given a value of its own. In particular, teachers expect it to motivate students, to provide variety, to simply attract attention, to do "what the students want nowadays". Many of the teachers see necessary to integrate the use of informatic tools into a coherent sequence of teaching across several lessons and integrating the use of informatic

[1] According to the usual guidelines for European research projects, each set of research studies has been referred to as a Work Package.

tools with other activities, within an overall guiding plan. There is a general perception of the computer as facilitating other learning but rarely as knowledge in itself. It is fairly common to find teachers in our case studies not giving the computer the primary place in a teaching sequence. A tool with new potential for things to learn is confined to a role of the servant of the usual.

It is noticeable how some teachers respectively transform the use of tools to get a closer fit with these expectations. Essentially every case study shows a teacher trying to achieve a fit between the use of the computer, and the habitual classroom practices with which that teacher and his or her class are familiar. There can be little doubt that 'habits die hard'. A common transformation of explicit didactic intentions, provided in teachers' guides, is to select only those which fit with the teacher's practices. The teacher's perception of the degree to which a proposed use of an informatic tool 'fits' (that is, has not too great a distance from) some accepted segment of required content knowledge in the subject, is critical to its acceptance. If the fit is not good enough, the use of the tool will be transformed to make the fit closer. Where the proposed use of the tool carries with it new content approach, which is far from that which the teacher expects, both use of the tool and the new approach are likely to be transformed.

Practices in the use of informatic tools are significantly shaped by what resources are available, by where they are located, by when they are available, and by who has charge of them. A very common pattern, in all the countries, is to concentrate computing resources in a central bookable suite (or suites). Practices, which require informal ad-hoc spontaneous use, do not then develop, except where students have computers at home.

Rather few case studies report significant interaction with other teacher colleagues. Rather, the picture at present is mainly of individuals developing their own ways of using informatic tools.

Further information can be found at: http://www.blues.uab.es/~idmc42 files RW1.1 and RW1.2: *The nature of use by science teachers of informatic tools*. Stylianidou, F., Ogborn, J. and Balzano, E., Giberti, G., Gutierrez, R., Kolst¯, S. D., Monroy, G., Pérez, O., Pintó, R., Quale, A., Rebmann, G., Sassi, E., 2000.

7. INVESTIGATION ON THE IMPLEMENTATION OF GRAPHIC REPRESENTATIONS IN SCIENCE CLASSES WP 2

The intention was to investigate students' difficulties in reading certain images and to determine teachers' awareness of them. Transformations carried out during the implementation of activities using images were also a relevant part of this set of research studies. The following specific research questions needed to be answered:

> What are some of the problems and opportunities of the use of graphic representations in science classrooms that can be anticipated on grounds of prior evidence and/or theory? Are these confirmed by students' readings of such representations?
>
> How do teachers understand these anticipated problems? How do they deal with such problems, in the context of innovative teaching involving the use of images and graphic representation?

What transforming mechanisms can one conjecture to be involved?

This set of research studies constituted the so-called **Work Package 2,** and was coordinated by Roser Pintó from the Universitat Autònoma de Barcelona. Four research teams (i.e. those from FR, IT, SP and the UK) were involved in four parallel research initiatives.

Each research team selected a set of documents[2] containing images. In order to identify problems of representation, an a-priori common scheme about possible students' difficulties was elaborated. The theoretical framework in which to root the analysis was drawn from social semiotics (Kress, G. & Van Leeuwent, T. 1996) and from physics. Here, we outline the conclusions on the above research questions:

A 'list' of features of the images to detect students' difficulties designed by the research teams, has shown its usefulness to categorise the students' difficulties.

1. Images requiring interpretation of the roles of elements representing both real world and schematic or symbolic entities. (R/S)
2. Images constituting on-screen graphs, representing data being collected in real-time experiments, where external observable processes and graphic representations need to be related to one another. (RT)
   a) Ideal versus Real (RT-IvsR
   b) Global versus Local (RT-GvsL
   c) Gestalt (RT- Gest)
3. Images whose interpretation requires giving importance to or highlighting certain elements, often in relation to textual/graphical features which make or do not make them salient. (SEL)
4. Images containing elements, which require appropriate readings of symbols, and which contain examples of synonymy, homonymy and/or polysemy of symbols. (SIM)
5. Documents containing verbal elements, included in the image or used as captions, that need to be read as an important part of the whole. (VE)
6. Documents containing more than one image, thereby requiring interpretation of the relationships between the different images. (INT)
7. Images whose compositional structures are significant and thereby require the reading of spatial distributions and different representational structures. (CS)

Some general trends of interpreting images were detected in students' readings:

- Tendency to make narrative readings of the images, i.e. to interpret them as if they had a story-like structure.
- Tendency to misread the documents when information necessary to correctly interpret them is missing.
- Tendency to make use of previous knowledge when facing iconic representations that do not fit to their interpretative capabilities. In such cases, the image does not help students to learn new concept or information
- Tendency to give detailed explanation referring only the image's graphical elements and not taking into account its conceptual meaning.

The teachers' awareness of observed students' difficulties in reading images has proved to be generally low but increases when they are called upon to interpret and justify prototypical examples. Teachers tend to justify students' difficulties in terms of disciplinary aspects, instead of in terms of the characteristics of the graphical elements. The features of the images detected by the teachers as being sources of

[2] The word 'document' is used in this report to refer to the combination of text and images on a page.

difficulties do not match completely with the features detected as sources of the students' difficulties in the analysis done for the STTIS researchers

When asked about suggestions for changing the images to overcome students' difficulties, some teachers suggest the addition of more verbal elements in the document evoking a view that visual representations are not by themselves appropriate tools to convey precise ideas.

Some transformations in the use of images have been observed: Changes of the graphic elements that affect the meaning conveyed by the elements of the image have been displayed. It has been observed a weak addressing of graphical features of images that reduce its effectiveness to teach science. Transformations related to the use of the visual language or to the didactical role assigned to the images have both been observed. Some changes of the didactical role of the images seem to evidence an insufficient awareness of didactic advantages of using good quality image.
More information can be found at http://www.blues.uab.es/~idmc42/document/index.html (file RW2): *Investigation on the difficulties in teaching and learning graphic representations and on their use in science classrooms*. RW2. Pintó, R., Ametller, J., & F. Chauvet, P. Colin, G. Giberti, G. Monroy, J. Ogborn, F. Ormerod, E. Sassi, F. Stylianidou, I. Testa, L. Viennot, 2000.

8. RESEARCH INTO THE IMPLEMENTATION OF INNOVATIVE TEACHING SEQUENCES.

The workpackage is addressed to analyse the difficulties arising when teachers are faced with an innovative teaching sequence. The goal was to analyse in detail how teachers' non-neutral interpretation of a proposed teaching sequence may result in transformations that have conceptual consequences. This set of research studies constitutes WorkPackage 3 and was co-ordinated by Laurence Viennot from the University of Paris VII. It represents an extension of Hirn's work (1998). The specific research questions were formulated as:

> What specific requirements are expected of teachers in certain selected well-defined curriculum innovations in science?
>
> How do teachers understand these expectations? How do they act on these understandings in the classroom? What transforming mechanisms can one conjecture to be involved?

Four parallel research initiatives from four research teams (those from FR, IT, SP and UK) were carried out by analysing the implementation of specific innovative teaching sequences for secondary school courses.

Actual teaching strategies employed relating to teachers' conceptual difficulties, to their epistemological views, and to their previous practice were analysed through the implementation of six innovative sequences. Data sources from 171 teachers, from four countries were obtained through interviews before and after implementation, from class observation and video recording.

An open attitude toward the implemented innovation is observed in all the consulted teachers. They often seem to be aware of the learners' previous ideas,

language problems and learning difficulties, and may willingly claim that these deserve attention. However, important transformations are observed in all the investigations concerning the respective emphasis on the taught concepts, their mutual relationships, and their linkage. There is a trend to conflate of old and new strategies giving cause to a hypertrophy of the contents or to incoherence. The recommendations proposing to make links between different approaches or languages, between concepts and activities, or among conceptual paths are often not taking into consideration. It has been common to observe the performance of traditional views as: "seeing is understanding" or "doing is understanding". The innovations referring new sequencing of contents often have not been implemented according the didactical intentions of their designers. Many key features or details could not be perceived or adopted by our teachers, affecting the effectiveness of the innovative strategy.
More information in: (http://www.blues.uab.es/~idmc42/document/index.html) file RW3, under the title: *Investigation on teacher transformations when implementing teaching strategies*. RW3, Viennot, L. and Balzano, E., Chauvet, F., Giberti, G., Gomez, R., Hirn, C., Monroy, G., Ogborn, J., Pintó, R., Sassi, E., Stylianidou, F., 1999.

9. NATIONAL AND TRANSVERSAL REPORTS

A relevant task for the project was the elaboration of the so-called *National reports,* where each research team presented the conclusions of empirical research, usually triangulating results from different areas of its research about the implementation of a specific innovation (for example, from the results of different subworkpackages).

It has also been possible, by cross checking the results obtained across countries clustered in the National reports, to elaborate Transversal reports referring one of the specific curricular innovation. This "second level" research, *Transversal report*, has allowed to detect robust phenomena that appear despite contextual differences. They have been elaborated by the co-ordinator of each of the three empirical research studies, namely Jon Ogborn, Roser Pintó and Laurence Viennot. This corresponded to Phase 2 of the STTIS project.

10. INFERRING TRANSFORMATION TRENDS. WORK PACKAGE 4.

Previous empirical research studies, in which five European research teams working in parallel were involved, have made possible the crosschecking cross-site analysis of the National reports and Transversal Reports. There was not only a cross comparison across countries but also a crossing of results from the three types of selected curricular innovations. The aim of this investigation was to detect commonly shared features in teachers' performance when faced with an innovation. Therefore, the following research question was tackled:

> Can general trends be inferred in the transformation undergone by teachers when facing different curricular innovations?

This was the aim of **Work Package 4,** co-ordinated by Roser Pintó from the Universitat Autònoma de Barcelona. The analysis followed a back-and-forth strategy (Miles and Huberman 1984, Denzin and Lincoln 1998) and provided data whose contrast and comparison led to general clusters.

Interesting tendencies have been noticed. Information can be found at http://blues.uab.es/~idmc42/document/index.html file RW4: *Teachers Implementing Innovations: Transformations Trends.* Pintó, R., Gutierrez, R. and Ametller, J. Andresen, O, Balzano, E. Boohan, R., Chauvet, F., Colin, P., Couso, D., Giberti, G., Hirn, C., Kolsto, S.D., Monroy, G., Ogborn, J., Quale, A., Rebmann, G., Sassi, E., Stylianidou, F., Testa, I., Viennot, L., 2001.

11. RESEARCH-BASED MATERIALS FOR TEACHER TRAINERS

Previous research tasks should give rise to an important goal of the STTIS project: to elaborate *research funded material.* Teacher training activities mainly based on the results of the STTIS project have been prepared to favour the up-take of some innovations. They constitute **Work package 5** co-ordinated by Elena Sassi from the University of Naples. Eleven workshops aimed at teacher trainers were prepared in order to make teacher and teacher trainers more aware of common problematic points, and to offer activities for overcoming them.

Information can be found at (http://blues.uab.es/~idmc42/document/index.html) file RW5. *Teacher training materials favouring the take-up of innovations.* Sassi, E., Monroy, G., Testa, I., Ametller, J., Andresen. O., Balzano, E., Boohan. R., Chauvet, F., Colin, P., Couso, D., Giberti, G., Gutierrez., R., Hirn-Chaine, C., Ogborn, J., Pintó, R., Quale, A., Rebmann, G., Stylianidou, F., Viennot, L., 2000.

12. APPLICATION

Since we consider a crucial point the extension and diffusion of the results amongst educational authorities, the most applicable STTIS results have been summarized, and a set of recommendations have been proposed *Guidelines for policy-makers of teacher training program.* This summary constitutes **Work package 7** co-ordinated by Elena Sassi from the University of Naples, RW7

As stated, the STTIS project intended to contribute towards increasing teachers' acceptance of particular innovations and towards improving the class intervention designs, in order to bridge the gap between innovation designers and teachers. Our view is that the STTIS research outcomes were able to achieve these objectives. It is now hoped that they will be useful tools for policy makers and for teacher trainers.

REFERENCES

Bell, B., (1998). Teacher development in Science Education. In: Fraser, B. J. and Tobin, K. J. (eds.), *International Handbook of Science Education* (pp. 681-693). Kluwer. London,

Black, P. & Atkin, J. M., (eds.), (1996). *Changing the subject: innovations in Science, Mathematics and Technology Education.* Routledge. London.

De Jong, O., Korthagen, F. & Wubbels, T., (1998). Research on Science Teacher Education in Europe: Teacher thinking and Conceptual Change. In: Fraser, B. J. y Tobin, K. J. (eds.), *International Handbook of Science Education* (pp. 745-758). Kluwer. London.

Denzin, N. K. & Lincoln, Y. S. , (eds.), (1998). *Collecting and Interpreting Qualitative Materials*. SAGE Publications. London.

Hirn, C., (1998). *Transformations d'intentions didactiques par les enseignants: le cas de l'optique élémentaire en classe de Quatrième*. Unpublished thesis, Université Denis Diderot (Paris 7).

Kress G. & Van Leeuwen T., (1996). *Reading Images: the grammar of visual design*. London: Routledge and Kegan Paul.

Miles, M. B. & Huberman, A. M., (1984). *Qualitative data analysis. A sourcebook of new methods*. SAGE Publications. London.

Pinto, R., Ametller, J., Chauvet, F., Colin, P., Giberti, G., Monroy, G., Ogborn, J., Ormerod, F., Sassi, E., Stylianidou, F., Testa, I., Viennot, L., (2000). *Investigation on the difficulties in teaching and learning graphic representations and on their use in science classrooms*. RW2. STTIS Project.

Pinto, R., Gutierrez, R. & Ametller, J. Andresen, O., Balzano, E., Boohan, R., Chauvet, F., Colin, P., Couso, D., Gibert I. G., Hirn, C., Kolsto, S.D., Monroy, G., Ogborn, J., Quale, A., Rebmann, G., Sassi, E., Stylianidou, F., Testa, I., Viennot, L., (2001). *Teachers Implementing Innovations: Transformations Trends*. RW4.

Pinto, R., Ametller, J., Boohan, R., Chauvet, F., Giberti, G., Gutierrez, R., Monroy, G., Ogborn, J., Ormerod, F., Quale, A., Sassi, E., Stylianidou, F., Testa, I., Viennot, L., (2000). Dissemination activities. *Presentation of the STTIS results at the International Conference Physics Teacher Education Beyond 2000*. RW6.

Sassi, E., Monroy, G., Testa, I., Ametller, J., Andresen. O., Balzano, E., Boohan, R., Chauvet, F., Colin, P., Couso, D., Giberti, G., Gutierrez, R., Hirn-Chaine, C., Ogborn, J., Pintó, R., Quale, A., Rebmann, G., Stylianidou, F., Viennot, L., (2000). *Teacher training materials favouring the take-up of innovations*. RW5

Sassi, E., Ogborn, J., Pintó, R., Quale, A., Viennot, L., (2000). *Guidelines for policy-makers of teacher training programs*. RW7

Sttis Project, (1998). *Outline and Justification of Research Methodology*. Internal Report RW0:

Stylianidou, F., Ogborn, J. & Contini, M., Gutierrez, R., Kolstoe, S. D., Ott, M., Perez, O., Pinto, R., Quale, A., Rebmann, G., Sassi, E., Viennot, L., (1999). *The State of Art in the Use and Value of Informatic Tools*. RW1.1

Stylianidou, F., Ogborn, J. AND Balzano, E., Giberti, G., Gutièrrez, R., Kolst¯, S. D., Monroy, G., Perez, O., Pinto, R., Quale, A., Rebmann, G., Sassi, E., Viennot, L., (2000). *The nature of use by science teachers of informatic tools*. RW1.2.

Van den Akker, J., (1998). The science curriculum: between ideals and outcomes. In: Fraser, B. J. y Tobin, K. J. (eds), *International Handbook of Science Education*, (pp. 421-447). Kluwer. London.

Viennot, L. & Balzano, E., Chauvet, F., Giberti, G., Gomez, R., Hirn, C., Monroy, G., Ogborn, J., Pinto, R., Sassi, E., Stylianidou, F., (1999). *Investigation on teacher transformations when implementing teaching strategies*. RW3

STTIS REPORTS

Balzano, E., Giberti, G., Monroy, G., & Sassi, E., (1999). *Transformations done by teachers when using an IT: the case of MBL (micro computer based laboratory)*. NR1.2-IT.

Contini, M., Ott, M., & Sassi, E., (1999). *The State of the Art in the Use/Value of InformaticTools*. NR1.1 -IT

Chauvet, F., Colin, P., & Viennot, L, (1999). *Reading images in optics. Students' difficulties, teachers views and practice*. NR2-FR.

Chauvet, F., Hirn, C., & Viennot, L., (1999). *Investigation on teacher transformations when implementing teaching strategies*. FR NR3.

Monroy, G., Testa, I., Giberti, G., & Sassi, E., (1999). *Transformation of didactic intentions: The Italian case of motion and force proposal*. NR3-IT.

Pinto, R., & Ametller, J., (2000). *Using images to teach Energy: students' difficulties, teachers' interpretations and teachers' transformations*. NR2-SP.

Pinto, R., Ametller, J., Chauvet, F., Colin, P., Giberti, G., Monroy, G., Ogborn, J., Ormerod, F., Sassi, E., Stylianidou, F., Testa, I., Viennot, L. (2000). *Investigation on the difficulties in teaching and learning graphic representations and on their use in science classrooms*. RW2. STTIS Project.

Pinto, R., Gutierrez, R. and Ametller, J. Andresen, O, Balzano, E. Boohan, R., Chauvet, F., Colin, P., Couso, D., Giberti, G., Hirn, C., Kolsto, S.D., Monroy, G., Ogborn, J., Quale, A., Rebmann, G.,

Sassi, E., Stylianidou, F., Testa, I., Viennot, L., (2001). *Teachers Implementing Innovations: Transformations Trends.* RW4.

Pinto, R., Ametller, J., Boohan, R., Chauvet, F., Giberti, G., Gutierrez, R., Monroy, G., Ogborn, J., Ormerod, F., Quale, A., Sassi, E., Stylianidou, F., Testa, I., Viennot, L., (2000). *Dissemination activities. Presentation of the STTIS results at the International Conference Physics Teacher Education Beyond 2000.* RW6.

Pinto, R., Perez, O., and Gutierrez, R., (1999), *Implementing MBL (Microcomputer Based Laboratory) technology for the laboratory work in Compulsory Secondary school Science classes.* NR1.2-SP.

Pinto, R., and Gomez, R., (1999), *Teaching about energy in Spanish secondary schools: teacher's transformations of innovations.* NR3-SP.

Quale, A., Andresen, O., & Kolstoe, S.D., (1999). *Using IT-tools in the teaching of physics in Norwegian secondary schools: intentions, expectations and practice.* NR1.2-NO.

Rebmann, G., (1999). *Investigation of actual use of informatic tools by science teachers: the French case.* NR1.2-FR.

Sassi, E., Monroy, G., Testa, I., & Giberti, G., (2000). *Reading and interpreting graphs from real-time experiments: students' difficulties, teachers' interpretations and class practice.* NR2-IT.

Sassi, E., Monroy, G., Testa, I., Ametller, J., Andresen. O., Balzano, E., Boohan. R., Chauvet, F., Colin, P., Couso, D., Giberti, G., Gutierrez., R., Hirn-Chaine, C., Ogborn, J., Pintó, R., Quale, A., Rebmann, G., Stylianidou, F., Viennot, L., (2000). *Teacher training materials favouring the take-up of innovations.* RW5

Sassi, E., Ogborn, J., Pintó, R., Quale, A., Viennot, L., (2000). *Guidelines for policy-makers of teacher training programs.* RW7

STTIS Project, (1998). *Internal Report RW0: Outline and Justification of Research Methodology*

Stylianidou, F., Ogborn, J. and Contini, M., Gutierrez, R., Kolstoe, S. D., Ott, M., Perez, O., Pinto, R., Quale, A., Rebmann, G., Sassi, E., Viennot, L., (1999). *The State of Art in the Use and Value of Informatic Tools.* RW1.1

Stylianidou, F., & Ogborn, J., (1999). *Teachers using computer modelling and simulation in the science classroom: the English case.* NR1.2-UK.

Stylianidou, F., & Ogborn, J., (1999). *Teachers' transformations of innovations: the case of teaching Energy in English secondary schools.* NR3-UK.

Stylianidou, F., Ogborn, J. and Balzano, E., Giberti, G., GutiÈrrez, R., Kolstoe, S. D., Monroy, G., Perez, O., Pinto, R., Quale, A., Rebmann, G., Sassi, E., Viennot, L., (2000*). The nature of use by science teachers of informatic tools.* RW1.2.

Stylianidou, F., Ormerod F., & Ogborn, J. (2000). *Difficulties in teaching and learning with pictorial representations.* NR2-UK.

Viennot, L. and Balzano, E., Chauvet, F., Giberti, G., Gomez, R., Hirn, C., Monroy, G., Ogborn, J., Pinto, R., Sassi, E., Stylianidou, F., (1999). *Investigation on teacher transformations when implementing teaching strategies.* RW3

QUALITY DEVELOPMENT PROJECTS IN SCIENCE EDUCATION

[1]M. Beeth, [2]R. Duit, M. Prenzel, [2]C. Ostermeier, [3]R. Tytler, [4]P.-O. Wickman

[1]*The Ohio State University, USA,* [2]*Institute for Science Education at the University of Kiel, Germany,* [3]*Deakin University, Melbourne, Australia,* [4]*Stockholm Institute of Education, Sweden*

ABSTRACT

This paper discusses four system wide development projects, on three continents, which have arisen as a response to increasing international concern about the quality of school science. The paper describes the characteristics of these projects concerning the context, the underlying principles, the implementation strategy, and the research and evaluative features that accompany them. A number of interesting convergences and divergences are uncovered in this comparison, and issues worthy of ongoing discussion identified.

1. BACKGROUND, AIMS AND FRAMEWORK

How best to promote the quality of schools in general and of science teaching and learning in particular has been an issue for a long time. However, more recently this issue appears to have been given increasing attention. There are a number of reasons for this. Among them a most significant motive seems to be thinking patterns that stem from quality management conceptions in industry. There are increasing demands that schools and universities have to become more effective. Further, in a number of countries disappointing results of students in TIMSS tests have alarmed a broader public, as well as governments intent on improving science provision and making schools more effective.

Quite frequently, discussion in the mass media has focused on the idea of making school more effective in a narrow sense. It is therefore most fortunate that actual projects, among them the projects discussed in this paper, have aimed at views of teaching and learning that meet recent student oriented conceptions of scientific literacy. These approaches take into consideration both the demand to recruit students for science and technology related careers and the necessity to provide basic skills for civic responsibility in an increasingly technological world.

Given these circumstances, it is important that attention is given to the characteristics of projects that represent contemporary research-based thinking about quality teaching and learning, and how best to support this in schools. This paper

D. Psillos et al. (eds.), Science Education Research in the Knowledge-Based Society, 447–457.

gathers together the experience of four such projects, from different continents, in order to identify the key issues and the strategies likely to lead to success.

Analysing the projects represented in this paper, it appears that quality development projects with such an orientation have the following key characteristics:

- Supporting schools and teachers to rethink the representation of science in the curriculum.
- Enlarging the repertoire of tasks, experiments, and teaching and learning strategies and resources.
- Promote strategies and resources that attend to student engagement and interest.
- Setting constructivist principles into practice. *On the teachers' side*: The reflective practitioner with a non-transmissive view of teaching and learning. *On the students' side*: The active, self-responsible, co-operative and self-reflective learner.
- Attention to principles of teacher development.
- Attention to principles of school and systemic change.
- Development of methods to monitor the change process for groups of teachers and the overall impact on student learning/scientific literacy.

The four projects discussed in this paper share these characteristics however with different emphases. All projects are large scale projects at the national or regional level. The school levels addressed vary ranging from Kindergarten to grade 13. The projects are from 3 different continents and sit within rather different educational traditions, and converge and diverge in interesting ways.

2. DESCRIPTION OF THE PROJECTS

2.1 A Quality Development Program to Improve Science and Mathematics Education in Germany

Overview Results of TIMSS clearly indicate that science and mathematics education in Germany is far less successful than expected and necessary to guarantee a minimum of scientific and mathematic literacy. As a reaction to these insufficient results a nation-wide five year program was started in the autumn of 1998 to increase the efficiency of mathematics and science teaching. The program conception was developed by a national group of science and mathematics educators on the one hand and educational psychologists on the other. 30 pilot schools, connected with another 150 network schools, work on selected modules which concern key problem areas in mathematics and science teaching. The program is school based. Input to support the teachers' work is provided by the institutions responsible for the program. The input provided includes seminars to make teachers familiar with the above modules and papers summarizing major findings of research concerning the particular modules as well as ideas and examples to improve the situation. These materials are available to all participants of the program on an internet server. Co-operation among teachers is a fundamental principle of the program. Teachers have to document their work plans and goals achieved to make the information available to their co-operation partners. The new approaches developed are tested in the individual schools and school networks and evaluated by the teachers (see Prenzel & Duit, 2000, for an overview of the program; see also: http://www.ipn.uni-kiel.de/projekte/Sinus_eng).

Guiding principles The program draws on school based approaches of quality development. The basic view of teaching and learning is in a broad sense constructivist. The teacher is seen as a reflective practitioner. Making teachers aware of the problems and then familiarizing them with ideas concerning how to solve these problems (based on findings of research) is seen as the key to the success of the program. According to the constructivist view the student is seen as an active self-reflective learner. Throughout the program this view also holds for teachers. Most modules not only describe guidelines for reflective and efficient learning for students but also for teachers.

Research and evaluation A study investigating teacher acceptance of the program was conducted after the program had been under way for two and a half years (Ostermeier & Prenzel, 2001). The goal of this study was to obtain indicators of how well the program conception supports teachers' development. Another study on the acceptance of the program by teachers, headmasters of the participating schools and parents will follow.

In summer 2000 baseline data collection was carried out to investigate the schools "starting conditions" with regard to school context and climate as well as science and mathematics performance drawing on the instruments used in PISA (1999). The program schools were part of the German PISA sample. Accordingly data will be available to identify characteristics of the schools in the program in comparison to the other participating German schools. In 2003 the schools in the program will also be integrated in the second round of PISA. There is further an option for participation in the third round of PISA in 2006.

2.2 The Science in Schools Research Project: Victoria, Australia

Overview The SiS Research Project is a major component of a suite of projects initiated by the Victorian Department of Education, Employment and Training, to improve science teaching and learning in schools. The project is one of a number in which research is a major feature of education initiatives, in which universities have worked with governments to develop and monitor reforms.

The project is working with 126 'study schools' from across the state of Victoria, and this number will be increased each year in a 'roll out' strategy. Each study school has a SiS coordinator who is working with staff in the school to implement the SiS Strategy, described below, and groups of schools are supported by a consultant who visits or otherwise communicates regularly. Each study school is funded to provide time release for the SiS coordinator, schools are given teacher release support, a resource fund, and some support for professional development activities. From the start, schools were given advice regarding specific professional development (PD), but increasingly the project has focused on the provision of tailored whole school PD focusing on core issues of teaching and learning. A project website (www.scienceinschools.org) delivers support material.

Project principles During the first year of the project, in 2000, we worked with 27 schools to develop, refine and test our model, which consists of two major features: The *SiS Components*, a framework for describing effective science teaching and learning; and *the SiS Strategy*, which is the process through which schools can improve their science teaching and learning.

The Model allows flexibility for schools and teachers to plan and implement initiatives based on the particular needs of the school, within an overall framework provided by the SiS Components. It is consistent with current thinking on teacher development and school change (e.g. Bell & Gilbert, 1996; Fullan, 1992; Hoban, 1997; Franke et al., 1998; Hall & Hord, 2001). The components were developed by interviewing identified effective teachers of science to extract the core elements of their practice, in conjunction with a careful reading of the literature. They are consistent with constructivist principles, extending beyond a conceptual focus to incorporate issues of student engagement, broad perspectives of the nature of science, and understandings of learning as a communal enterprise. The Strategy involves an audit procedure and the development of an action plan which focuses attention on specific initiatives, associated PD, and monitoring of outcomes. All aspects of the Strategy, including the support processes, were rated highly by the study school teachers in an evaluative survey. A major issue the project had to address was how to ensure school and teacher control of their processes, and at the same time ensure a clear focus on quality teaching and learning. The Strategy and Components have been successful, we feel, in this.

Research and evaluation The Project has a major research component. During 2000, the components were used to develop a mapping procedure, based on an interview between each teacher and the SiS coordinator, which constructs a profile of teacher classroom practice that enables change to be tracked. A powerful outcome of this process is the development of shared language concerning teaching and learning which has facilitated change in schools. Tests of student learning outcomes, attitudes and perceptions are used to monitor progress and outcomes. A major issue for the project has been the large size of the monitoring exercise involving the 126 schools, and this has necessitated some compromises in the nature of the testing we have undertaken. We have also collected a substantial amount of qualitative data which has provided insights into how the project has progressed in schools, the processes of teacher and school change and factors which affect this, and many stories of science initiatives that point to significant differences between primary and secondary school cultures. The project is gaining improved understandings of school science practice, and the conditions under which change can occur.

A significant issue currently facing the project, concerns how to extend the model to all schools, without the input of the substantial funding that has characterised the project in its developmental phase.

2.3 *The Science and Technology for All Project: Sweden*

Overview Science and Technology for All (NTA–Naturvetenskap och Teknik för Alla) is a school development project run by the Royal Swedish Academy of Sciences and the Royal Swedish Academy of Engineering Sciences in cooperation with municipalities throughout Sweden. The project is currently, primarily aimed at grades K-6 (children 6-12 years old), but will be expanded to grades 7-9. It was started 1997 and is financed by the Ministry of Education and Science and by different private funds.

In Sweden the municipalities manage most schools. The project currently involves 17 municipalities and about 100 schools. However, each year new municipalities are recruited, and in five years more than 25% of the municipalities in Sweden are expected to be involved. The Swedish NTA project is developed from a project run by the National Science Resources Center (NSRC) founded by the U.S. National Academy of Sciences in cooperation with the Smithsonian Institution, although the details of the original concept have been changed and adapted to Swedish circumstances and the Swedish National Syllabus. In the participating municipalities NTA makes it possible to create a long-term plan for school development involving also the local industry and institutions for higher education and research. In each participating municipality a local coordinator for the project is appointed.

Guiding principles The project is organized around a number of mainstays that were inspired by NSRC. One concerns *curriculum development* and is centered on 14 thematic experimental inquiry based units taking about 10 weeks to complete. All units are tested in several versions in multiple classrooms. Another mainstay is *organized material support* that supplies the teacher with all the material needed for a unit, something that is badly needed because most Swedish schools for grades K-6 lack equipment to do science. Finally teachers participate in *continuous competence development*. Every teacher using a unit completes one day of training in that thematic area. Teachers are also given repeated opportunities to share their classroom experiences. Competence development is also carried out in collaboration with local educators (experienced teachers), industry, colleges and universities. Special courses in science curriculum theory are organized for the local educators.

A starting point for NTA was that the units and organization developed by NSRC represent well-tried knowledge based on science education research and practical experience on how to achieve increased interest and improved learning in science and technology. The first trials with the American units in Swedish schools were also favourably received both among teachers and students. However, there was a concern that the units did not leave much space for teacher or student initiatives and that the American material was not adapted to the Swedish National Syllabus. Early discussions in the project thus dealt with how detailed, ready-made units better can take advantage of the prior experience of teachers and students and develop them beyond the units and give them more participatory roles. This discussion dealt with how the constructivist perspective could be expanded, so that the project included perspectives from research also related to how a community of shared understanding can be developed (Wickman & Östman, 2001). The project therefore agreed on a number of pedagogical aims stressing participation, which means that the purposes and contexts of a certain curricular design must be discussed and made clear among all participants–with their different backgrounds and at all levels in the project–to avoid mechanical copying and also to allow alternative routes to be taken.

Research and evaluation The NTA project is continually evaluated. Teachers generally are very positive to the aid they are given through the project. Teachers and students become more interested in science and more time is allocated to the science subjects. NTA helps teachers to teach more extended courses on the same theme. Teachers can also easily change and adapt the units to local circumstances.

2.4 Systemic Reform of Science and Mathematics Education in Ohio (USA)

Overview Ohio was one of ten states in the USA funded under the State Systemic Initiative by the National Science Foundation. From 1991 to 1995 Ohio's Systemic Initiative-Project Discovery (OSI-Discovery) received additional financial support from the Ohio Board of Regents and the Ohio Department of Education. The mission of OSI-Discovery was to facilitate systemic reforms that produced improved learning by K-12 students in mathematics and science. The main thrust of this effort (see Kahle, et al. 1995) focused on sustained professional development institutes for individual middle school teachers through research-based, inquiry teaching models such as Physics by Inquiry (Arons, 1990). By 1999, approximately 5000 teachers had participated in OSI-Discovery institutes taught by scientists, mathematicians, teacher educators, and master teachers. Another 5,400 teachers had participated in regional or local institutes taught by OSI-Discovery-trained leaders.

From 1995 to the present, OSI-Discovery continued its systemic improvement mission with a focus on post-secondary mathematics and science education. The focus of reform expanded from teachers, schools, and districts grades K-12 to teacher education projects through a higher education component - Project SUSTAIN. Evaluations of efforts associated with Project SUSTAIN also show striking changes in how teacher educators are experiencing mathematics and science learning throughout the state of Ohio (Wagner, Meiring & Costner, 2000). Specifically these efforts have led to inquiry-based mathematics and science courses offered to pre and inservice teachers, educating university faculty to teach mathematics and science content courses through inquiry methods, and strengthening communication among college/university faculties in Education and the Arts and Sciences.

Research and evaluation Overall, Ohio's systemic reforms have been successful in: reshaping of teacher participants' beliefs about learning and their pedagogical practices, raising achievement scores of students - particularly in buildings with a critical mass (51-100%) of Discovery teachers, preparing a cadre of teacher leaders, instilling inquiry-based curricula in mathematics and science courses, supporting inquiry-based teacher education projects at state-assisted universities, and promoting collaboration among Arts & Science and Education faculties (Kahle & Meece, 2000; Wagner, Meiring & Costner, 2000). Ohio's systemic initiative have yet to: obtain resources necessary to impact at least 50% of all K-12 teachers in Ohio, solve the infrastructure and school improvement support needs of K-12 teachers and administrators - particularly as they relate to project improvement rather than growth by an individual teacher or building, provide adequate professional development to significant numbers of higher education faculties, and become an integral part of state agency planning relative to mathematics and science education (e.g., the Ohio Department of Education).

3. ANALYSIS: A FRAMEWORK FOR DESCRIBING KEY FEATURES OF SYSTEM-WIDE PROJECTS

Across these diverse projects there are interesting divergences that flow from the different project contexts and purposes, but also convergences that show evidence of broad agreement on the nature of a problem with science in schools, fundamental principles of teaching and learning in science, and the nature of teacher change. The discussion below offers a framework for analysis of such projects.

3.1 The Context and Purposes for the Projects

All four projects have been spawned by a perception that science in our schools is failing both in terms of the quality of learning that is happening, and the engagement of students in science as evidenced by falling participation in post compulsory science pathways. The perception of poor quality learning was in each case strongly influenced by national testing programs, and also in the case of the Victorian, Ohio and German projects also by the TIMSS (1995) results. In the Ohio project attention stems from the National Commission on Excellence in Education's report - A Nation At Risk (NCEE, 1983). One conclusion expressed in this report is that there is "a rising tide of mediocrity that threatens our very future as a nation and a people" (p.5). National assessments of learning have confirmed this assertion (NEGP, 1994). In each case the testing results have given increased status to documents that call for reform in almost every aspect of teaching and learning, and teacher training and development. Science educators sometimes decry the trend towards large scale comparative testing, but its political usefulness is apparent in these four cases.

The national analyses and international comparisons have generated a sense of urgency for politicians and the scientific community in each of the four countries. In the Victorian project this has been explicitly linked with a science and technology innovation whole government initiative, but the sub text in each project has been the need to involve more students in science and science careers. Three of the projects have been initiated by government. The Swedish project was started by scientists and engineers and not by the government or by science education researchers. Only subsequently, as the project produced results, did it receive government funding.

It is interesting, given these imperatives, that each project is focusing explicitly on scientific literacy as a major aim, rather than the narrower aim of producing a technological workforce. Perhaps this is a recognition by all players of the complex ways in which citizens in contemporary technological societies engage with science.

3.2 *Project Control and Direction: Who Makes the Key Decisions?*

The management of each of the projects is devolved across a number of key groups, and in some cases involves a high degree of local control. Projects of this size are complex, represent the interests of many players, and inevitably need to fulfil a range of purposes within the broad aims. Each project incorporates structures to deal with the tension between maintaining project integrity and allowing local autonomy.

Ohio's systemic initiative currently supports 13 different projects at state universities. Each project is administered by the project staff to accomplish goals they set for themselves. Annual conferences allow each project to share the results of their projects with all other state universities involved in Project SUSTAIN. The Victorian SiS project team comprises science educators, seconded teachers, and regional science project officers. A government reference group controls the overall direction and funding, but the project operates with a high degree of autonomy, and there is a high degree of trust, and healthy discussion and debate, between the various players. The structure allows schools considerable control over their initiatives. The Swedish project is run by a large team of representatives from the academies, of school teachers, administrators from municipalities and by science education researchers. A number of decisions are made directly by municipalities and schools, as a major aim is to encourage them to produce their own long-term structures that will support school development in science and technology. The German program is run out of a consortium of institutions, and managed by science educators and educational psychologists. Again, a good deal of local control is given to schools in the way the program proceeds.

3.3 *The View of Teaching and Learning Underpinning the Projects*

Each of the projects takes a broad scientific literacy perspective on the purposes of science education, and subscribes to a view of teaching and learning that is consistent with constructivist principles. The emphases are different, however, for various practical and contextual reasons. In the German program, constructivist principles in a broad sense are explicitly referred to in framing both the view of teaching and learning, and of teacher development. The Ohio approach is consistent with general tenets of constructivist learning, especially those that call for full engagement of the learner in his or her learning, but the approach sits strongly within a rhetoric of inquiry-based pedagogy.

The Victorian SiS Components are strongly student centred, and focus strongly on conceptual understanding and engagement with the processes and conceptual structures of science. Concern to build autonomous learners, and the belief that learning should acknowledge the realities of students lives out of school and the communities within which learning occurs, reflect broader concerns in Victoria with a perceived disengagement of students across the middle years of schooling in general. Similarly, the Swedish characterisation has been expanded to a more inclusive view of learning emphasizing the situatedness of learning and learning as participation. This has become decisive in that the people involved have different backgrounds and different conceptions of learning and teaching practices. The project has agreed on a number of pedagogical aims in more general terms that not only emphasise the learner as an active subject but also the pragmatic, communicative and shared aspects of knowledge.

Thus, while constructivist principles are strongly represented in all of the projects, the focus has broadened in each case to respond to the perspectives of different participating groups, and wider conceptions of schooling issues

3.4 The Model of Teacher Change Underlying the Projects

Each of the projects has adopted a whole school approach to change, to some degree, but they differ in the details of resources and processes that support teachers and schools. Each project has worked out the tension between central resource and process provision, and local control, in different ways. The Victorian and German projects are the most obviously based on a whole school change model in that each school decides the particular agenda they will follow, and the processes they will use. The role of the school community is emphasised, and communication between schools is also a feature. The German schools are guided by a series of modules that focus on key issues. Victorian schools undergo a strategic planning process that must be based in the SiS Components. In both cases there are professional development and collegial support, and reporting procedures, that ensure the integrity of the project.

The Swedish project is based on existing unit materials, but as the project has developed, more attention has been paid to the way teachers interact within the project, and adapt the materials, and to the different ways participating groups might adapt the project to suit local needs. Control has thus progressively devolved.

In the Ohio project a number of teachers from each school attends a training program, but there is attention paid to how the innovation is applied in each school. Broader conceptions of participating groups is reflected in the systemic reform aim to involve everyone connected to the education of a student and everything that a student might experience formally or informally during their years at school. Targets for systemic reform include those related to schooling, the professional knowledge of teachers, the pre-service and professional development of teachers, state and national standards for learning and teaching, and partnerships with parents, community members and care givers.

3.5 Project Evaluation: Processes and Targets

There is scope in the projects such as these for many features to be evaluated, using a range of methodologies. Possible targets for evaluation, each allowing a variety of possible dimensions, are student outcomes, teacher and school change outcomes, and features of the implementation strategy. In each of the projects choices have been made as to whether to use large scale, quantitative monitoring, or smaller scale, more fine grained evaluative methods, or a combination.

The German and Ohio projects both use existing large scale testing to evaluate student achievement. The Ohio project is further down the track and has posted impressive results. The Victorian project has produced its own tests, including attitude and perceptions surveys, and is also developing closer monitoring procedures. The Swedish project includes student outcomes monitoring as a smaller project with a subset of the schools. In each case analysis is occurring at both the classroom and school level.

School and teacher change, and the change process, is monitored in each case using school or regional reports and questionnaires. The German program collected comprehensive baseline data from its schools. The Swedish project is using interviews and classroom observations in a subset of schools as part of a linked project. The Victorian project has developed an interview based instrument to monitor change in teachers' classroom practice quantitatively, and is using interviews with coordinators and principals to track school change.

The implementation process is monitored in each case using questionnaire and school report data, and in some cases more in depth case study methods.

4. DISCUSSION: ONGOING ISSUES

The discussion during the symposium at ESERA, on which this paper is based, and also comments from the various projects, raised a number of issues and problems faced by the projects, which have been solved to varying degrees. These issues include:

- The need for large scale and complex projects such as these to develop explicit views about teaching and learning science, and the role of the teacher and schooling;
- The need to match evaluation of student outcomes with the core beliefs about teaching and learning promoted by the project;
- Possible tensions between government policy, the views of various stakeholders, and beliefs about teaching and learning and purposes held by science educators, and teachers;
- Possible discordance between results on system wide science tests that drive these projects, and more fundamental issues addressed by the projects concerning teaching and learning, and participation in school science;
- Tension between the need for a coherent vision that focuses on core issues, and the need to devolve ownership of innovation to schools and teachers;
- How to balance the need to utilise pre-existing materials, including assessment materials, with the demands for matching the innovation to local or regional needs;
- How to harmonise the different levels at which such projects operate; individual teachers, science departments, schools, regional groups, and the broader community; and
- How to balance a need for research and evaluation, against the need for effective and responsive implementation of initiatives.

REFERENCES

Arons, A. (1990). *A Guide to introductory physics teaching.* New York: John Wiley & Sons.

Bell, B. & Gilbert, J. (1996). *Teacher development: A model from Science Education*. London: Falmer Press.

Franke, M., Carpenter, T., Fennema, E., Ansel, E., & Behrend, J. (1998). Understanding teachers' self-sustaining generative change in the context of professional development. *Teaching and Teacher Education*, 14(1), 67-80.

Fullan, M. (1992). *Successful school improvement.* Buckingham: Open University Press.

Hall, G. & Hord, S. (2001). *Implementing change: Patterns, principles, and potholes*. Boston: Allyn & Bacon.

Hoban, G. (1997). Theories and models of professional development. In R. King, D. Hill & J. Retallick (Eds), *Exploring professional development in education* (pp. 1-20). Australia: Social Science Press.

Kahle, J. B. & Meece, J. L. (2000). *Reform that works* [Booklet]. Oxford, OH: Miami University.

Kahle, J. B. et al. (1995). *Project Discovery, The Ohio statewide systemic initiative in mathematics and science education*. Annual Reports, 1992 to 1995. Oxford, OH: Miami University.

National Commission on Excellence in Education (1983). *A Nation at Risk*. On-line. [Available from: http://www.ed.gov/pubs/NatAtRisk/title.html]

National Education Goals Panel (1994). *Data for the National Education Goals Report*, Volume II: State Data. Washington, DC: U. S. Government Printing Office.

Ostermeier, C., & Prenzel, M. (2001, August). Implementation of a teacher-based quality development program: Findings from a study on teachers' acceptance and requests for future assistance. *Poster presented at the 9th Conference of the European Association for Research on Learning and Instruction, Fribourg (Switzerland).*

PISA – Programme for International Student Assessment (1999). *Measuring student knowledge and skills. Paris*: OECD.

Prenzel, M. & Duit, R. (2000, April). Increasing the efficiency of science and mathematics instruction: Report of a national quality development program. *Paper presented at the Annual Meeting of the National Association for Research in Science Teaching (NARST).* New Orleans [Available from: http://www.ipn.uni-kiel.de/projekte/blk_sinus.pdf].

Third International Mathematics and Science Study (1995). *TIMSS Report*. On-line. [Available from: http://timss.bc.edu/timss1995.html].

Wagner, S., Meiring, S. P. & Costner, K. M. (2000). *Milestones 2000: Reform efforts in mathematics and science teacher education programs at SUSTAIN Collaboratives in Ohio*. Columbus, OH: The Ohio State University.

Wickman, P.-O. & Östman, L. (2001). University students during practical work: can we make the learning process intelligible? In H. Behrendt, H. Dahncke, R. Duit, W. Gräber, M. Komorek, A. Kross, & P. Reiska (Eds), *Research in science education - Past, present, and future* (pp.319-324). Dordrecht, The Netherlands: Kluwer Academic Publishers.

VIDEO-BASED STUDIES ON INVESTIGATING DEFICIENCIES OF SCHOOL SCIENCE TEACHING

Reinders Duit, Helmut Fischler, Hans Fischer, Elke Sumfleth

IPN Kiel, Free University of Berlin, University of Dortmund, University of Essen

[1]ABSTRACT

Video based research has become a powerful tool to investigate teaching and learning processes in science education. These studies have been employed to reveal the microstructure of teaching and learning processes and to investigate the interplay of instructional scripts and various variables of the particular learning environments. In the following synopsis of a poster symposium at the ESERA Conference in Thessaloniki we present four studies that employ video-analysis of science instruction in various ways. All studies were initiated by the rather disappointing results German students achieved in TIMSS. It is the aim of these studies to investigate means of improving science teaching and learning.

SYNOPSIS

German students did not well in TIMSS. Their results in science were just mediocre. What is even more worrying, however, is the fact that large numbers of German students had problems solving the more demanding tasks, especially those requiring conceptual understanding. In addition science instruction obviously is only able to awaken and maintain interest in learning science to a limited degree. It appears that these deficits have to do with the way science is taught in Germany and the limited support provided to take learning science seriously in school and out of school contexts. A priority program on investigating the quality of science instruction and literacy has been launched by the German Science Foundation in order to support manifold initiatives to improve the situation (http://www.ipn.uni-kiel.de/projekte/biqua/biqua_eng.htm). Since April 2000, 23 projects are funded. Science educators closely cooperate with colleagues from psychology and pedagogy. It is a key characteristic of the priority program that the different projects cooperate and investigate science teaching and learning from various perspectives. The priority program will be funded for six years.

The four studies presented here are among the 23 projects funded. They share the method of video-analysis of real physics and chemistry instruction, in order to investigate various aspects of the scripts dominating physics and chemistry instruction. The concept of script as introduced by Schank and Abelson (1977) was

[1] The authors listed here are responsible for this synopsis. The names of the colleagues participating in the four groups is given when their work is presented.

D. Psillos et al. (eds.), Science Education Research in the Knowledge-Based Society, 459–469.

adopted for research on patterns of instruction by the TIMS video-study comparing mathematics instruction in three countries (Stigler et. al. 1999). The concept of script employed in our studies includes the level of what may be observed (surface structure) as well as the level of teachers' *and* students' subjective theories (deep structure). The interpretation of the observed instructional patterns is also informed by the framework of a "choreography" introduced by Oser and Patry (1990). Script and choreography are metaphors denoting our assumption that instruction is based on certain (at least implicit) plans of instruction.

It is the key aim of the studies to identify the kind of physics and chemistry instruction predominant in Germany, to understand the teaching and learning processes observed on the background of scripts shared by teachers and students, and to investigate various means of improving teachers' subjective theories of "good" physics and chemistry instruction. In all studies analyses of video data are seen as powerful means for studying the microstructure of teaching and learning processes. Video documents allow seeing the same lessons as often as necessary from various perspectives. Manifold coding systems may be used to investigate a broad spectrum of variables of the teaching and learning processes. Although there are many features that are shared by the four projects video-analysis is employed in different ways and for different purposes. Therefore, the following brief overviews of the studies may provide insight in different means to use this powerful method.

In the first study presented video technique is employed to identify characteristic patterns of physics instruction. Various coding systems are used to allow multi-perspective analysis of these patterns. It is also investigated which patterns are affiliated with the most sufficient development of affective variables like interests and self-concept and cognitive variables. The second study investigates how an improvement of science teaching and learning can be achieved through a modification of teachers' pedagogical and psychological expertise. Videotaped lessons of the teachers involved in this project make it possible to analyse their classroom behaviour prior to and after special coaching procedures. The third study intends to investigate the structures of typical physics lessons in order to identify ways of achieving more successful "choreographies". Various reference theories are used to develop sets of categories to code the video taped instruction. In the last study teaching-learning situations at school and at home are videotaped for monitoring the teachers' behaviour and for analysing students' discourses. Here video analysis is embedded in a formal 2X2 design to investigate the efficiency of a constructivist learning environment.

1. STUDIES ON THE INTERPLAY OF TEACHING AND LEARNING PROCESSES IN PHYSICS INSTRUCTION

IPN-Physics-Video-Study-Group: Manfred Prenzel, Reinders Duit, Manfred Euler, Helmut Geiser, Lore Hoffmann, Manfred Lehrke, Christoph Müller, Rolf Rimmele, Tina Seidel, Ari Widodo

1.1 Overview

It is the aim of the video study presented here to investigate the scripts and the teaching and learning processes characteristic for German physics instruction. The role experiments play in instruction will be given particular attention. There are three phases of two years each. In the first phase (May 2000 to May 2002) physics instruction in two German states is investigated. 13 teachers and classes are involved. The first phase also serves to develop appropriate methods of data collection and interpretation. In the second phase physics lessons in Germany are compared to physics lessons in the German speaking part of Switzerland. In the final phase results gained will be used in intervention studies on improving the quality of teaching and learning physics.

1.2 Aims and theoretical frameworks

The aim of the video based studies is not merely to identify patterns of instructional phases and interactions but also to detect key indicators for conditions and processes that characterize the quality of physics instruction. Theoretical perspectives guide interpretation of video data and results of teachers' and students' questionnaires and interviews. Of specific significance are teachers' and students' conceptions and beliefs of teaching and learning in general and teaching and learning physics in particular. These conceptions and beliefs are viewed as culturally mediated.

1.3 Sample and methods

In this first phase eleven grade 8 classes and two grade 7 classes participated in the studies. Every class was taught by a different teacher. Two sequences of three lessons on the topics of "Introduction into the Electric Circuit" and "Introduction into the Force Concept" were video documented. A basic coding follows the categories used in TIMSS Video (Stigler et al. 1999). Further, the various research questions call for special and rather elaborated coding categories. A coding system investigating whether instruction meets characteristics of constructivist oriented science instruction is in the final state of development. The same holds for a coding system addressing the different roles of experiments. Further systems are under construction. For every lesson the content structure is reconstructed using a certain kind of flow diagrams.

At the beginning and at the end of the school year students were given a questionnaire on affective variables like interests and self-concepts, on their meta-cognitive views and their physics knowledge in the two topics investigated. The students also filled in small questionnaires at the end of every lesson video-taped, especially on their mental activities and learning motivation during instruction. The teachers were given only a small questionnaire in the beginning, mostly on technical issues and on their meta-cognitive views. After documenting the second topic teachers were interviewed about various facets of their views of teaching and learning physics. They were further confronted with episodes from their instruction.

1.4 Results

TIMSS Video (Stigler et al., 1999) showed that German mathematics instruction is dominated by a single script. In contrast our analyses revealed a greater variety of teaching. However, a somewhat narrow "questioning-developing" approach, which is characteristic for German mathematics teaching, is also prevailing here.

Two different kinds of scripts may be distinguished. In some classes, the lessons are characterised by a high level of student activity, minimum control by the teacher and a high frequency of student experiments. In other classes we find a more traditional script that is characterised by high teacher control and the preference of classroom activities that employ variants of a "questioning-developing" format. Furthermore, experiments are usually carried out by the teacher. The amount of content covered in the first case is significantly smaller than in the second one. As a consequence the content structure in the second script is much richer and more complex. A key variable with regard to affective and cognitive outcomes appears to be the extent opportunities for learning are provided (such as explicit cognitive activation). A significant finding is also that most of our teachers' thinking about physics instruction is rather content oriented. Considerations about the content in question predominate planning; reflections about students' perspectives play a rather minor role.

There are other preliminary findings on relations between characteristics of physics instruction and affective and cognitive student variables and on relations between teachers' subjective theories of teaching and learning physics and their actual classroom behaviour (Seidel et al. 2002, Prenzel et al. 2001, Widodo; Duit & Müller 2001) which are promising and show the particular power of combining more traditional research methods (employing questionnaires and interviews) and video-analyses.

2. TEACHING EXPERTISE AND TEACHING SCRIPTS: CONDITIONS FOR THEIR MODIFICATION

Members of the project group: Peter Zedler, Claudia Tonhäuser, University of Erfurt; Hans-Joachim Schröder, Helmut Fischler, Free University of Berlin

In many countries the improvement of science education is a widely accepted goal. Often the enhancement of teachers' competencies is regarded as an appropriate means to improve science teaching. However, the effects of in-service training courses have proven to be less effective than anticipated. Subsequently, a research project was designed and implemented to systematically identify the conditions for modifying teachers' pedagogical and psychological expertise. The research project concentrates on physics teachers in secondary schools. The main research methods are: interviews and maps for identifying teachers' conceptions of teaching and learning, videotaped lessons for identifying teaching scripts, and three different treatments for changing teachers' conceptions.

2.1 Background, Aims and Framework

Traditional in-service training courses have often proven to be less effective than anticipated (Tillema & Knol 1997, Yerrick et al. 1997) because teachers very often show a fundamental discrepancy between their conceptions on the one hand and their actions on the other (Fischler 1994). The main research question of this project, therefore, is how teachers' competencies can be enhanced in such a way that, as a result of this intended development, their pedagogical and psychological expertise improves on students' achievements. The conditions necessary for modifying teachers' pedagogical and psychological expertise are to be identified systematically.

According to the aims of the research project the design of the project includes the following steps:

1. Initial analysis: Identification of the teachers' expertise (conceptions about teaching and learning, conceptions about the nature of science, pedagogical content knowledge), identification of teaching scripts; identification of students' specific attitudes towards science education.
2. Treatments to change teachers' expertise: Reframing, reflecting-team, process consultation.
3. Final analysis: Analysis of the differential effects of the treatments, identification of teachers' expertise and their teaching scripts; questioning of the students.

2.2 Methods and samples

Identification of teachers' expertise: Teachers' conceptions about teaching and learning (science, physics) are being investigated by means of interviews in which the presentation of videotaped teaching scenes plays an essential role. The scenes represent situations, which are typical for physics lessons and contain problematic actions of teachers. In the interview, the teacher participating in the research project is asked questions like these: Do you know this situation, does this occur in your lessons very often? How do you decide in these situations? Are there alternatives to the teacher's acting in the videotaped scene? In addition to the interview, lessons of the teachers are videotaped in order to get to know which orientations are guiding their actions in classrooms and to contrast these orientations with their conceptions previously stated.

Treatments: Three different treatments were applied to change teachers' conceptions about teaching and learning. The first treatment, called reframing, concentrates on individual teachers so that they restructure their cognitive frame to interpret situations and processes of actions in classrooms. The second treatment involves a reflecting-team where different teachers describe their views of classroom processes by means of group sessions that identify alternatives for each teacher. The third treatment, called process consultation, focuses not on teachers' conceptions but on altering situation specific chunks of teachers' actions.

Sample: About 35 teachers are taking part in a first phase of this investigation, which is designed for six years, including the development of research instruments and the investigation of changes in students' attitudes towards physics (teaching) and of changes in their achievement.

2.3 Results

In case studies the interrelationship between teachers' conceptions and their orientations of acting in lessons was investigated and analysed so far. In a first treatment phase, the behaviour of an exemplary teacher was observed and related to her conceptions and her teaching scripts. One of the main results describes the stability of conceptions and scripts, which are grounded in her professional biography. Process consultation seems to be the most difficult but most effective treatment. Besides this general result the research delivers a lot of information about how practicable and in details how effective this treatment is.

In order to improve science instruction it appears to be necessary to enhance teachers' competencies by means of special treatments, which aim at changing teachers' acting in classrooms. It is not sufficient to improve teachers' knowledge or their conceptions of teaching and learning. It is important that teachers experience themselves their new orientation under the guidance of experts and in co-operation with colleagues.

3. TEACHERS' CONCEPTS AND TEACHING STRATEGIES IN PHYSICS INSTRUCTION AND THEIR EFFECTS ON INSTRUCTIONAL DESIGN AND LEARNING OUTCOMES

Hans E. Fischer, Wilfried Bos, Thomas Reyer, Nicole Höllrich,*
*University of Dortmund, *University of Hamburg*

3.1 Background, aims and framework

The first study presented above revealed that physics lessons in Germany rarely support students to learn the abilities they are supposed to achieve. The results of the TIMS video study show that a Germany mathematics lesson basically follows a single script which is dominated by a narrow use of the "questioning-developing" approach (Baumert & Lehmann 1997, Stigler et al. 1999). Obviously, that lack of variety of lesson "choreography" does not represent the whole scope of teaching and learning. As a consequence we assume that instruction has not only to be analysed on the level of the surface structure but on the level of the deep structure as well. Oser and Patry (1990) developed an instructional theory, which is based on a small number of specific teaching aims, concept types, and related learning processes relevant for classroom learning. Modifying their theory for physics lessons, we propose ten different concept types. These are learning by experience, conceptual change, problem-solving, theoretical knowledge, contemplation, training routines, transformation of affective excitement, social learning, constructing values and personality and analysing outlines. In order to analyse the lessons accordingly we identify four different levels: (1) The level of conceptual structure of the teachers represents the teachers' pedagogical knowledge. (2) The level of concept types as representation of the teachers' pedagogical knowledge in the classroom. (3) The level of the students' content related action sequences, as the students' answer to the teachers' activities. (4) The students' results concerning performance, skills and attitudes.

3.2 *Methods and Samples*

The four different levels shown above require different methods for analysis. Data sources for investigating conceptual structure are records of the teacher's views, attitudes, and behaviour investigated by interviews. These are partly oriented at scales used by Labudde (2000). The action sequences can be surmised from what the teachers express, and should, as well as the students' content related operations, be re-discovered in the video recordings as deep and surface structure of the lessons. The learning outcome and the attitudes are accessible by using a set of performance and attitude inventories. We are observing and analysing the concepts and the instruction of six physics teachers starting in grade eight of six different classes in two different schools (Gymnasium in the city of Dortmund). Three different topics of physics, electricity, optics (both in grade 8), and finally Newton's force concept (in grade 9) are video-documented. The observations took place from September 2000 to February 2002. The teachers' concepts of teaching and learning, their ideas of physics education, and their estimations of their own role as teachers will be derived from a qualitative analysis of interviews and questionnaires.

The investigation of particular lessons includes the analysis of video-recordings of the lessons, guided by rules and categories similar to those used in the TIMS Videotape Classroom Study (cf. Stigler et al. 1999). The categories for the deep structure have been developed as operational expression of the concept types of the teachers and the expected action sequences of the students.

3.3 *Results*

Interpretation of data is still in progress. As we can already see from the interviews and the video recordings of the lessons, the teachers know little about their students' learning processes. Moreover, they are hardly aware of how they influence them by their classroom activities and by instructional design; they know little about influencing the students' learning by intending specific types of education aims. Regarding these results we come to the hypothesis that instruction is more efficient if a broader repertoire of teaching methods is used. Evidently, effective learning only occurs if the cognitive needs of the students are taken into account.

We expect to explain how teachers' concepts of teaching influence the structure of the lessons and how the surface structure affects learning processes and learning outcomes. In a subsequent phase of the project the results will be used to construct a model of instructional design as a contribution to (science) teacher education.

4. CO-OPERATIVE LEARNING OF CHEMISTRY – INFLUENCES ON COGNITION AND INTEREST

Elke Sumfleth, Stefan Rumann, Nina Buttler, Chemistry Department, Essen University and Elke Wild, Judith Gerber, Josef Exeler, Psychology Department, Bielefeld University

4.1 Background

One possible reason for the deficiencies of German students in conceptual understanding of science is seen in the dominant teaching practice in Germany, which is determined by narrow questions aiming at specific terms to be used. Such a direct instruction prevents the development of a flexible knowledge structure (Gerstenmaier & Mandl 1995) and the maintenance of interest in science as well as self-directed motivation (Duit & Häußler 1997). Constructivist approaches appear to allow improvement of this situation (Klieme & Clausen 1999). The characteristic features of learning in these approaches are students' activities and situated problem oriented learning. A main point is the optimal degree of self-control versus external control because learners achieve rather badly in complex and totally open learning environments. In addition to these cognitive effects constructivist approaches also appear to support self-directed learning motivation (Krapp 1999).

4.2 Aims and framework

This investigation focuses on improving students' co-operative learning in chemistry classes and in homework situations in order to develop their daily-life conceptions, to support self-directed learning motivation and to promote a positive attitude towards scientific phenomena. From a cognitive point of view we start with the idea that learners who get the chance to become active and self-regulated in working on complex problems in structured co-operative teaching-learning-environments (at school and at home), will gain flexible and useful knowledge. In addition scientific knowledge should be improved by intensive discussions in classes and at home. From a motivational point of view we suppose a qualitative shift in students' motivational orientations towards a stronger emphasis on self-directed motivation. Furthermore the interventions should foster the development of personal interest in chemistry as well as in chemistry lessons. With regard to chemistry content we investigate the efficiency of learning environments planned in a constructivist way pursuing cognitive and motivational goals. But these environments have to be well-structured regarding time economy and learning efficiency. The main unanswered question is how to achieve the right proportion of construction and instruction concerning the above properties of an efficient learning environment.

4.3 Mode of inquiry

The investigation is conducted as a quasi-experimental study with a pre-, post and follow-up test design to assess variables such as personal interest and motivational orientations as well as specific pre-knowledge. In order to study the influence of the kind of homework and of the co-operative group-work at school on interest and knowledge a 2x2-design is employed. The instruction of the experimental group is oriented towards self-regulated group investigation and co-operative working on the homework. The instruction of the control group is oriented towards direct instruction and control of the homework by the parents. The interventions last six lessons and homework situations. To control effects caused by the behaviour of the teacher or by the type of school the investigation is carried out at two different schools (Gymnasium) with four parallel classes at each school which are taught by the same teacher. The lessons are video-taped to monitor the treatment and to assess if the teachers have special difficulties in playing a more advising role. With regard to instructional and motivational aspects they offer relevant insights into mechanisms, which are important for the co-construction of meanings and for conditions, which influence the effects of co-operative group work. The qualitative analysis of interactions is focused on the influences of students' conceptions on the course of discussion in the small working groups, in class and in homework situations. It is expected that dealing with students' conceptions will differ in these three kinds of communication (considering, ignoring, non-recognising, taking up explicitly, re-interpreting, inquiring, accepting as right answer). The sequences are chosen according to the students' use of special terms. This analysis allows identifying the reasons why students occasionally develop conceptions, which are wrong from the scientific point of view.

4.4 Results

Among other methods a multiple choice test was used to investigate the main relationships between concepts in the field of acids and bases. Figure 1 shows that knowledge gain is best for the group with both interventions (S/H) and lowest for the group with no intervention at all (-/-). The analysis of the data, especially of the video data is ongoing.

From a social psychological point of view the analysis of the interactions leads to the identification of such parameters, which support an effective co-operation, and of other parameters, which allow explaining insufficient results of group work. There are totally different patterns of communication, ranging from groups dominated by statements of a single student with nearly no influence of others to well distributed actions which are complimentary to each other. The respective pattern seems to influence the outcome of the group-work. From the point of view of chemistry education the content specific progression of solving complex problems is analysed. Finally, video data serve as quality criteria of student adequate co-operative task setting.

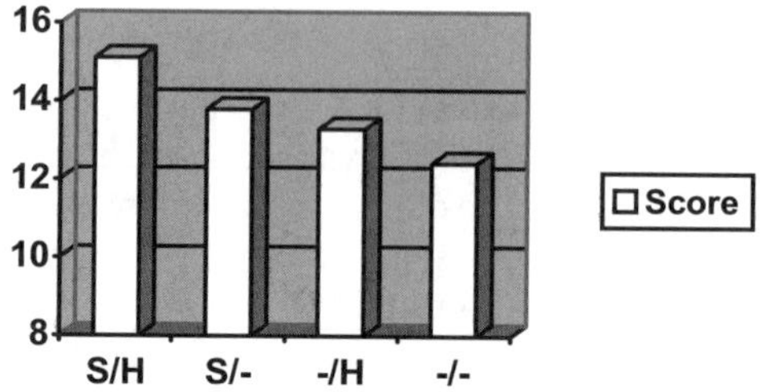

Figure 1: Results of the multiple choice test

REFERENCES

Baumert, J. & Lehmann, R. et al. (1997). *TIMSS - Mathematisch- naturwissenschaftlicher Unterricht im internationalen Vergleich [TIMSS – Mathematics and science instruction in international comparison]*. Opladen: Leske + Budrich.

Duit, R. & Häußler, P. (1997). Physik und andere naturwissenschaftliche Lernbereiche [Physics and other science learning domains]. In: F.E. Weinert (Eds.), *Psychologie des Unterrichts und der Schule* (Enzyklopädie der Psychologie, D, Serie I, Bd. 3, pp. 427-460). Göttingen: Hogrefe.

Fischler, H. (1994). Concerning the difference between intention and action. Teachers' conceptions and actions in physics teaching. In: I. Carlgren, G. Handal, & S. Vaage (Eds.), *Teachers' minds and actions. Research on teachers thinking and practice* (pp. 165-180). London: Falmer Press.

Gerstenmaier, J.& Mandl, H. (1995). Wissenserwerb unter konstruktivistischer Perspektive [Knowledge achievement from a constructivist perspective]. *Zeitschrift für Pädagogik*, 6, 867-888.

Klieme, E. & Clausen, M. (1999, April). Identifying facets of problem solving in mathematics instruction. Paper presented at the AERA annual meeting in Montreal, Canada.

Krapp, A. (1999). Interest, motivation and learning: An educational-psychological perspective. *European Journal of Psychology of Education*, 14, 23-40.

Labudde, P. (2000). *Konstruktivismus im Physikunterricht der Sekundarstufe [Constructivism in upper secondary physics instruction]*. Vienna: Haupt.

Oser, F. & Patry, J.-L. (1990). Choreographien unterrichtlichen Lernens, Basismodelle des Unterrichts [Choreographies of school learning, basic models of instruction]. In: *Berichte zur Erziehungswissenschaft*, Nr. 89. Freiburg, Switzerland: Pädagogisches Institut der Universität Freiburg.

Prenzel, R., Duit, R., Euler, M. , Geiser, H., Hoffmann, L., Lehrke, M., Müller, C. Rimmele, R., Seidel, T., Widodo, A. (2001). *Studies on the Interplay of Teaching and Learning Processes in Physics Instruction*. Kiel, Germany: Institute for Science Education at the University of Kiel (http://www.ipn.uni-kiel.de/projekte/video/main16.htm).

Schank, R.C. & Abelson, P.R. (1977). *Scripts, plans goals and understanding. An inquiry into human knowledge structures*. Hillsdale: Erlbaum.

Seidel, T., Prenzel, M., Duit, R., Euler, M., Lehrke, M., Geiser, H., Hoffmann, L., Müller, C., Rimmele, R. & Widodo, A. (2002). "Jetzt bitte alle nach vorne schauen!" - Lehr-Lernskripts im Physikunterricht und damit verbundene Bedingungen für individuelle Lernprozesse ["Please turn your attention to the front of the classroom" – Teaching and learning scripts in physics instruction and related conditions for individual learning processes]. *Unterrichtswissenschaften*, 30(1), 52-77.

Stigler, J.W., Gonzales, P., Kawanaka, T., Knoll, S. & Serrano, A.. (1999). *The TIMSS Videotape Classroom Study: Methods and Findings from an Exploratory Research Project on Eighth-Grade*

Mathematics Instruction in Germany, Japan, and the United States. National Centre for Education Statistics (NCES 99-074); Washington, D.C.: U.S. Government Printing Office.

Tillema, H. H. & Knol, W. E. (1997). Promoting student teacher learning through conceptual change or direct instruction. *Teaching and Teacher Education*, 13, 579-595.

Widodo, A., Duit, R. & Müller, C. (2001, April). Constructivist views of teaching and learning in practice: Teachers' views and classroom behaviour. Paper presented at the annual meeting of the National Association for Research in Science Teaching, New Orleans.

Yerrick, R., Parke, H. & Nugent, J. (1997). Struggling to promote deeply rooted change: The "Filtering effect" of teachers' beliefs on understanding transformational views of teaching science. *Science Education,* 81, 137-159.

AUTHOR INDEX

SUBJECT INDEX